Environmental Impact Assessment

NATO ASI Series

Advanced Science Institutes Series

A series presenting the results of activities sponsored by the NATO Science Committee, which aims at the dissemination of advanced scientific and technological knowledge, with a view to strengthening links between scientific communities

The series is published by an international board of publishers in conjunction with NATO Scientific Affairs Division

A	**Life Sciences**	Plenum Publishing Corporation
B	**Physics**	London and New York
C	**Mathematical and Physical Sciences**	D. Reidel Publishing Company Dordrecht and Boston
D	**Behavioural and Social Sciences**	Martinus Nijhoff Publishers Boston/The Hague/Dordrecht/Lancaster
E	**Applied Sciences**	
F	**Computer and Systems Sciences**	Springer Verlag Berlin/Heidelberg/New York
G	**Ecological Sciences**	

Series D: Behavioural and Social Sciences – No. 14

Environmental Impact Assessment

edited by

PADC Environmental Impact Assessment and Planning Unit

University of Aberdeen
United Kingdom

1983 **Martinus Nijhoff Publishers**
A member of the Kluwer Academic Publishers Group
Boston / The Hague / Dordrecht / Lancaster
Published in cooperation with NATO Scientific Affairs Division

Proceedings of the NATO Advanced Study Institute on Environmental Impact Assessment, Chateau de Bonas, Toulouse, France, August 30 – September 12, 1981

Library of Congress Cataloging in Publication Data

NATO Advanced Study Institute on Environmental Impact
Assessment (1981 : Toulouse, France)
Environmental impact assessment.

(NATO advanced study institutes series. Series D,
Behavioural and social sciences ; no. 14)
Includes index.
1. Environmental impact analysis--Congresses.
I. University of Aberdeen. PADC Environmental Impact
Assessment and Planning Unit. II. Title. III. Series.
TD194.6.N37 1981 333.7'1 82-22579

ISBN-13: 978-94-009-6797-7 *e-ISBN*-13: 978-94-009-6795-3
DOI: 10.1007/978-94-009-6795-3

Distributors for the United States and Canada: Kluwer Boston, Inc., 190 Old Derby Street, Hingham, MA 02043, USA

Distributors for all other countries: Kluwer Academic Publishers Group, Distribution Center, P.O. Box 322, 3300 AH Dordrecht, The Netherlands

Softcover reprint of the hardcover 1st edition 1983

TABLE OF CONTENTS

PREFACE

Brian D. Clark

PADC Environmental Impact Assessment and Planning
Unit Project Director

Events throughout the world substantiate the view that planning and decision-making systems need a better integration of environmental, economic and social considerations. Many organizations are showing considerable interest in Environmental Impact Assessment (EIA) and its role in project planning and policy evaluation and as an aid to decision-making. Consequently, it was decided to hold a NATO Advanced Study Institute on EIA for the following reasons. First there is evidence of uncertainty, particularly amongst many scientists and decision-makers, as to the nature, scope and objectives of EIA. Secondly, there is much confusion over the objectives and utility of certain EIA methods. Third, there appears to be a gulf developing between decision-makers and what they require from EIA, and the ability of the scientist to provide information which is scientifically rigorous. Finally, there appears to be little concern as to the relationship between "impact prediction" and the actual consequences of a development activity, suggesting that if EIA is not to become both politically and scientifically disreputable greater emphasis should be placed on prediction, monitoring and post-audit studies.

As will be seen from the contents of this volume the ASI attempted to address all of the above topics and indeed many more. It was perhaps inevitable that the ASI raised more questions than were answered but this is indicative of the vigorous debate that is now taking place about the role and utility of EIA.

At the time of writing a number of major international gatherings have been held such as the UNEP Session of a Special Character convened in Nairobi, Kenya, to evaluate environmental successes and failures in the ten years after the 1972 Stockholm Conference.

Whilst such meetings reflect the increasing concern throughout the world for environmental issues it would appear that there is a lack of both political commitment and financial resources to introduce measures to maintain and enhance the quality of the environment on the scale which is now necessary. It appears that we are strong on description of environmental problems but weak on prescriptions for improvement.

This volume is published in the belief that the concepts, procedures and methods of EIA are a mechanism that can be of utility both to the decision-maker and the scientist. It brings together the views of an international faculty of academics, scientists and decision-makers who contributed to the NATO ASI held in Bonas, France in 1981. It is hoped that it will not only make a contribution to the debate on EIA but also provide positive ideas as to what EIA can achieve and how it can be implemented.

Brian D. Clark

on behalf of the
Course Directors

ACKNOWLEDGEMENTS

As we discovered, the running of a NATO ASI was a complex business, and the fact that it was a relative success according to participants' evaluation, was due to the commitment and dedication if a great many people. Initially I would like to thank Dr Andrew Sors of the Monitoring and Assessment Research Centre, Chelsea College, University of London (now at the E.E.C.), for mentioning NATO ASIs which we must admit had never been heard of in the far north of Scotland! Once committed to run the ASI on Environmental Impact Assessment (EIA) we received constant support and advice from Dr Mario Di Lullo (NATO), Dr Tilo Kester and Dr Craig Sinclair (NATO), who was able to join us with his wife and make a positive contribution.

At Bonas there can be no doubt that the academic success of the ASI was as much due to the physical and social environment which has been created by Professeur Jean-Claude and Madame Françoise Simon. Their painstaking conversion of the Château de Bonas to an international conference centre, with the development of the estate, make it an ideal venue both for formal sessions and social relaxation, when the real discussions take place. Throughout the course Françoise Simon gave advice and assistance and she was assisted only by Maryse, the Château secretary. The Course Directors would also wish to acknowledge a number of French Institutions that contributed to the success of the meeting and these include, Gascogne Floc, Armagnac, "Vin Plastique", the Mayor of Bonas and Xaver Monbailliu who did so much to arrange the local programme and bring the French nuclear industry to a halt!

Many people have contributed to the production of this volume. We would like to thank the faculty for this written contribution - unfortunately not all were available in time for publication. Pearl Allan and Kathleen Brebner have borne the brunt of the typing and Janet Schucksmith has undertaken the proofreading and indexing. All three have done these tedious tasks with dedication and good humour. To our publishers we are particularly indebted for their patience and forebearance.

Without Sandra Ralston, our conference organizer, the course could not have been held. Her stamina, and ability to remain calm against all the odds must be recorded. Julie Martin has been responsible for co-ordinating the volume and quietly and efficiently has handled its preparation through all stages of production.

To all those listed, and many who are not, such as Marion who held the fort in Aberdeen, and to the faculty and participants who made the seminar possible, we extend our grateful thanks. We hope that the volume will bring back happy memories of the seminar but more importantly hope that it will make a useful contribution to the development and understanding of EIA.

Brian D. Clark

on behalf of the
Course Directors

OVERVIEW

THE AIMS AND OBJECTIVES OF ENVIRONMENTAL IMPACT ASSESSMENT

Brian D. Clark

University of Aberdeen,
Aberdeen.

INTRODUCTION

Since the 1950s growing environmental awareness is increasingly focussing attention upon the interactions between development actions and their environmental consequences. In developed countries this has led to the public demanding that environmental factors be explicitly considered in the decision-making process and a similar situation is now occurring in developing countries.

Early attempts at project assessment were crude and often based upon Technical Feasibility Studies and Cost Benefit Analysis (CBA). CBA was developed as a means of expressing all impacts in terms of resource costs valued in monetary terms. Up to the 1960s, a number of major developments such as the proposed third airport at London and the Aswan Dam, which were assessed using CBA techniques, caused considerable public disquiet. Flaws in CBA became more apparent and one consequence was the development of a new evaluation approach which came to be known as Environmental Impact Assessment (EIA). It has evolved as a comprehensive approach to evaluation, in which environmental considerations, as well as economic and technical considerations, are given their proper weight in the decision-making process.

When EIA was first conceived, it was regarded as an "add on" component to CBA, and was designed to incorporate all those potential impacts that had proved troublesome in CBA (1). As a consequence, early practitioners used the system as a means of collecting information but often failed to comprehend the policy environment in which the development was proposed. Today, more fundamental questions are being posed such as:

a) Is the development required?
b) What are the relevant alternatives which could provide the same benefits and how do respective EIAs compare?
c) What is the appropriate level of public safety in relation to hazardous technologies?
d) What degree of environmental protection should be guaranteed for areas of significant ecological and landscape value?

EIA can, therefore, be thought of as a basic tool for the sound assessment of development proposals. Whilst EIA can play an important role in formulating environmentally sound policies and plans, and in their evaluation, this paper will concentrate on the application of EIA at a project level. Aspects of EIA such as procedural mechanisms, including selection of projects for EIA (screening), terms of reference (scoping), EIA methods and EIA reviews, have also been excluded from the discussion.

THE RATIONALE FOR EIA

Projects until recently were often formulated and assessed according to technical, economic and political criteria, and the potential environmental, health and social impacts of projects were rarely considered in a vigorous manner. Even when considered, such assessment usually took the form of cost-benefit analysis which crudely attempted to place a monetary value upon non-economic variables such as the destruction of marine ecosystems or the social and health impacts of air pollution. As a consequence of such restricted assessment, many developments resulted in unforeseen harmful impacts which reduced predicted benefits. An example is the Aswan Dam which created deleterious secondary effects such as a loss of agricultural productivity and a reduction in the importance of the Mediterranean fishing industry.

The purpose of an EIA is to determine the potential environmental, social and health effects of a proposed development. It attempts to define and assess the physical, biological and socio-economic effects in a form that permits a logical and rational decision to be made. Attempts can be made to reduce potential adverse impacts through the identification of possible alternative sites and/or processes.

There is no general and universally accepted definition of EIA. The following examples illustrate the great diversity of definitions:

a) "... an activity designed to identify and predict the impact on the biogeophysical environment and on man's health and well-being of legislative proposals, policies, programmes, projects and operational procedures, and to interpret and communicate information about the impacts" (2).

b) "... to identify, predict and to describe in appropriate terms the pros and cons (penalties and benefits) of a proposed development. To be useful, the assessment needs to be communicated in terms understandable by the community and decision-makers and the pros and cons should be identified on the basis of criteria relevant to the countries affected" (3).

c) "... an assessment of all relevant environmental and resulting social effects which would result from a project" (4).

d) "... assessment consists in establishing quantitative values for selected parameters which indicate the quality of the environment before, during and after the action" (5).

Such definitions provide a broad indication of the objectives of EIA, but illustrate differing concepts of EIA. The scope of EIA is clearly defined in only the first definition. Three definitions include socio-economic impacts, but in the last definition only the environment is mentioned. The definition of the United Nations Environment Programme (3) in b) above, implies that decision-making on the relative importance, based on local conditions, of beneficial and adverse impacts, should be part of EIA. The other definitions merely indicate that EIA is an "objective", technical and predictive exercise, with no decision-making component. For the purposes of this paper it is assumed that the main objective of EIA is to provide decision-makers with an account of the implications of proposed courses of action before a decision is made. Decision-makers are not usually involved in the assessment process, as this is normally considered to be a technical exercise. The results of the assessment are assembled into a document called an Environmental Impact Statement (EIS) which contains a discussion of beneficial and adverse impacts considered to be relevant to the project, plan or policy. The completed report, or EIS,is one component of the information upon which the decision-maker ultimately makes a choice. At this stage, other factors such as unemployment, energy requirements or national policies may influence the outcome of the decision. A final decision can be made with due regard being paid to the likely consequences of adopting a particular course of action, and where necessary introducing appropriate monitoring and auditing programmes to reduce any deleterious impacts that may have been identified in the assessment process.

THE ADVANTAGES OF USING EIA

EIA is a mechanism which aids the efficient use of natural and human resources and which has proved valuable to both those promoting developments and those responsible for their authorization. EIA may reduce costs and the time taken to reach a decision by ensuring that subjectivity and duplication of effort are minimized, as well as identifying and attempting to quantify the primary and secondary consequences which might necessitate the introduction of expensive pollution control equipment, compensation or other costs at a later date.

There are a number of ways in which EIA can improve the efficiency of decision-making, but to be effective EIA should be implemented at an early stage of project planning and design. It must be an integral component in the design of projects, rather than something utilized after the design phase is complete. Preferably, EIAs should be part of an incremental decision-making process which has a number of decision-points in the project planning procedure. This means that there can be a continuous feedback between EIA findings, project design and locations. EIAs can be implemented to test alternative project designs at an early stage, to help choose the project design which emphasizes benefits and minimizes harmful effects. EIA therefore can be used not only to investigate and avoid harmful impacts but also to increase likely benefits.

The emergence of an optimum alternative in terms of the objectives or goals relevant to a proposed project means that EIAs may have significant long-term financial advantages. If a potential problem is identified early in project planning it may allow considerable financial savings to be achieved. At the crudest level the abandonment of a project may be required if all alternative designs or locations are considered unsuitable in terms of likely detrimental effects. This would save capital costs. It is more likely, however, that design modifications may obviate the need for expensive ameliorating action once a project becomes operational. If a development is not assessed for its likely impacts, it may cause serious social and health problems. For example, a proposed dam and reservoir may have health effects which may require an expensive health care programme. The wrong location for a resettled population may result in agricultural failure and the need for food supplies to be sent to the relocated people from other areas.

The incorporation of EIA into decision-making may create a number of benefits. If a forecast of the likely impacts of developments is available, allowances can be made and the infrastructure can be provided in a manner whereby impacts are minimized. Where uncertainty exists as to future development, EIA can identify those areas most susceptible to adverse impacts and so guide site selection. To be effective EIAs can only be used when the alternative sites are few in number; otherwise EIA can be time-consuming and very expensive. EIA can, nevertheless, aid the identification of the most suitable site in terms of benefit maximization and reduction of harmful effects. Should no site be considered suitable, then the results of EIA aid the determination of broad environmental, social or health criteria to be used when a large number of sites are screened for their suitability. The relevance and importance of EIA for site selection has been recognized in a report entitled *Draft Guidelines for Assessing Industrial Environmental Impact and the Siting of Industry* (6).

THE SCOPE OF EIA

In principle, an EIA procedure should apply to all actions likely to have a significant environmental effect. The potential scope of a comprehensive EIA system is, therefore, considerable and could include the appraisal of policies, plans, programmes and projects. Lee and Wood (7) identified the benefits of a "tiered" EIA structure, in which the environmental implications of policies at different levels of government, were regarded as the "top level" of environmental decision-making. Such policies need not, in themselves, be environmental but they may have environmental consequences such as the subsidization of a fishing fleet. The potential advantages of a tiered approach over a procedure which is restricted to development projects are:

a) At the development project stage, the available options are often severely limited by earlier decisions made at a higher level. Mis-specification of a project assessment (e.g. a road scheme) may occur if the higher levels (e.g. transport policies) were not subject to such evaluation. The issue may be where environmental impacts will occur, and not whether they ought to.

b) Assessment of individual projects can only be conducted once proposals have been made. It cannot guarantee optimum site selection, and a thorough assessment of all alternative actions may be prohibitively expensive and time consuming (8).

c) The scope of viable alternatives decreases at the project level; also the willingness to contemplate alternatives may decline.

d) The available time for collection and analysis of environmental data will become increasingly restricted at the lower tier unless a programme of establishing environmental baseline data is undertaken independently of individual project EIAs.

e) When projects are individually small in size, but collectively large in number (e.g. housing), an EIA at the plan and programme stages may lead to a reduction in the time required for evaluation.

If plans are based on a sound environmental assessment, specific project proposals which fail to conform are likely to require detailed environmental assessment. The preparation of plans must be based on adequate data relating to the existing environment and the implications of possible changes. The systematic collection, analysis, storage and regular updating of such data must, therefore, greatly improve the quality of subsequent project EIAs as well as reducing time and costs. EIA and landuse planning ought, therefore, to be seen as complementary to each other.

THE RELATIONSHIP OF EIA TO THE DECISION-MAKING PROCESS

The purpose of EIA is to generate and provide information on the environmental consequences of a proposal. By itself it does not assure that adverse impacts are averted or minimized and beneficial impacts increased. If EIA is to be effective, it must be related in form and timing to the decision-making process in order that optimum strategies for averting or reducing adverse consequences can be formulated and evaluated (8).

Whilst the overall objective of an EIA is to safeguard the environment, it has often been regarded as a tool of advocacy by environmental interest groups. Such interests are, however, rarely the only or dominant criteria governing decision. The conclusions of an EIA must be presented in such a manner that they can be taken into account alongside other relevant economic and social factors and be as scientifically objective as possible.

A frequent objection to EIA in countries with well developed decision-making procedures, is that it imposes an additional stage in the process and may duplicate the existing system. Evidence exists, however, to suggest that EIA may be beneficial even in countries with a well developed decision-making procedure. By the systematic examination of environmental constraints and the development of measures designed to protect or conserve the environment, EIA prevents such considerations becoming a late addition to the decision-making process, thereby saving financial resources and time and also ensuring a more environmentally-sound project.

Responsibility for undertaking an EIA can be allocated in a number of ways, depending upon the particular EIA system planned or operating and the nature of the project. Four major alternatives that have been used are: the authorizing agency; the developer; shared responsibility between the authorizing agency and the developer; or an independent specialist body. Impartiality may be achieved in a number of ways:

a) Guidelines or minimum standards for the form and content of an EIA.
b) Supervision by a reviewing or controlling body with no vested interest in the project.
c) Mandatory consultation with relevant and competent organizations.
d) Publication and provision for public discussion of the impact statements.

PUBLIC INFORMATION AND PARTICIPATION IN EIA

Public involvement should be an integral part of any EIA system. Efforts should be made to obtain the views of, and to inform, the public and other interests who may be directly or indirectly affected by the project. The authorizing agencies may not always identify the environmental issues which the public perceives to be important and they may also lack the detailed local knowledge that the public possesses. Advantages of participation may lead to the provision of information about local environmental, economic and social systems; the possible identification of alternative actions; an increase in the acceptability of the project as the public will better understand the reasons for the project; and a minimization of conflict and delay. Problems may nevertheless arise. Public participation may, in the short-term, be time-consuming, and increase costs, and participants may be unrepresentative of the community. In spite of these potential problems, many countries are actively encouraging public involvement in EIA.

THE FINANCIAL ASPECTS OF EIA

Initially, EIAs may be expensive to implement, particularly in areas where little is known about existing environmental and social conditions. Design changes produced as a result of EIA findings may also increase capital costs, but it can be argued that savings to local, regional and national economies arising from the avoidance of deleterious impacts and from the maximization of beneficial impacts will outweigh the costs of an EIA system in the long term. The cost of an EIA system will decline once procedures and the techniques have been established and assessment personnel have become accustomed to their tasks. Indeed, it can be further argued that thorough investigation of impacts at an early stage of project planning may save money by speeding up the process of implementing a proposal. The costs of EIA are commensurate with the complexity and significance of the problem and the level of detail required. In many countries, the cost is borne by the proponent of the development, while in others it is borne by the authorizing authority. In those countries with considerable experience in EIA, the costs vary between 0.05% and 2% of the project value. It is misleading, however, to regard the "actual costs" of EIAs as being saved if an EIA is not undertaken, for if the information is required it will have to be collected by some means.

The benefits represented by EIA have usually not been determined, because it is difficult to assign monetary values to such benefits. Many of the environmental amenities that would otherwise have been degraded or destroyed have a unique value, which over time will far outweight EIA costs. Many cases show that the use of EIA has allowed the choice of an option which is both environmentally and economically superior to the original choice - the Alaskan Pipeline being a good example.

CONCLUSIONS

There is now a general recognition that environmental considerations ought to be integrated into the planning and decision-making framework, but differences exist as to the exact form that such integration should take. The administrative structures of the EIA process will also vary. Some countries will implement EIAs through legislative or administrative regulations whilst others will integrate it with planning or other authorization systems.

EIA can ensure that environmental aspects are given equal status with economic, technical and social considerations during the evaluation of development proposals. Attention must also be given not only to immediate impacts, but also to indirect, secondary and long term effects. It is necessary to stress the importance of a proper framework for deciding which project activities ought to be subject to EIA and that project EIAs are limited by decisions made at a policy or plan level. The evaluation of higher order decisions is important and this is where EIA is now increasingly focussing attention.

REFERENCES

1. O'Riordan, T. *Environmental Impact Assessment and Policy Review*, Paper presented at the IBG Conference, Lancaster University, January 1980.

2. Munn, R.E., ed. *Environmental Impact Assessment: Principles and Procedures*, SCOPE Report 5, 2nd edition (Chichester, England, John Wiley, 1979).

3. United Nations Environment Programme. *Draft Guidelines for Assessing Industrial Environmental Impact and Environmental Criteria for the Siting of Industry* (Paris: UNEP Industry and Environment Office, 1978).

4. Battelle Institute. *The Selection of Projects for EIA* (Brussels: Commission of the European Communities Environment and Consumer Protection Service, 1978).

5. Heer, J.E. and D.J. Hagerty. *Environmental Assessments and Statements* (New York, New York: Van Nostrand Reinhold, 1977).

6. United Nations Environment Programme. *Guidelines for Assessing Industrial Environmental Impact and Environmental Criteria for the Siting of Industry*, UNEP Industry and Environment Guideline Series, Vol. 1 (Paris: UNEP Industry and Environment Office, 1980).

7. Lee, N. and C. Wood. "EIA - A European Perspective" *Built Environment* 4(2)(1978)101-110.

8. Organization for Economic Cooperation and Development *Environmental Impact Assessment* (Paris: OECD, 1979).

ENVIRONMENTAL IMPACT ASSESSMENT FROM THE POINT OF VIEW OF A DEVELOPING COUNTRY

Kriton Curi

Boğaziçi University
Institute of Environmental Sciences
Istanbul, Turkey

INTRODUCTION

"Environmental Impact Assessment" although it became popular in the last decade, in reality was an activity performed under different names even from Bibilical times. Eve, for example, made a very rapid EIA after the snake recommended her to convince Adam to eat the "Apple of Paradise". She immediately considered the beneficial points of the proposal, namely the satisfaction which they were going to feel by eating the apple, and decided to put the plan into action. Her EIA, however, was incomplete. She considered only the short term, beneficial effects, not the long term, adverse effects. She did not foresee, for example, that the human race might become addicted to eating apples.

The failure of Eve was the first error made in predicting the effects of an activity on the environment, and has been followed by many other similar mistakes. The main reason for these failures has been the lack of an objective and scientific approach. The desire to systematize this type of evaluation gave rise to the modern EIA concept.

WHAT IS ENVIRONMENTAL IMPACT ASSESSMENT?

Although the term "EIA" is now widely used, it is still difficult to define. In particular, it is difficult to provide a single answer acceptable to everybody. Furthermore, most of the definitions given in relation to EIA use terms like "environment" and "impact", without making clear what is meant by them. Some of these definitions are given below.

EIA is ...

-"... an activity designed to identify and predict the impact on the biogeophysical environment and on man's health and well-being of legislative proposals, policies, programmes, projects and operational procedures, and to interpret and communicate information about the impacts" (1).

-"... the process of actually calculating what effect the proposed action will have on environmental quality" (2).

-"... an assessment of all relevant environmental and resulting social effects which would result from a project" (3).

-"... the systematic examination of the environmental, social and economic consequences of projects, plans and policies" (4).

As can be seen, the above definitions are far from "scientific". This is to be expected, given that EIA is in some respects an "art" as well as a "science". There are, however, some conditions that an EIA should fulfil. These are:

a) Objectivity. The information given should be as unbiased as possible.
b) Completeness. The information given should attempt to cover all the possible effects of an activity, and not only some of them.

The U.S. Environmental Protection Agency (5), for example, recommends that the following considerations be kept in mind:

a) The need for a description which is broad enough to provide a meaningful basis for evaluating both direct and induced impacts (which may be social and economic, as well as environmental).

b) The need to incorporate a basic understanding of the interrelationships between social, economic, and environmental elements.

c) The need to incorporate a basic understanding of the chemical, physical, biological, and other properties of all process discharges that determine behavior and impact, in sufficient detail for the determination of impacts.

d) The need to describe the baseline environment as it exists at present, and as it would exist in the near-term future (10 to 15 years) if the proposed alternative were not implemented.

e) The importance of considering the effects of cumulative impacts. The proposed project must be weighed against not only its own set of impacts, but also against those additional impacts which will arise because of interaction with other environmental stresses which cumulatively cause much greater impact. For example, a stream of a given flow and gradient has a certain ability to assimilate degradable materials discharged to it, by virtue of its capacity to reaerate itself. If the stream course is altered by dams, channels, diversions, etc., this capacity is greatly altered, with a likely outcome of severe damage to stream ecosystems and man's ability to use it.

f) The need to provide sufficient detail to allow independent impact assessments by public reviewers and to minimize delays in EIA review.

EIA AND DEVELOPING COUNTRIES

Developing countries, in comparison with the developed ones, have a basic advantage, the advantage of making use of the experience of the developed countries. In particular, in work aimed at environmental protection, they should be able to avoid repeating the mistakes of the developed countries. However, the protection of the environment often conflicts with short-term economic benefits. Developing countries are sometimes reluctant to request the preparation of an environmental impact statement (EIS) for the following reasons:

a) The preparation of an EIS may cause unacceptable delay of the development, due to lack of baseline data.

b) The cost of examining the environmental impact of a development properly may be too high, or may be considered an unbearable burden on the economy of an underdeveloped or developing country.

c) The politicians, who are also the decision-makers in most developing countries, may not be interested or may not even be aware of the possible detrimental effects of an activity. Alternatively, the belief that a proposed development will improve their chances of re-election may cause them to disregard the long-term effects of the development, and not to favour the preparation of an EIS.

d) There is an awareness, in developing countries, of the risks of bringing in "foreign experts", usually from an international organization which is supporting the development financially, to conduct the EIA. Often these "experts", not knowing the local conditions, and unable to properly understand or evaluate the needs of the local people, prepare reports which are wrong or misleading.

The reluctance to study environmental impact sometimes has undesirable and even catastrophic consequences.

AN EXAMPLE OF FAILURE TO ASSESS ENVIRONMENTAL IMPACTS

Ignorance, combined sometimes with wrong decisions, as well as an absence of any legislative obligation to conduct a detailed study of the possible impacts of a development or an activity, may have disastrous results. An example of such a case is the following event which took place in Istanbul.

Baruthane Valley is a valley located close to a densely populated district of Istanbul. Below the bottom of this valley a huge sewer line (2.4 m in diameter) carried the sewage of the nearby residential-commercial area (Figure 1a). The Highways Department decided to make a boulevard along the valley, and to do so filled one side of the valley (Figure 1b). This was followed by filling the other side in order to reclaim some land to enlarge an adjacent cemetery (Figure 1c). Finally, the Municipality of Istanbul, which was in search of land on which to dispose of its solid wastes, decided to fill the remaining part of the valley (Figure 1d). In the meantime some hundreds of *gecekondu* 1) were built in the surrounding area. The inhabitants of these slums suffered the first unfavourable impact of the development. In the summer of 1978, several explosions, caused by methane gas generated in the landfill, took place and killed a number of slum-dwellers. This was followed by another unpredicted impact. The sewer line below the valley, unable to carry the load imposed on it by the fill material, which had reached a depth of 30 m, collapsed and clogged. Thus sewage started to accumulate, unnoticed at first, behind the artificial dam of fill (Figure 1e). Finally, in December 1979, after heavy rain, the area was filled entirely with a mixture of sewage and rainwater, to a depth of 20 m. Several slums were flooded, some of them disappearing entirely below the water. (6, 7)

THE NEED FOR EIA IN DEVELOPING COUNTRIES

The incident examined above is enough to indicate that EIA studies are necessary in developing countries. Wherever possible, these studies should not interfere with or delay developments. They are essential in order to make decision-makers aware of the beneficial and adverse effects of developments on the environment, and thus to assist them in taking informed decisions. If potentially detrimental effects are known in advance, preventive measures are

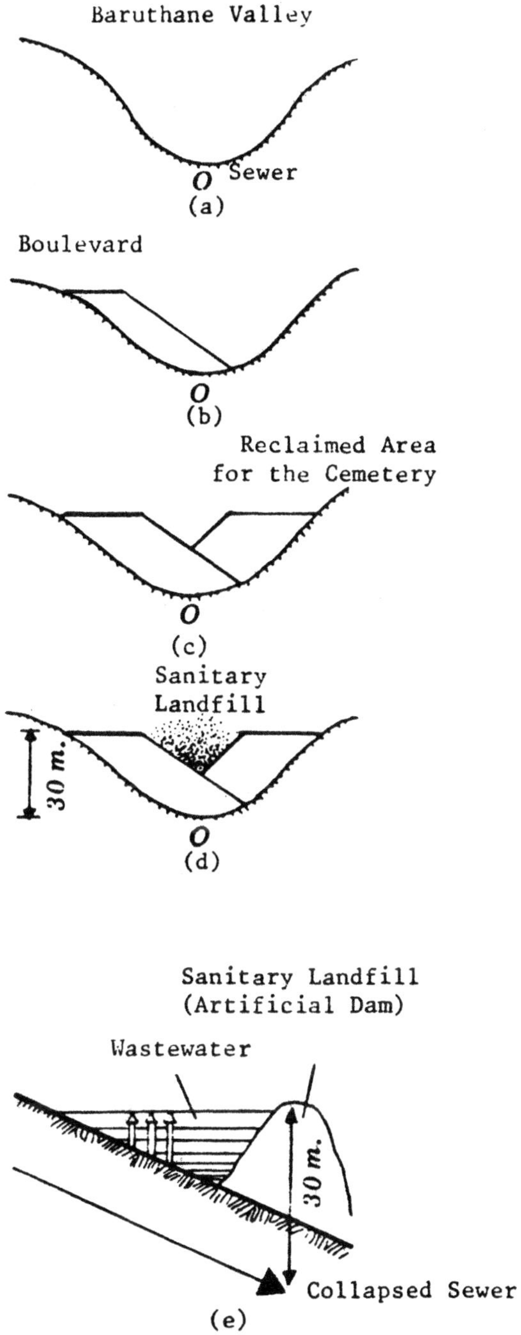

Figure 1: Development in Baruthane Valley
(a) - (d) are cross sections;
(e) is a longitudinal section.

possible. Something which should be kept in mind, however, is that existing EIA methods are sometimes incomplete and unscientific. Developing countries, instead of accepting and using EIA as an "imported technology", should try to form their own interdisciplinary teams of scientists to develop the most appropriate technology for their own conditions.

FOOTNOTE

1. *Gecekondu* is the name given to slums in Turkey, meaning "built in a night".

REFERENCES

1. Munn, R.E. *Environmental Impact Assessment,* SCOPE Report 5, 2nd edition (Chichester, England: John Wiley, 1979).

2. Vesilind, P.A. *Environmental Pollution and Control* (Ann Arbor, Michigan: Ann Arbor Science Publishers Inc., 1979).

3. Battelle Institute. *The Selection of Projects for Environmental Impact Assessment* (Brussels: Environment and Consumer Protection Division, Commission of the European Communities, 1978).

4. Clark, B.D. *Environmental Impact Assessment: Background and Objectives,* paper presented at WHO Regional Seminar on Environmental Impact Assessment, New Delhi, 1979.

5. U.S. Environmental Protection Agency. *Environmental Impact Assessment Guidelines for Selected New Source Industries* (Washington, D.C.: U.S. Environmental Protection Agency, 1975).

6. "Sel Değil, Kanalizasyon Suyu", in *Cumhuriyet* (daily newspaper), Istanbul, 15 December, 1979.

7. "Baruthane Deresi Taşti, Halk Evlerini Boşatti", in *Hürriyet* (daily newspaper), Istanbul, 12 December, 1979.

SELECTED NATIONAL PROCEDURES

THE U.S. MODEL OF ENVIRONMENTAL IMPACT ASSESSMENT

C. P. Wolf

Department of Social Sciences
Polytechnic Institute of New York
Brooklyn, New York

Along with the invention of the future, as something we can know and do better about, and the invention of the public, as distinct from the electorate, the invention of the environment as an object of generalized social concern distinguishes the intellectual history of the 1960s. These three converge in the analytical and institutional system of environmental impact assessment (EIA), a major social innovation of the 1970s usually thought to have originated in the United States. Futurism enters in the prediction of future environments and environmental problems, and the conditions which will likely result from a proposed environmental modification. The public (or publics) enter as active participants in the formulation of environmental issues and the expression of environmental concerns. EIA is a rational means of systematically relating these conditions and concerns in a process of collective decision and action.

This chapter reviews why EIA was introduced, when it is undertaken, and whether it is achieving its objectives. The social impacts of EIA on state and local political processes are also examined, including public awareness of environmental issues and the effectiveness of EIAs in influencing decisions on intended actions. Finally, the unanticipated consequences of EIA procedures--beneficial and deleterious--are considered.

THE ORIGINS OF EIA

Just why EIA arose when it did is an unsolved problem in the sociology of knowledge. No precipitating "crisis event" such as the Santa Barbara Channel oil spill of 1969 (1) suffices to explain the upsurge of concern. The history of environmentalism in the U.S. goes back at least to the Progressive Era, around the turn of the century. The conservation movement which emerged at that time was instrumental in securing the establishment of our system of national parks and national forests. Its intellectual ancestry was the "frontier hypothesis" of historian Frederick Jackson Turner, who perceived that America was ceasing to be a land of boundless expanse and limitless resources. Despite the environmental disaster of the 1930s (2) and the eloquent warnings of Barry Commoner in the 1950s, not until the decade of the 1960s did concern over environmental pollution become widespread.

The modern environmental movement can be dated from the public response to Rachel Carson's Silent Spring (3) in the early 1960s. As its title suggests, that book made the connection between pesticide use and biological effects on wildlife. More important, it aroused fears of systematic contamination of food and water supplies destined for human consumption. The linkage between environmental hazards and personal health has been potent in raising public apprehension and mobilizing action. An early precursor was Upton Sinclair's novel, The Jungle, set in the stockyards of Chicago. While Sinclair intended it as an indictment of the whole capitalist system, the popular revulsion fastened on passage of the Pure Food and Drug Act. As he later complained, "I aimed for their hearts and hit their stomachs." It is this same linkage which perhaps accounts for the current popularity of "diet" books--in a recent week, four of the five nonfiction best sellers were of this genus--and for other manifestations of "personal fitness" such as natural foods and running.

The culmination of the environmental movement was Earth Day 1970, an event which carried "counterculture" hedonism of the Woodstock variety into the realm of social activism. The ground had been prepared by the anti-war movement, protesting U.S. involvement in Vietnam; organizational skills learned in that effort were put to good account by environmentalists. Indeed, the "environmental crisis," as it was sometimes pictured, appeared to be a repudiation of the same adult values, legitimating the attack by the youth generation. A complicating factor was the civil rights movement, in which black militants displaced middle class white activists. Added to this was the cumulative exposure to environmental damage caused or threatened by development projects of all descriptions--dams, highways, nuclear reactors, urban renewal, and the like--in localities across the entire country (4).

Movement dynamics had their counterpart in the policy arena. A ten-year legislative struggle led to enactment in 1969 of the National Environmental Policy Act (NEPA) which authorizes EIA for "major federal actions." This landmark legislation does not stand alone, of course; it is extended by other significant acts, such as the Clean Air Act, the Clean Water Act, and the Toxic Substances

Control Act. Amplifying this legislative framework are the enabling regulations and guidelines of executive agencies; their legality and applicability has been repeatedly contested and attested in litigation and judicial decision (5).

Although NEPA pertains only to major federal actions, a broad construction shifts its weight to non-federal actions, e.g. through revenue-sharing in Community Development Block Grants, and to non-governmental actions as well, through the complicated tangle of public-private sector relations. Many states and even a few municipalities have enacted "little NEPA" laws as well--the California Environmental Quality Act, for example (6, 7). Nor does the institutionalization process end there; besides the legal authority for EIA, there must also be the intellectual resources to fulfill its requirements. Since the early 1970s, when no one knew what an environmental impact statement (EIS) should contain, much less what was "adequate," a small industry of consultants and other environmental professionals has formed to provide the requisite expertise. Completing the process is the involvement of potentially affected publics as mandated in revised Council on Environmental Quality (CEQ) guidelines and other authority (8).

THE "ENVIRONMENT" OF EIA

The U.S. model of EIA partakes of the peculiarities of the "environment" from which it arose, a decidedly mixed system of compromises and contradictions, conservatism and innovation. It is a confusing, even chaotic, picture to foreigners, and to not a few Americans themselves. It is also a situation with great potentiality for creativity and experimentation. In characterizing the institutional system of EIA in the U.S., these terms come to mind: pluralism, legalism, capitalism, pragmatism, and scientism. Each will be taken up in turn.

Pluralism

Although the impetus for EIA came at the federal level, many other actors and interests are involved in its implementation. For example, "legalism" denotes the legal system as a leading institutional actor; "capitalism" indicates the market orientation conspicuous in much public policy, especially under the Reagan administration; and "scientism" implies the role of professional communities. Within the federal system itself are overlapping and often conflicing jurisdictions and constituencies. Typical of this diversity is the problem of "federal consistency" between federal and non-federal (state and local, for the most part) interests. Even on the federal level conflicts of agency interest occur, say between the economic development and environmental protection of barrier islands, or the resource extraction and environmental preservation of public lands. The application of uniform federal standards to diverse local conditions guarantees variability in administration; extralocal and interstate sources of environmental pollution dictate a federal presence.

Legalism

The ultimate arbitrer of conflict in U.S. society is the legal system. Because of great and growing diversity, legal recourse is taken in the absence of other mechanisms for conflict resolution. While our planning models appear to require consensus in order to operate, situations to which they are applied often contain irreconcilable differences. Environmental litigation has expanded in proportion to environmental legislation, so much so that a system of environmental courts was seriously proposed. There are contradictions within the legal system itself; the same jurisdictional fragmentation that besets legislative assemblies and executive agencies at all levels of government is reflected in the courts.

Litigation often touches on technical matters the legal system cannot easily process. While procedural questions are readily decidable, their substantive adequacy is often beyond legal review (9). Were this not the case, the courts would be placed in the untenable position of prescribing what "good science," say the proper method of computing a benefit-cost ratio, might be. A system of "science courts" has been proposed to handle such technical questions, but in any case the problem of relating them to value (policy) questions would remain. Because of these inherent difficulties, legal recourse is increasingly viewed as a "no win" situation for all parties at interest. The alternative of environmental mediation (10) has become correspondingly attractive, though as yet not widely practiced.

Capitalism

Market solutions are often sought in the implementation and enforcement of environmental regulations whose basis is in NEPA or related legislation, e.g. in the "bubble concept" for auctioning rights to air pollution among industry bidders (11). Mainly these consist of "incentives" (subsidies) offered in exchange for regulatory compliance. Of course, many industries and firms make their profits precisely by not "internalizing externalities." In some states--Illinois, for instance--impact statements are prepared on the burden of environmental regulation; the other side of the equation, benefit estimation, has been more difficult. Transfer effects have raised questions of equity across class and political boundaries. A bargaining game to negotiate the timing of industry's compliance with regulatory standards has often resulted in "stetched-out" deadlines. None of this is meant to imply that enforcement is superior in centrally planned economies (12).

Pragmatism

The essentially non-ideological character of American politics has long puzzled foreign observers. Our politics serve more to unite than to divide us; as the "art of the possible," politics means compromise--"rising above principle" to secure even "half a loaf." When it comes to environmental matters, "un-politics" (13) seems an apter description. There is no Ecology Party in the U.S. and environmental issues have never been seriously in dispute between the two major parties. On occasion, state and local issues have received political attention, but again seldom as a matter of party politics. NEPA and related laws enjoy substantial bipartisan support.

Although there seems to be a selective affinity between liberal Democrats and environmental values (14), the identity of interest between conservatism and conservation is not infrequent--Senator Barry Goldwater and the late Charles A. Lindbergh, for two. Neuhaus (15) and other populist critics would contend the common denominator is elitism, a charge levelled against environmentalists generally (16). Conversely, the success of the environmental movement did not produce disaffection and alienation on the part of any significant minority; indeed, environmental values continue to receive a high level of popular endorsement (17).

Scientism

The final term to be considered in trying to account for the advent of EIA is "scientism," in the sense of developing a science-based system of environmental standards. "Empiricism" might be a more descriptive term. The advancement of environmental science and of pollution control technology have made this position available; the inability to gain value consensus has made it acceptable. Despite the legislative accomplishments of the past decade and the strong popular support for environmental values, it cannot be said that the political system has been effective in articulating and aggregating environmental issues. When it comes to a choice between "jobs versus environment," for example, public opinion divides more evenly. No satisfactory mechanism for making "trade-offs" has been devised.

Because of our seeming inability to reach value consensus on environmental issues, there is a pronounced tendency to convert value questions into scientific ones. No amount of measurement of environmental indicators will yield an answer to the question of what is "adequate" environmental protection, however. That is a value question, and it must be answered in value terms, consensually validated. Failing that consensus, we are left with a patchwork of environmental standards that are science-based to some extent--there can never be enough scientific research to establish them conclusively--but not values-based. A viable and defensible system would, of course, require both.

It would seem then that EIA is a "distinctively" American product, and that is sometimes held against its adoption in other developed and developing countries. While the environmental science may be the same, though applied to diverse environmental settings, the institutional systems may be quite divergent. For example, how might one do EIA under conditions of martial law? The answer to that might be the same as to the question of how porcupines make love: carefully. Moreover, the enactment of NEPA must be viewed against a background of U.S. national planning efforts that is quite weak compared to other countries. Our achievement may appear less imposing when viewed from this perspective. Nevertheless, EIA has evoked worldwide interest, just as technology assessment on the American plan is widely admired, though so far little emulated. That is not to say the U.S. model of EIA is without faults and flaws; we will come to an examination of its structural defects shortly. In any case, EIA is being adopted by growing numbers of countries and adapted to their particular circumstances. Not so apparent is what "appropriate" EIA procedures and practices might comprise under these varying environmental and institutional conditions.

WHY EIAS ARE UNDERTAKEN

The simple answer to why EIAs are undertaken is compliance with environmental laws and their enabling regulations. For example, deciding what constitutes a "major" federal action usually involves such criteria as the size of public investment, whether any population displacement and relocation is likely, and the degree of public controversy a proposal incurs. Local opposition, often joined by national "public interest" groups, may be instrumental in forcing EIA by appeal to judicial review. Legal threat is often a sufficient forcing action.

Underlying these administrative and social criteria are the environmental perceptions and values of which NEPA is a tangible expression and practical application. Whereas compliance is simply a procedural matter, the values which NEPA and related legislation embody are of the essence. In principle, EIA should be undertaken to inform and to help resolve <u>any</u> environmental policy issue, from ozone depletion to farmland preservation. In practice, environmental values must compete and accommodate with others, particularly economic, and no societal consensus exists on what is "acceptable" environmental quality, what is the proper relation between environmental information and decision-making, or what the balance should be between conflicting values.

The Social Orientation of NEPA

The manifest intent of NEPA and similar legislation is protection and maintenance of the natural environment. While NEPA is not omnibus social welfare legislation, and EISs are not required for programs such as manpower training (except insofar as they involve a physical alteration of the natural environment, such as facility construction), there is a human dimension both in the social interpretation of environmental conditions and in the value of human activities which impinge upon them.

The social orientation of NEPA is found in its stated purpose "to encourage productive and enjoyable harmony between man and his environment and stimulate the health and welfare of man. . . NEPA further recognizes "the profound impact of man's activity on the interrelations of all components of the natural environments" and prescribes the use of "all practicable means and measures . . . to create and maintain conditions under which man and nature can exist in productive harmony and fulfill the social, economic, and other requirements of present and future generations of Americans." The means and measures cited, including those "which will insure that presently unquantified environmental amenities and values may be given appropriate consideration," entail an interdisciplinary approach "to insure the integrated use of natural and social sciences . . . in decision-making which may have an impact on man's environment." Although partial to environmental concerns, NEPA is clearly not purist in promoting them ahead of all other social concerns. Rather than a pristine environment, the objective is one of advancing towards social goals in ways that are compatible with environmental values. McHarg's (18) admonition to "Design With Nature" captures this concept of balance between natural and human environments.

EIA is nothing if it is not an effort to engage the human--that is, social--consequences of environmental modifications. Those consequences are themselves the products of actions by human agents. Quoting Robert Sommer, Harvey (19) observes, "The sensibilities of mankind cannot remain permanently immune from the environmental changes man is bringing about through his own actions. It is, therefore, salutary to remind ourselves occasionally that 'the long-range question is not so much what sort of environment we want, but what sort of man we want.'" The social interpretation of EIA is cogently argued by Pattison (20):

> "Any assessment of environmental impact--to be meaningful--must necessarily be built upon assumptions as to public attitudes as well as on technical findings. Our benchmarks for clean air, clean water, or clean streets have less to do with a definable "degree above zero" than with what various members of the public

consider acceptable or attainable. No matter how accurately the probability and magnitude of, say, certain fish kills are quantified, the impact of such kills is essentially subjective, loaded with emotional factors not amenable to cost-value analysis. It is the impact of a particular set of findings or predictions on the minds of men--not the impact of the pollution on the environment *per se*--that is our key unknown.

"Thus it would appear that those responsible for environmental impact studies really should begin with attitude measurement in the affected communities. This might well call for a scientific opinion profile of a representative sample of citizenry to determine their levels of concern for changes in the environment, good or bad, from the standpoints of health, economics, recreation, and aesthetics.

"Only with such a public opinion study at hand can the significant impact of probable environmental changes be forecast. . . . Without such an analysis to set a baseline for an impact study, the findings of scientists and engineers will never satisfy the ecologists, the anti-ecologists, or the folks who pay the bills."

The social definition of environmental quality underscores a reciprocity of human environment and human experience. It is the *social* environment which is our experience and expression. Environment is the carrier of human value; it is shaped to human purpose. "Environmental quality" must therefore resolve to "quality of the human environment." Correspondingly, the aims and the achievements of environmental planning must be to enlarge and enhance the quality of life. A good deal of the intellectual work required of social scientists is to effectively formulate these "quality of life" criteria and to accurately gauge their indicators. Both are needed, to deepen its conceptual understanding and to refine its operational measurement. NEPA provides the institutional context in which this social knowledge and social research can make a positive contribution. The EIA process it establishes is a powerful social technology for determining and regulating the social causes, conditions and consequences of environmental modification.

WHETHER EIA IS ACHIEVING ITS OBJECTIVES

EIA would appear to hold considerable promise for improving environmental quality as an integral part of the overall quality of life. Does its performance fulfill this promise? In answering this question, we must address some others:

1. Is EIA effective in focusing environmental concerns?

2. Is EIA a means of achieving environmental policy objectives?

3. Are future environmental problems being identified and forestalled?

4. Is the quality of environmental planning, design, decision-making, and management improving through the use of EIA?

5. Is the environment getting better in ways that can be attributed to EIA?

In brief answer to each:

1. In general, EIA has been a powerful environmental problem focuser, so much so that the EIS format has been adopted or urged in a wide variety of other policy contexts: inflation, regulation, arms control, energy, the judiciary, mental health, the family, urban development, and so on. EIA requirements themselves have been extended to international development proposals and even to weapon systems (e.g. the MX missile mobile basing system), as well as to state and local actions and to private development proposals which intersect governmental operations. Of course, there are still a number of areas in which "impact statements" are not mandatory, such as the equity impacts of budget programs--disinvestments as well as expenditures. While EIA procedures in other countries may not involve the same kind or amount of documentation, it has become customary in the U.S. to place heavy reliance on EIS preparation. The adequacy and quality of these documents is eminently debatable however.

The appeal of "impact statements" is partly due to the forcing action inherent in NEPA. Not only does NEPA provide--it demands--an occasion on which formal accounting is made of proposals for environmental modifications and their possible consequences. In doing so, it creates the expectation that these findings will weigh heavily both in decisions to proceed and in procedures for carrying out approved proposals. Another reason for the popularity of the EIA format is that whenever the American political system discovers something that halfway works, the temptation is to try to make it work for everything. Perhaps the best example is--or was, before "Reaganomics"--the tax system. Partly, too, the reason for wanting a proliferation of "impact statements" is to offer a rational, or rationalized, means for competing with the concerns which NEPA promotes. All this presupposes that EIA is "working" effectively. That impression must now be qualified.

2. In the U.S. tradition, there is but weak connection between environmental policies and environmental actions. For example, Senate Document 97 (21) proclaims that "Well-being of all the people shall be the overriding determinant in considering the best use of water and related land resources." The result of water project developments such as Garrison Diversion resembles more a scramble for special advantage than the public interest, however. The disjunction of means and ends long predated NEPA, and perhaps it is unfair to hold EIA accountable for its correction. Nevertheless, this separation of public goals and actions not only impedes progress toward meeting national needs but actively undermines public trust.

The underlying difficulty is that EIA is normally conducted at a project level, far downstream from the point at which policy questions are settled. In the case of nuisance facility siting, for instance, the environmental issue is not so much where to put it as whether we want it at all (and we don't). How local concerns can engage policy questions decided elsewhere, and whether their interest is entitled to prevail over some collective definition of regional or national interest, have yet to be resolved. Moreover, project development tends to be highly site-specific, and such policy guidance as may be available, say from CEQ, is usually too broad for easy translation to local conditions. Appeals to "move up" EIA in the planning process--the National Association of Environmental Professionals makes that a tenet in its Code of Ethics--are primarily aimed at approaching closer to the level of policy decision. Proposals for "tiering" are likewise calculated to increase the policy influence of EIA procedures. It is also at the policy level that competing agendas, such as economic development, must be negotiated; again the result is a dilution of environmental policy objectives.

A related problem is the range of policy questions that can be successfully engaged through the EIA process. Does fossil fuel combustion portend a catastrophic increase in atmospheric CO_2? At present there is no direct way in which the National Energy Plan can be subjected to EIA review. While the CO_2 problem is still highly speculative, the SO_2 one is not; acid precipitation has become not only an interregional problem but also an international one. Yet coal-fired plants are typically assessed on a case by case basis, a condition which in highway planning is called "segmentation" (assessing isolated stretches of road while disregarding their connections). Efforts have been made to raise the level of assessment by means of "programmatic" or "generic" impact statements on regional and national scales. Clearly though, improvement is needed in systematic EIA review across policy, program and project levels of assessment (22). Although project modifications have sometimes resulted from EIA, never have environmental policies been formulated or revised on the strength of these findings.

3. Future environments and environmental problems have been previewed on both national and international scales (23, 24). Until placed in the context of EIA, however, the systematic relation of proposed actions to future conditions will remain indefinite. Partly this is a matter of policy failure or inattention; long range environmental policies are elusive even when they have serious near-term consequences, e.g. the disposal and management of radioactive wastes. The same forces that militate against far-futures planning operate in this policy arena as well. This is understandable in light of constitutionally limited political authority, but the potential of EIA as a tool for constructing alternative futures is correspondingly diminished.

The EIA process is future-oriented in respect to impact prediction, but that can occur only within the scope of specific proposals. Although NEPA calls for the assessment of cumulative impacts, rarely has this been achieved in EIA practice. Similarly, NEPA's emphasis on the "irreversible and irretrievable commitment of resources" has been narrowly construed in most project appraisals. EIA is not even accurate in tracking the changes that often accompany project development between the proposal and implementation stages, much less monitoring the actual as against the predicted impacts and imposing mitigation measures. Far from being a kind of general "foresight provision," EIA's time horizon is near-term, limited by the expected lifetime of the project. In the case of nuclear development, this is arbitrarily projected over a forty-year time frame. The effects--and economics--of decommissioning have scarcely been considered to date, except for the accident at Three Mile Island. As a form of applied futuristics, EIA so far is woefully deficient.

4. The general complaint that EIA lacks timeliness, occurring as it does too late in the decision process to influence the shaping of environmental policy, has already been voiced. This is but one aspect of the institutional system of environmental control. Ideally, EIA would enter at all phases, from planning through management. The relation to planning might be to conceive initiatives and alternatives for achieving environmental quality goals, either directly through programs for environmental enhancement or indirectly through avoidance or alleviation of environmental damage from other project development. In the design phase, EIA could suggest alternatives (sites, structures, technologies, etc.) to moderate environmental damage.

Only through consideration of these objectives and alternatives could informed policy decisions be reached. Yet there is no clear proof that EIA has had such influence on decision-making. Partly this is due to the lack of a definite relationship between environmental information, such as the EIA process might generate, and the decisions logically based upon it. Even the criteria for evaluating decision quality are for the most part lacking. Finally, the relation of EIA to environmental management lies in the realm of mitigating adverse impacts and monitoring the environmental soundness of project development. One difficulty here is that frequently there is a shift in agency responsibility between the construction and operations stages of a project, e.g. from a federal to a state level.

In mitigation of EIA, it does seem that environmental information is taken seriously as a constraint--as a potential barrier to project development that must be surmounted. Its influence in this respect is more negative than constructive. In general, it is difficult to demonstrate that EIA is producing information on which to base environmentally sound decisions, or that EIA is an effective tool for environmental planning, design, decision-making and management. The trouble may be, as G. K. Chesterton said of Christianity, "not that it was tried and found difficult, but that it was found difficult and not tried."

5. The "bottom line" for evaluating the effectiveness of EIA is in environmental quality improvements that can be ascribed to its utilization. To confirm this would require a time series of environmental indicator measures and periodic "state of the environment" audits of indicator performance compared to environmental standards such as "swimmable, drinkable" water. For many years, the only consistent environmental reporting was done by the National Wildlife Federation, a voluntary association. In recent years CEQ has worked toward constructing a system of environmental indicators, and on these measures some perceptible gains in environmental quality have been registered, especially in some areas of water and air quality. That incipient system has now been dismantled, with the reorganization of CEQ and the downgrading of its responsibilities by the Reagan Administration. In any case, the contribution of EIA to improving environmental quality could only have been inferential. Nevertheless, there is reason to believe that EIA has curbed the worst abuses of callous bureaucrats and rapacious developers. It is widely recognized today that environmental quality can no longer be taken for granted. EIA has played an essential part in transforming that realization into practical action.

Structural Weaknesses of the U.S. Model

The discussion to this point reveals a number of obvious weaknesses in the way EIA business is done in the U.S. These have been commented upon repeatedly, but as yet without much relief or remedy.

Given NEPA's stress on environmental quality maintenance, an expectation that EIA results would prove decisive seems reasonable. The indefinite relation of environmental information to decisionmaking is predictable however from the standpoint of competing interests and agendas which necessarily enter into the policy process. The same may be said regarding the "timeliness" problem--that EIA comes after the fact of policy decision rather than serving as a sound basis for reaching it. The type of environmental information EIA generates is also questionable in terms of its adequacy and quality (25). At present, EIA is a one-time, "once through" procedure, whereas the proposals under review continuously evolve. Seldom is an environmental monitoring system coupled with project development to provide an accurate reflection of ongoing change and a check on the accuracy of impact prediction.

Another main criticism of EIA under NEPA is the advocacy role of proponents for a development proposal under review. The proponent is normally designated as the "lead agency" responsible for EIS preparation, immediately casting doubts on the impartiality of the process. This built-in agency bias prompts frequent appeals for "third party" or "disinterested" assessment, patterned, say, on the review panel model in Canada. Equalizing the resources available to various parties at interest, e.g. the funding of public intervener groups, is an alternative solution. If the adversary system is to function equitably, there must be at least a rough parity among the parties. Disinterested assessment incurs its own structural problems, however. Who is to pay for adequate assessment work if not the proponent? Or, if the proponent pays, what guarantees can there be that this will not distort the process? Whether the burden of proof lies with proponents or opponents is a related issue.

Based on U.S. experience, the chief problems of EIA appear to be more institutional and political than methodological. While I have emphasized the difficulties and deficiences surrounding EIA it should be realized that before NEPA there was nothing even approximating a coherent system of environmental controls. If EIA has its defects--and certainly it does, in abundance--they are the defects of its virtues. It is scarcely conceivable that the process could be dispensed with without risking irreparable harm.

THE SOCIAL IMPACT OF EIA ON LOCAL POLITICAL PROCESSES

Has EIA heightened public awareness of social and environmental issues? Certainly it has provided a focus, a forum, and a format for addressing and expressing local concerns. The focus is provided by the "forcing" action of NEPA; proponents must prepare and defend a case that proposed developments will not seriously degrade the environment. The EIA process allows for public comment on these arguments. The EIS identifies the issues and assembles the evidence on which evaluations of predicted impacts are presumably based. So far so good, but as we have seen, EIA incurs the liabilities of the U.S. political process just as it draws from its strengths. Environmental concerns are not determining factors in all public policy decisions; the extent of public involvement in the EIA process is usually quite modest; and the EIS, as a "full disclosure" document, bears only a tenuous relation to the eventual outcome.

Perhaps the greatest social impact of EIA has been in the area of public participation itself. Under revised CEQ guidelines lead agencies are explicitly charged with involving "the public." This implies going beyond the 45-day "public review" period for receiving comments on a draft EIS, which in any case is too abbreviated for the careful examination of voluminous and highly technical reports. It means transcending the often acrimonious "public hearing," at which fixed positions are rehearsed in an adversary setting. First of all, it means finding out who "the public" themselves are and what their concerns and greviences might be.

In the past, local concerns were relegated to the "normal politics" of elected officials and representative assemblies. Current provisions for public involvement clearly surpass these traditional structures and functions. Whoever "the public" may now be, they are no longer the electorate. This momentous change has come about in no small part because it has been unclear whom the representatives were representing. This is a radical departure from conventional wisdom, and perhaps raises as many questions as it settles. It turns out, for instance, that the public are not singular but plural, and that the decision situation faced is one of multiple publics pursuing multiple purposes and applying multiple criteria of evaluation.

From this diversity and complexity, a unitary decision is supposed to emerge. As the Chief of Engineers so rightly has said, however, "You can't please everyone." Since "pleasing everyone" is a practical impossibility in most impact situations, a richer bargaining game is required in which everyone--though not entirely pleased--at least gets something.

As matters now stand, it is unclear <u>which</u> expression of public preference should be accorded the greatest weight and how its representativeness should be judged. In cases such as predator control and wildlands preservation, do local interests outweigh regional and national ones? Where does <u>the</u> public interest--which all federal agencies are charged with upholding--reside? Sensitivity to local conditions is urged on agency officials, yet how considerations of equity and fairness can be honored under such conditions is far from obvious. Doing equal justice does not necessarily mean treating everyone the same.

Confronted with these dilemmas, public officials have sometimes been inclined to dismiss the feasibility and desirability of genuine public involvement. The publics' interests are discounted as ignorant and erroneous, despite a long history of effective citizen efforts in environmental affairs (26). For there to be a realistic expectation of effective public involvement, however, there must be a basis for participation. Failure to create such a basis has led to misguided allegations of public "apathy" and to rejection of the "irrational fears" of an aroused citizenry. Above all, inviting in the public implies their involvement will somehow be "meaningful." That can only mean the sharing of power affecting public choice. Since the basis of representation has not been clarified, this amounts to an abrogation of political responsibility on the part of officials who at best are not overly anxious to relinquish their own discretionary power. Even when successful, public involvement incurs the equal and opposite charge of "cooptation" and "public relations."

Despite these evident difficulties, public participation has become an indispensable instrument of democratic governance. It is not as though elected officials are eager to assume responsibility for environmental policy choices and their implementation in development projects. Evasion of accountability is the more likely expedient, just as political rationality dictates the benefits of a resource allocation decision be captured and its disbenefits be "externalized," where equity considerations would imply their internalization. The necessity for public involvement has arisen because of the erosion of political support at the local level, where project impacts are likely to be most severe. At the same time, public involvement also must contend with the structural contradiction that its episodic and unsystematic attention to issues cannot be corrected without undermining its voluntary character.

UNANTICIPATED CONSEQUENCES OF EIA

Whatever the congressional intent, EIA has become unexpectedly potent in asserting environmental interests (27). In turn, this has generated countervailing pressures from competing interests, who tend to view the relative success of EIA as an inversion of national priorities. One form these counterpressures have taken is an attack on "overregulation" and a demand for "regulatory reform." This environmental "backlash" has so far not materialized in a rollback of national legislation, although there has been a marked dilution of regulatory enforcement under the Reagan administration. A crucial variable is the independence of the judiciary in affirming environmental law. Although Attorney General Smith has opined that the courts are overzealous in their desire to enforce the laws of the land, notably NEPA, previous efforts to subject the judicial system to political control have proved disastrous. Environmentalists are waging an active counterattack, e.g. the petition to "recall" Secretary of the Interior Watt, and it seems likely that the political sensitivity on environmental issues will at least restrain efforts to "streamline" EIA in areas such as nuclear development.

One topic on which virtual unanimity exists is that of "balance" in national priorities. While the demand for balance, e.g. between environmental protection and economic development, is ubiquitous, no plausible means have yet been found to determine where that point lies or how to reach it. In some policy areas economic considerations have been deliberately subordinated, however, as in the Rare and Endangered Species Act. On the other hand, the Tellicoe Dam example suggests that an unglamorous endangered species such as the snail darter has its price. (Parenthetically, the economic justification of this project has been condemned by several disinterested observers, suggesting that in this as in other water resource projects, national economic efficiency is still vulnerable to special interests.)

Another consequence of EIA, which in retrospect does not appear so unexpected, has been its "perversion" to obstruct projects on grounds seemingly extraneous to NEPA. One such case was the proposed construction of a large prison facility. According to one informant, a penologist,

> "Ever since the 1969 environmental protection legislation, we have become increasingly involved in formulating statements regarding social, cultural, and physical impact of proposed correctional institutions. Our most recent work pertains to the proposed correctional institution at Lorton, Virginia. While construction of this facility would have been a grave error from a correctional point of view, it has been largely our environmental impact study which has brought the project to a halt."

In this instance, penology and not ecology was the point at issue. Is this barely hidden agenda to be lightly regarded as a perversion of NEPA? Or does it rather place a proper emphasis on social impacts and their involvement in physical planning projects? While no definitive answer is possible, it would seem that if genuine concerns are denied effective legal representation they will resurface in extralegal forms.

Also unsurprising is the emphasis in EIA on quantification.

Although NEPA explicitly admits consideration of "intangibles," in reality EIA turns out to be something of a numbers game. What those numbers might mean, and what their weight in environmental decision-making might be, appear secondary. Bisset (28) has observed that preoccupation with quantitative analysis may actually serve to obscure contentious social issues. The case of quantification underscores the scientistic tendency mentioned earlier, of converting value questions into technical ones. Perhaps that stems from value insecurity and the fragility of value consensus over such questions as "balance," but the result is misplaced concreteness in the extreme. Not only does it lead to a distortion in the allocation of scientific resources, it places environmental issues on a false basis.

In most policy decisions, the degree of analytic precision required is more likely in the range of orders of magnitude than of parts per million. Scientific precision--although EIA is scarcely regarded as "good science" in virtue of its policy relevance and use (19)--has a part to play, but it sometimes appears that has been allowed to occupy the whole policy space. Public policy issues nowadays do often have a large technical component, as well as a value component; they are of the kind that Weinberg (30) has called "trans-scientific." No amount of science, however good, will provide answers to the value question by itself. If sought for there, we are looking in the wrong place. Insofar as EIA has encouraged this delusion, it is disserviceable and mischievous. More likely, EIA itself has been victimized by a common cultural fallacy to ask only those questions for which we can find easy answers. Whereas public discussion and resource commitment have emphasized science-based EIA, our real problems remain more on the side of values. If we cannot get the "right" answers, let us at least get the questions right.

REFERENCES

1. Easton, R. *Black Tide: The Santa Barbara Oil Spill and Its Consequences* (New York, New York: Delacorte, 1972).

2. Worster, D. *Dust Bowl: The Southern Plains in the 1930s* (New York, New York: Oxford University Press, 1979).

3. Carson, R. *Silent Spring* (Boston, Massachusetts: Houghton Mifflin, 1962).

4. Talbot, A.R. *Power Along the Hudson: The Storm King Case ... and the Birth of Environmentalism* (New York, New York: E.P. Dutton, 1972).

5. Anderson, F.R. *NEPA in the Courts: A Legal Analysis of the National Environmental Policy Act* (Baltimore, Maryland: Johns Hopkins University Press, 1973).

6. Wandesforde-Smith, G. "The evolution of environmental impact assessment in California", pp. 47-76 in T. O'Riordan and W.R.D. Sewell, eds. *Project Appraisal and Policy Review* (New York, New York: John Wiley, 1981).

7. Catalano, R. and R. Reich. "Local government and the environmental impact assessment process". *Urban Lawyer* 9(1977)185-207.

8. Lucas, A.R. "Legal foundations for public participation in environmental decision-making". *Natural Resources Journal* 16(1976) 73-102.

9. McDonald, T.M. "The relationship between substantive and procedural review under NEPA: A case study of SCRAP v. U.S." *Environmental Affairs* 4 (1975)157-178.

10. Lake, L.M., ed. *Environmental Mediation: The Search for Consensus* (Boulder, Colorado: Westview, 1980).

11. Liroff, R.A. *Air Pollution Offsets: Trading, Selling and Banking* (Washington, D.C. : The Conservation Foundation, 1980).

12. Komarov, B. *The Destruction of Nature in the Soviet Union* (White Plains, New York: M.E. Sharpe, 1980).

13. Crenson, Matthew A. *The Un-Politics of Air Pollution: A Study of Non-Decisionmaking in the Cities* (Baltimore, Maryland: Johns Hopkins University Press, 1971).

14. Buttel, F.H. and W.L. Flinn. "The politics of environmental concern: The impacts of party identification and political

ideology on environmental attitudes". *Environment and Behaviour* 10(1978)17-36.

15. Neuhaus, R. *In Defense of People* (New York, New York: Macmillan, 1971).

16. Sills, D.L. "The environmental movement and its critics". *Human Ecology* 3(1975)1-41.

17. Council on Environmental Quality. *Public Opinion on Environmental Issues: Results of a National Public Opinion Survey* (Washington, D.C.: U.S. Government Printing Office, 1980).

18. McHarg, I.L. *Design With Nature* (Garden City, New York: Doubleday, 1971).

19. Harvey, D. "Social processes, spatial form and the redistribution of real income in an urban system", pp. 296-337 in M. Steward, ed. *The City: Problems of Planning* (Baltimore, Maryland: Penguin, 1972).

20. Pattison, E.S. "The key ecological unknown". *Technology Review* 76(1974)4.

21. U.S. Senate. *Policies, standards and procedures in the formulation, evaluation and review of plans for use and development of water and related land resources,* Document No. 97 (Washington, D.C.: U.S. Government Printing Office, 1962).

22. Wolf, C.P. "Getting social impact assessment into the policy arena". *Environmental Impact Assessment Review* 1(1980)27-36.

23. Darling, F.F. and J.P. Milton, eds. *Future Environments of North America: Transformation of a Continent* (Garden City, New York: Natural History Press, 1966).

24. Council on Environmental Quality and Department of State. *The Global 2000 Report,* 3 volumes. (Washington, D.C.: U.S. Government Printing Office, 1980).

25. Schindler, D.W. "The impact statement boondoggle". *Science* 192(4239)(1976)509.

26. Caldwell, L.K. *et al. Citizens and the Environment: Case Studies in Popular Action* (Bloomington, Indiana: Indiana University Press, 1976).

27. Fairfax, S.K. and H.M. Ingram. "The United States experience", pp. 29-45 in T. O'Riordan and W.R.D. Sewell, eds. *Project*

Appraisal and Policy Review (New York, New York: John Wiley, 1981).

28. Bisset, R. "Quantification, decision-making and environmental impact assessment in the United Kingdom". *Journal of Environmental Management* 7(1978)43-58.

29. Matzke, G. *et al. An Examination of the Moral Dilemmas of University Scientists Participating in the Preparation of Environmental Impact Statements* (Stillwater, Oklahoma: Department of Geography, Oklahoma State University, 1977).

30. Weinberg, A. "Science and trans-science". *Minerva* 10(1972) 209-222.

ENVIRONMENTAL IMPACT ASSESSMENT IN CANADA

W. J. Couch*, J. F. Herity* and R. E. Munn**

*Federal Environmental Assessment Review Office, Ottawa, Ontario, Canada
**Institute for Environmental Studies, the University of Toronto, Toronto, Ontario, Canada

1 INTRODUCTION

This paper presents an examination of Canadian approaches to environmental impact assessment (EIA). First the factors existing in the late 1960s that led to the formal adoption of an EIA process in Canada in 1973 are examined. Then the nature of Canadian government structures is discussed so as to understand how the various procedures have evolved. This leads into a discussion of the key features of Canadian EIA processes, from project screening to formal public review, the relationship of EIA to planning and regulatory activity, the status of scientific method, and the related issues of social impact assessment and public participation.

2 HISTORICAL REVIEW

Although the term EIA was not used until the 1970's, environmental assessments have been undertaken in Canada for at least 50 years. The famous Trail smelter investigation of the 1930s is an early example (1). Vegetation damage caused by SO_2 had been occurring in the U.S. part of the Columbia River Valley, and an international tribunal was set up to investigate. Other examples include the major studies in the 1960s for siting an industrial complex on the north shore of Lake Erie (2) and for siting an oil-fired power station in New Brunswick (3). The results of these three investigations became available as public documents.

In the case of Canadian nuclear power development, baseline

studies and assessments were required by the Atomic Energy Control Board from the outset in the late 1950s. Public hearings were not held but local information offices were opened and selected persons such as doctors were invited to take one-week courses in nuclear energy and radiological health. In the early 1960s, nuclear power was believed to be a great technological achievement, and possible environmental, health and social impacts were not prominent in public debate.

What happened in the late 1960s? What forces led to the institutionalization of EIA, first in the U.S. with the passage of National Environmental Policy Act (NEPA) in 1969, and later in Canada and other countries? This is an interesting question deserving attention by social scientists. Several possible reasons can be mentioned in the Canadian context:

a) A series of books containing alarming predictions of future environmental states was published in the 1960s, for example, Carson (4) and Ward (5).

b) The UN Stockholm Conference on the Human Environment was held in 1972. For several years prior to Stockholm, national statements and positions were being prepared. In Canada this included a series of public meetings which generated much interest and was widely covered by the media.

c) In the late 1960s, there was a rather sudden drop in public confidence in technology and in the opinions of scientists.

d) Several worrying environmental accidents took place in the 1960s, including the Torrey Canyon oil spill and Minamatol mercury poisoning.

e) In Canada, as in many other countries, the scale and influence of development projects increased dramatically. Huge energy transportation projects were being contemplated, having the potential for immense public benefit as well as widespread environmental disruption.

f) Canada, like other industrialized countries, experienced a period of economic expansion and prosperity in the 1960s and early 1970s. It was the decade of cheap fuel and the two-car family. Governments and industry had the money required to undertake the pollution control programmes that the public wanted.

g) These social forces were reflected in the appearance of new private and public institutions, whose activities were directed towards environmental matters. Many new citizen organizations such as the Friends of the Earth and, in Canada, Pollution Probe were created, which heightened public awareness of environmental issues. In turn, governments at all levels responded by creating new agencies whose main responsibility was the environment. (The Canadian federal government and all 10 provinces now have such agencies.)

The Canadian Federal Department of the Environment was created in 1971. Wishing to promote activities with high public visibility and aware of the NEPA example, Canadians began exploring the idea of an EIA process as early as 1972, leading to the establishment of the Environmental Assessment and Review Process on December 20, 1973. The exact procedures to be used were still being discussed at that time, and the Department of the Environment co-sponsored the SCOPE 1974 Workshop at Victoria Harbour, Canada which led to the publication of a book on EIA Principles and procedures (6). That Workshop helped clarify a large number of ideas that had been circulating in the early 1970s.

3 A FEDERATION OF DIVERSITY

Canada is a diverse nation. The Atlantic provinces of Newfoundland, Prince Edward Island, Novia Scotia and New Brunswick were settled in the eighteenth century by people from the British Isles. Their laws, as with most provinces, are derived from English Common Law. They are compact provinces whose traditions have been affected by proximity to the sea.

By contrast, Manitoba, Saskatchewan and Alberta are large, land-locked prairie provinces. Most settlers came during this century from eastern Canada, continental Europe, the U.S. and the British Isles. Agriculture has historically dominated their economies.

About 60 per cent of Canada's total population of 24 million live within the Windsor-Toronto-Montreal-Quebec City corridor. It is about 1100 km in length and its economy is based on industry and intensive agriculture. The corridor is divided politically and culturally: the eastern half is French-speaking Quebec. Most Quebecois are descendants of French settlers who came prior to 1763. Also, Quebec's civil law system is unique in Canada because it is not derived from English Common Law, but the Napoleonic Code. The western half of the corridor is English-speaking Ontario. Ontario was first settled by people of British origin who came either directly from the British Isles or from the U.S. The majority of immigrants who came to Canada since World War I from countries other than Britain have settled in Ontario.

The Yukon Territory and Northwest Territories present a sharp contrast. These federally administered territories constitute almost 40 per cent of Canada's total area (4.9 million km^2). However, their population is less than 1/3 of 1% of Canada's total, or about 75,000. Approximately 40 per cent of the Territories' inhabitants are native people with very special problems and concerns. The Territories have assumed a new importance during the past decade because they are rich in hydrocarbons and other untapped natural resources now in declining supply.

Finally, there is British Columbia which is an amalgam of the old and new colonial traditions, but which is rich in both renewable and non-renewable resources.

It is understandable therefore, that approaches to EIA vary in such diverse political and geographic settings. The federal government has not assumed a role of standardizing procedures. The national diversity of Canada has demanded different approaches in different circumstances.

Since the British Parliament passed the British North America Act (1867), which created Canada from formerly separate colonies, the division of powers between the federal government and the provinces (the former colonies) has been the subject of continuing discussion in Canadian public life. The responsibility for EIA is also divided, with each jurisdiction conducting EIA activities within its own area of responsibility. Decisions in each jurisdiction are taken by elected politicians in accordance with British parliamentary democracy.

The federal responsibility for EIA applies to the activities of federal departments and agencies. Further, projects involving federal lands or federal monies are also subject to the federal process. Some examples of federal undertakings are proposed airports or harbours built by the federal Department of Transport; construction done by the federal government's builder, the federal Department of Public Works; energy-related projects sponsored by, or subject to the federal Department of Energy, Mines and Resources; activities involving Indian Inuit groups or national parks; and provincial or private undertakings involving federal loans and grants.

Particular mention should be made of the Yukon Territory and Northwest Territories because of major plans to develop their hydrocarbon and mineral resources. The land and offshore area is under federal jurisdiction. Therefore, virtually all public and private developers in these territories are subject to the federal EIA process.

In contrast, activities carried out by provincial agencies

are subject to the respective provincial process, as are activities involving provincial monies or provincial lands. Roads and highways, the production and supply of electricity, health and education, natural resources, and activities of municipalities, are areas of provincial jurisdiction.

To avoid confusion or duplication in EIA activities among the various jurisdictions it is essential to have good communication. One formal mechanism for this is the Canadian Council of Resource and Environment Ministers (CCREM) which consists of the minister responsible for environmental matters from the federal and each provincial jurisdiction. Within the council, senior environmental assessment administrators from each government meet regularly to discuss issues of mutual concern and to exchange new ideas. This group has sponsored the publication of a fairly detailed description of the main characteristics of the various Canadian EIA processes (7). As a result of good communication between jurisdictions, there is an excellent degree of cooperation. Many assessments are carried out jointly so a developer is spared the need to meet the requirements of two or more separate systems.

4 EIA: GENERAL COMMENTS AND COMMON CHARACTERISTICS

Although there are differences in detail in the application of EIA in Canada, there is a large degree of similarity in philosophy and method. EIA evolved autonomously at the federal level and within each province. From the early stages of EIA development, political leaders and the public across the country were generally aware of events in Canada relating to the introduction of a federal EIA system. They also cast their eye to the south and to U.S. experience. The emerging Canadian philosophy held that EIA should not become a tangled legal battle.

The Honourable Jack Davis (then Minister of the Environment) struck this key in the development of EIA in Canada when announcing the establishment of the federal EIA Process to Parliament (March 14, 1974):

> "I hope ... that we can avoid the delays and other pitfalls which a strictly legalistic approach would cause this country ... We will not hold up developments which are clean from an environmental point of view and, ... we will not bring the environmental assessment process into disrepute". (8)

The principle of keeping EIA out of the courts was supported by Davis' advisers within the federal public service, by all subsequent ministers and by provincial politicians and public servants. They agreed that all decisions in EIA would ultimately rest with the Cabinet of the relevant political jurisdiction (federal or provincial) 2). Cabinet (or members of Cabinet) could delegate routine matters to public officials.

This fundamental position must be viewed within the realities of Canadian political and public life. Canadian courts do not exercise the quasi-legislative powers of courts in other countries. Furthermore, both federal and provincial Cabinets and parliamentarians are jealous about protecting their prerogatives to be the ultimate decision-makers within their own jurisdictions. The tendency to establish independent tribunals as the final authority is usually resisted.

Davis' announcement also touched another key element of EIA in Canada - public consultation. He said:

> "Public disclosure is important. Written assessment made by the panel will, therefore, be published. Public participation is also vital ..." (8)

Canadians were initiated to public participation in EIA in 1976 during the hearings of the Mackenzie Valley Pipeline Inquiry - the Berger Inquiry. Mr Justice Thomas Berger was asked by the federal government, before the official federal process came into effect, to assess the impacts of a gas pipeline down the Mackenzie River Valley and the subsequent development of the Valley as a transport corridor. Berger met not only with senior government officials, leaders of industry and scientific experts in southern Canada, but also went directly to the affected public in the north. In the informal surroundings of their own communities he heard their concerns about the proposed pipeline's potential impact - technical, environmental, social and economic. The media gave the Inquiry national coverage. Berger's report, submitted to the Federal cabinet on April 15, 1977, recommended a 10-year moratorium on major pipeline development in the Valley (9). It appeared to reflect what he was told by the people of the area. He had introduced a new mode of public involvement into the affairs of government. The Canadian public took notice. More importantly, governments took notice and increased their commitment to public involvement in government decision-making.

Public consultation and participation are now a part of EIA in Canada. Social commentators may speculate over why public participation grew and matured with EIA. It is clear that during the last five years the public has become a *de facto* watch-dog on matters pertaining to EIA. It is more difficult today for a public or private developer to force through any project involving sensitive topics, such as an activity having the potential for a major oil spill or long-distance pipeline, without an EIA including a public review.

Public participation in Canadian EIA processes has both formal and informal functions. The public is informed of the proponents' plans. The proponent and government officials learn about public concerns that pertain to the proposal. National, provincial and local political leaders have a forum in which to become more aware of public concerns about specific projects or general policy. The result is more informed citizens and better informed decisions. Some of the activities associated with EIA processes are examined below.

5 SCREENING IN EIA

Developers can be federal and provincial departments, agencies and Crown corporations, municipal agencies and private corporations or individuals. The usual situation in Canada is that developers are responsible for screening their proposed actions for potentially significant adverse environmental impacts early in project planning.

Some must screen because they have been explicitly directed under statute, or regulation. The regulations issued in 1980 under the Environmental Quality Act of the province of Quebec require that all pulp and paper plants, petrochemical industrial parks plants and nuclear fission or fusion plants must be subjected to an EIA before the developer will be issued a certificate of authorization by the provincial authority. Similarly, power stations, linear transmission lines, public roads and dams, above a specified minimum size, are also subject to an EIA before the developer will be issued a certificate of authorization. In Quebec, a list of projects provides the screening mechanism which enables developers to decide whether a particular action needs an EIA.

In the province of Ontario, all development actions by provincial agencies and municipalities, unless specifically exempted, require an EIA under the provincial Environmental Assessment Act. In addition, undertakings from the private sector will be subject to the Act's provisions when designated by regulation.

The Ontario act provides a screening technique through the "Class" EIA system. Proposed actions may be put forward for assessment either on a "Specific" or "Class" basis. Projects which a developer is likely to propose, on a regular basis, can be divided into either of these categories. A "Specific" development action is a proposal which requires an individual EIA before a decision is made. Such projects are likely to have significant environmental impacts. A "Class" development is one of a category of similar proposals which have a generally predictable range of impacts which require assessment, but which in most cases are unlikely to be significant. Examples of such projects are highway widenings, electricity transformer stations and extensions to sewage treatment works. Such projects can be grouped together in a "Class" and a general EIA carried out for projects within a "Class". Therefore, a single EIA is undertaken for projects defined within a particular "Class" rather than a series of individual EIAs for each "Class" project. "Class" EIAs are submitted and processed like other EIAs. It is likely, however, that most projects will be classed as requiring "Specific" EIAs (10). The aim of the procedure is to reduce the need to implement repetitive EIAs. By allocating certain types of projects to a "Class" on the basis of agreed criteria, proponents are undertaking a self-assessment screening procedure.

At the federal level, there is a different approach to screening. All federal departments and agencies have to screen their own proposals for potentially significant adverse environmental impacts. The screening must take place before commitments or irrevocable decisions are made. If the proposal is deemed by the sponsoring agency to have the potential for significant environmental effects, it is subjected to a formal public EIA review by an independent office.

This self-assessment approach to screening avoids the cost, administrative dislocation and delay which would follow from the unwelcomed imposition and policing of yet another step in all decisions. It also generates an environmental consciousness by placing the basic decision-making with the project initiator. Most federal and provincial agencies and private developers are accepting environmental screening with development. The formal authority which requires screening varies among and within jurisdictions. At present, much activity is being devoted to the design of specific guidelines, and other administrative mechanisms to aid screening.

6 THE INSTITUTIONALIZATION OF EIA

Special offices, agencies or positions have been created to administer EIA in the federal and provincial governments. For illustrative purposes, the systems used in two provinces (Alberta and Nova Scotia), and for the federal government are discussed here.

The Alberta Environmental Assessment (EA) Division is a part of the Department of Environment of Alberta. It is the responsibility of the developer to seek guidance from the EA Division on whether an EIA Report is required for a project. The provincial Minister of the Environment reviews advice from the EA Division, and from external or public expressions of concern, before deciding whether the project may proceed or whether an EIA Report is required. The EA Division co-ordinates the review of the Report for the Minister by other involved provincial departments and agencies. It also ensures public participation in the EIA system by both communities and proponents through regionally-based officials who explain the issues to the public and act as a local point of government contact.

The Environmental Assessment (EA) Division of the Nova Scotia Department of the Environment screens all projects subject to the permitting processes which the Department administers. It advises the provincial Minister of the Environment on projects which may have a significant adverse impact on the natural environment. If the Minister decides that a formal EIA is required, the EA Division develops the EIA Guidelines for the review. When it receives the EIA from the developer, it co-ordinates the document's review by Nova Scotia Department of Environment, and other involved provincial and federal agencies. It prepares the recommendations of the Department of Environment for the Minister of the Environment stating whether the project should be permitted, with or without conditions, or refused. The final decision rests with the Minister.

Although the federal process is more decentralized, it also has an administrative agency - the Federal Environmental Assessment Review Office (FEARO) (11). Unlike its provincial counterparts, FEARO maintains a distant relationship with the federal Department of the Environment and, instead, reports directly to the Minister of the Environment. FEARO is responsible for liaison and co-ordination between federal departments and agencies in matters pertaining to EIA. It aids and advises in the screening of proposals and is responsible for the preparation and submission of evaluation reports on the application of the Environmental Assessment and Review Process within federal departments and agencies. Major projects are referred to FEARO for formal public review by EIA panels. During a review FEARO provides the secretariat which supports the Panel formed for the review. It organizes the public participation programme, co-ordinates the technical review of the environmental impact statement (EIS) and other documents and implements other support tasks. As the federal government's EIA agency, FEARO is usually a point of contact for the EIA agencies of foreign governments and international organizations.

7 FORMAL PUBLIC REVIEWS

Most Canadian EIA processes provide for a public review of proposed actions which might cause public concern. The review is usually carried out by an independent panel. In some jurisdictions the review organization is termed a "board", however, in this paper they will be referred to by the term "panel".

The review panels derive their authority from various sources. In about half the provinces, panels receive their power from EIA legislation. They follow quasi-judicial procedures also set out in statute. Other provincial and federal processes are not established in law. Federally, the panel chairman establishes procedures for each review before public meetings are held. Although panels do not have the power to subpoena witnesses or permit cross-examination by counsel as done under some provincial systems, panel members, proponents and members of the general public may, under the control of the chairman, question or challenge the statements of all participants.

The developer prepares an EIS at his own expense. The required contents of an EIS are fairly standard: need for the project and reasonable alternatives, a description of the environment without the proposal, a description of the proposal, an analysis of possible impacts, description of mitigation measures, and an account of residual impacts. The EIS is made available to the public before the hearings and distributed to appropriate government agencies for expert review. The public have an opportunity to comment on the proposal, the EIS, and technical reviews of the EIS.

In all jurisdictions, after the panel has studied the EIS and other documents, and has heard the public's concerns a final report is prepared. The report recommends whether the proposal should proceed, with or without conditions, or be refused. The report is submitted either to the minister responsible for the environment or to the cabinet. The minister or cabinet then decides and announces a decision without delay. (There is a variation in Ontario of this decision-making sequence. Under the Environmental Assessment Act, the panel decides, but the cabinet, within 28 days of the panel's decision, may vary or rescind the decision or cause a re-hearing by the panel.)

8 EIA IN LAND AND RESOURCE-USE PLANNING

EIA is becoming an important first element in land-use and resource-use planning in Canada. Until recently, EIA processes have focused almost entirely on project-specific issues. All too often, major development activity is proposed and assessed for its environmental and social effects in isolation and outside an overall planning framework. Perceived economic necessities, or the aggressiveness of free enterprise have preceded the development of comprehensive regional and national social and environmental policies. This has become recognized as not only socially dangerous, but also potentially financially disruptive. It is disruptive for developers when they have to abandon, postpone or radically alter their development plans because they have not accurately predicted government's response to their initiative. If there were an established planning framework, this would be less likely to occur and environmental issues might be more fully considered.

There appears to be a shift in the emphasis of EIA. It is becoming a useful mechanism by which plan-making and policy formulation can occur either generically or regionally, in advance of specific developments. In many instances, specific, long-range development proposals will initiate this type of activity, in which case there can be a common understanding and focus for the planners.

Using EIA in this way tends to avoid one of the most common and frustrating features of most EIA processes - late reviews. Late reviews occur when developers wait until the last possible moment to initiate a public hearing. The last possible moment often coincides with a decision that the project is vital and must be approved with a minimum of delay.

However, the conduct of a non-project-specific assessment aimed at planning and policy-formulation in advance of firm development plans, provides a "breathing space" needed to determine the most suitable land and resource uses for the area in question. When specific developments are proposed, the need for further, detailed assessment will be determined by the extent to which the proposal matches the plans. Two examples in the Canadian federal context will illustrate these ideas.

In 1979, a developer proposed to drill one exploratory well in Lancaster Sound in Canada's eastern arctic, a uniquely rich biological area. The Department of Indian Affairs and Northern Development requested a regional EIA based on a single well to avoid a series of EIAs, one for each well. After reviewing the EIS and hearing the public's concern, the EIA panel concluded it could not assess the proposal in isolation, and that the federal government should consider the broader issues that would affect all uses of the area. It submitted these recommendations to the federal Minister of the Environment in February, 1979. The Minister and cabinet accepted the panel's recommendations. The Minister of Indian Affairs and Northern Development established, in conjunction with the Government of the Northwest Territories, and other concerned federal departments, the Lancaster Sound Regional Study to initiate comprehensive planning for the future of the region. In December 1980, the Working Group on the Lancaster Sound Region Study issued a draft planning document for public comment. Eventually, this will result in a government policy statement, following public consultation on the permissible uses of this unique area.

The Beaufort Sea Hydrocarbon Development Project entails frontier regional resource-use and land-use planning on a large scale in the western artic. It is potentially one of the most complex and costly industrial undertakings in Canadian history. It would include oil and gas production from the Beaufort Sea and subsequent transportation of hydrocarbons to southern markets possibly by ice-breaking tankers through the Northwest Passage or overland by the region of the Mackenzie River Valley, or both. If development were to proceed, the total cost of the project would be an estimated $40 billion (1981 dollars).

It should be noted that the decision to refer the project for a formal public review was made in advance of hydrocarbon production plans being developed. Government, the public and the proponents, led by three private oil companies, are now taking the opportunity to institute comprehensive planning activities. During the early phases of the review, government departments and agencies have the opportunity, without being unduly rushed, to develop and enunciate the policies that would eventually affect any development. All this is being done with thorough public involvement and is generating a sense of confidence among participants that the right decisions will eventually be taken. In the meantime, exploration drilling continues in the Beaufort Sea to determine if reserves will justify the enormous costs involved in development.

9 EIA AND OTHER REGULATORY PROCESSES

Both federal and provincial governments have a number of key regulatory authorities whose permission must be obtained before new projects may be undertaken. In a growing number of jurisdictions, EIA takes precedence over permitting and licensing requirements. For example, four provinces require provisions of the EIA processes to be met before permitting may proceed. Two other provinces have systems in which regulatory agencies play integral roles in reviewing EIAs prior to referral of the project to the permitting process. In one province, the EIA process takes precedence when there is a conflict.

There now appears to be a trend to expedite licensing and permitting procedures by ensuring that environmental and social problems have been sorted out before the granting stage. Two examples, New Brunswick and Alberta, will be discussed to show how EIA and the regulatory process can work together and speed government decisions.

In New Brunswick, regulatory agencies are part of the EIA process. For projects which may have significant adverse environmental impacts, the New Brunswick Department of the Environment in conjunction with concerned regulatory agencies prepares the

Guidelines for the EIA Report and reviews the EIA Report at each step until a decision is made. Although the EIA Process does not replace the regulatory process, the EIA Report is used to collect information required in regulatory processes.

Similarly, in Alberta, regulatory bodies participate through advice and direction given at each stage in the preparation of the EIA Report. After the Minister of the Environment is satisfied with the final EIA Report, he accepts the document and advises the proponent to proceed to the permitting stage. The proponent then may prepare details, plans and specifications for approvals for the various statutes that apply to the proposed development.

The Alberta Energy Resources Conservation Board (ERCB) is a provincial body which grants licences to developers in the petroleum and coal industries. An EIA must be completed to the satisfaction of the provincial Minister of the Environment before a proponent may proceed to the ERCB. The Board may also arrange for representatives of Alberta Environment to assist it at a public hearing. Depending on the issues under review, Alberta Environment may appoint a member of the ERCB.

10 THE SCIENCE OF EIA

A critical element of EIA is the scientific rigour characterizing the prediction of the effects of developments. In all Canadian jurisdictions, there are departments and agencies of government responsible for the preservation and enhancement of natural resources. With regard to EIA, it is the task of these agencies to provide expert advice to proponents, the public and the decision makers. In most jurisdictions, these agencies also formally judge the adequacy of EIAs carried out by developers.

There is, however, concern in the scientific community that the procedural aspect of EIA is more advanced than the scientific component. In this connection, Canadians have been in the forefront, both in criticising current EIA practices and in proposing better predictive methods (12, 13, 14). An important current example of the latter is a major study begun in Canada in 1980 which it is hoped will improve existing ability to apply ecological principles to EIA.

The study is being conducted by Dr Gordon Beanlands at Dalhousie University and is jointly funded by the federal government, industry and the university. The principle objectives are:

a) To determine the extent to which ecological principles and techniques have been applied to EIAs in Canada and to document areas where such applications would have significantly improved the quality of impact statements.

b) To provide guidance on the application of ecological principles and techniques in the formulation of impact assessment objectives, adoption of study designs, the collection and analysis of data and the interpretation of such data for the purposes of assessing and evaluating environmental impacts.

c) To make specific recommendations regarding the application of the above guidelines in related programs and activities including environmental baseline studies and post-project monitoring requirements.

d) To evaluate the potential for incorporating such guidelines within a legal framework related to impact assessment procedures.

The approach being used is a series of ten workshops (lasting for 3 days) held in all regions of the country and involving approximately twelve leading experts in the various aspects of EIA. Participants have been asked to focus on such questions as:

a) What basic principles or approaches underlie your particular disciplinary studies relative to environmental impact assessment requirements as you understand them?

b) How do you see the above principles or approaches, or the results of studies based thereon, contributing to the development of a set of overall ecological standards for environmental impact assessment?

c) Based on your experience with environmental impact assessments in general, could you determine the least "critical mass" of information (in space and time) you require from baseline studies and post-project monitoring programs?

It is still too early to determine whether a consensus is emerging. However, the workshops are proving to be of great value in their own right. The experts are asking themselves questions and discussing issues in a manner they had not previously taken an opportunity to do. It is hoped that this study will indicate how a streamlining of information collection can be obtained accompanied by a greater degree of confidence in the accuracy of ecological predictions.

Another research topic in which Canadians have made significant contributions is in the field of environmental risk assessment. The work of Burton and Whyte at the University of Toronto is well known (15, 16). The main message of this work is that different groups may perceive benefits and disbenefits quite differently, even when they have used the same data.

11 SOCIAL IMPACT ASSESSMENT (SIA)

Besides raising questions about the proper methods to be used in predicting the response of the natural environment, EIA has also posed new problems in assessing social impacts. In Canada, public participation has been the major catalyst in generating SIA. Members of the public are usually concerned about how a proposal will affect them or their community. Public hearings are a natural focal point around which citizens can organize, and in which they can express their concerns. They expect "The Government" to look into any difficulties they may have about a proposed project.

The Ontario Environmental Assessment Act (1975) defines "environment" to include, among other factors: "the social, economic and cultural conditions that influence the life of man or a community". This definition of environment is widely used, at least implicitly, in EIA in Canada. There are inherent difficulties in accepting such a broad definition, for example, what spatial or temporal parameters should be placed on a definition of environment which includes social impacts? When there are social impacts which cut across jurisdictional boundaries, and across areas of federal and provincial responsibility, who has the authority in a complex federal state to decide which jurisdiction and which agencies have the powers to deal with the impacts? Where do the powers of specially appointed bodies end, and the powers of elected legislatures begin?

These are difficult and delicate questions and officials in different jurisdictions are grappling with them. An EIA Task Force established by CCREM has published a general study entitled "Community Impact Assessment" which addresses the general problems and opportunities inherent in community impact assessment in Canada (17). Many examples of SIA guidelines are available in Canada as well as a thoughtful report of its potential application in the federal EIA process (18).

SIA concepts are being applied under the banner of EIA in every jurisdiction that institutes public reviews. It is inevitable that this should occur, no matter how narrowly the term "environment" is defined. Public reviews, by their very nature demand it. One problem is that SIA is a rather pliable, subjective science (or is it really an art?). It has become clear that much more thought and work are necessary before SIA, as it is presently perceived, becomes a streamlined administrative mechanism for effective decision-making.

12 FUNDING PUBLIC PARTICIPATION

A second major policy question which grew from public participation in EIA is funding - making grants or awards to individuals or private groups who wish to intervene effectively in public hearings. Advocates of funding point out that developers usually have large resources to prepare and defend EIAs. Private citizens and voluntary groups feel they need financial aid to improve their participation in the review process. Critics reply by asking whether funding would improve the effectiveness of interventions. Also they argue that during a period of financial restraint the country cannot afford such innovations and that technical experts in governmental departments should be relied on.

There are precedents for funding public participation in Canada. During the Berger Inquiry, the Federal Department of Indian and Northern Affairs made $1.77 million available for funding those wishing to participate in public hearings. The Ontario Royal Commission on the Northern Environment also funds such activities. Regulatory organizations such as the Canadian Radio-Television and Telecommunications Commission and the Alberta Public Utilities Board may make grants to members of the public or groups after the hearings if their intervention has made a contribution to the review.

The first policy statement supporting the funding of public participation in an EIA is contained in the Environmental Assessment Act of Saskatchewan (Section 6). The Act permits the Saskatchewan Minister of the Environment to grant up to $10,000 to persons preparing and presenting briefs related to an EIA.

Various public interest groups have requested funding from federal politicians and policy-makers. Federal panels have often included a recommendation for funding public participation in their reports to the Federal Minister of the Environment. In response to these requests and recommendations, the federal government is providing public funding in the review of the Beaufort Sea Hydrocarbon Development Project and using this as a case to study the effectiveness of funding federal reviews. In brief, although there seems to be a trend towards funding public participation, it is still impossible to discern whether or not funding will become generally accepted in all jurisdictions as an integral part of the public participation aspect of EIA.

13 CONCLUSIONS

EIA is being applied across Canada with varying degrees of rigour and authority. The institutional structures affecting EIA are similar in most jurisdictions, although there are variations in EIA procedure to fit the specific factors of each jurisdiction. Developers usually screen their own activities. There is an administering agency to oversee and coordinate EIA within each jurisdiction. Public participation is considered an integral part of all EIAs within each jurisdiction. When directed by the authorized elected officials, independent panels may hold public reviews of major projects. These review bodies submit recommendations to an elected minister or the Cabinet. This executive arm of government has maintained its prerogative as the final decision-maker in these matters.

Now that the main elements to EIA systems are either in place or being put in place, Canadian policy-makers are looking at a series of new questions which have been derived from EIA and its interaction with other governmental activities. What should be the role of EIA in land and resource-use planning? How can and should EIA work within and improve existing regulatory processes? How can the method for ecological impact prediction be improved so that impact information can improve the quality of EIAs and the validity of decisions based on them? How can SIA be most effectively built into the existing EIA systems? Is the funding of public participation necessary for improving reviews? The fact that such questions are being asked is a strong indication that the EIA process is alive and well in Canada.

FOOTNOTES

1. Throughout this paper reference will be made to EIA. It should be noted, however, that in certain provinces (for example Ontario) the term used will be "environmental assessment".
2. The Cabinet is the term given in British parliamentary democracy to the group including the Prime Minister and important ministers, which meets to determine government policies.

REFERENCES

1. Dean, R.S. and R.E. Swain. *Report submitted to the Trail Smelter Arbitral Tribunal* (Washington, D.C.: Department of the Interior, 1944).

2. Chanasyk, V., ed. *The Haldemand-Norfolk Environmental Appraisal,* 2 volumes (Toronto, Ontario: Ontario Ministry of Treasury, Economics and Intergovernmental Affairs, 1970).

3. Hall, H.A. *Lorneville Impact Assessment,* 2 volumes (Ottawa, Ontario: Environment Canada; and Fredericton, New Brunswick: New Brunswick Department of Fisheries and Environment, 1973).

4. Carson, R.L. *Silent Spring* (London: Hamish Hamilton, 1963).

5. Ward, B.M. and R.J. Dubos, eds. *Only one earth: the care and maintenance of a small planet* (London: Deutsch, 1972).

6. Munn, R.E., ed. *Environmental Impact Assessment: Principles Procedures,* SCOPE Report 5, 2nd edition (Chichester, England: John Wiley, 1979).

7. Couch, W.J., ed. *Guide to Environmental Assessment in Canada* (Ottawa, Ontario: Canadian Council of Resource and Environment Ministers, in press).

8. Canada, Parliament. *Parlimentary Debates (Commons),* 2nd session - 29th Parliament, vol. 1 (Ottawa, Ontario: Canada, Parliament, 1974), p. 499.

9. Berger, Justice T.R. *The Report of the Mackenzie Valley Pipeline Inquiry*, 2 vols. (Ottawa, Ontario: Ministry of Supply and Services, 1977).

10. Ministry of Environment. *General Guidelines for the Preparation of Environmental Assessments* (Toronto, Ontario: Ministry of the Environment, 1980).

11. Federal Environmental Assessment Review Office. *Revised Guide to the Federal Environmental Assessment and Review Process* (Ottawa, Ontario: Ministry of Supply and Services, 1979).

12. Schindler, W. "The impact statement boondoggle". *Science* 192(4239)(1976)509.

13. Holling, C.S., ed. *Adaptive Environmental Assessment and Management* (Chichester, England: John Wiley, 1978).

14. Andrews, W.A. and J.L. Cranmer-Byng. *Urban Natural Areas: Ecology and Preservation* (Toronto, Ontario: Institute for Environmental Studies, University of Toronto, 1979).

15. Whyte, A.V. and I. Burton. *Environmental Risk Assessment,* SCOPE Report 15 (Chichester, England: John Wiley, 1980).

16. Burton, I. *et al. Living with Risk: Environmental Risk Management in Canada* (Toronto, Ontario: Institute for Environmental Studies, University of Toronto, in press).

17. Canadian Council of Resource and Environment Ministers. *Community Impact Assessment* (Ottawa, Ontario: Canadian Council of Resource and Environment Ministers, in press).

18. Lang, R. and A. Armour. *The Assessment and Review of Social Impacts* (Ottawa, Ontario: Federal Environmental Assessment Review Office, 1981).

ENVIRONMENTAL IMPACT ASSESSMENT AND THE PLANNING PROCESS

ENVIRONMENTAL IMPACT ASSESSMENT AT THE PLANNING LEVEL: TWO SYSTEMS

Murray G. Jones

Senior Engineer, Environmental Affairs, Shell Canada Ltd.

PREFACE

Environmental Impact Assessment (EIA) or Environmental Assessment (EA), as it is known in some jurisdictions, is an instrument of planning developed to improve the way in which the environment is addressed and included in decision-making. Under the title EA the instrument has been evolving in a number of countries for 5, 10 or more years, and in certain other countries under different names a number of the principles of EIA have also been applied.

EIA has been and continues to be applied most often to activities on the project level, but there has been considerable interest in the possibilities of using EIA on, or adapting EIA to, the planning and policy levels. In this paper the possibility of using EIA within these higher levels is explored. To keep the topic in focus, Part 1 provides descriptive statements on certain principles of EIA followed by particular consideration of the necessity for the differences in applying EIA at the planning level. Part 2 then describes the proposed approach in the Netherlands, a country which already has an extensive system of policy and planning requirements, but will be supplementing it by the new possibilities brought about by a law on EIA. The Dutch have not moved into this field without considerable thought. They have undertaken much research into the use and application of EIA. As to how it might be applied on the planning side, two examples are described: one which looks at how an EIA was prepared on an experiemental basis for a new industrial estate, and second, a more theoretical examination of the application to regional planning.

Finally, Part 3 looks at a comparative approach in use in Ontario, Canada. A recently submitted plan-stage EA is examined to demonstrate a system which has particular advantages for those countries having a less developed physical planning or land use planning approach in existence.

1 EIA AND PLANNING

1.1 Introduction

Environmental Impact Assessment (EIA) is a planning instrument originally created for two purposes:

a) to ensure that environmental factors are included in decision-making; and
b) to show how environmental factors have been included.

It has become more sophisticated as it has been recognized that EIA should be undertaken as part of the normal planning process for a particular activity starting at an early stage, and that the statement or document produced should be a demonstration to the decision-maker of the potential ramifications of the decision to be taken based on a choice between a number of real alternatives. In this way environment becomes integrated in both the planning and decision-making.

Also important is the fact that the statement, prepared at a certain point in the EIA process, is a public document which is a means of conveying to the public information about the activity being proposed. There are, therefore, two fundamental terms to keep in mind EIA and EIS. EIA represents the overall planning process which can continue even after the submission to decision-makers of the EIS - the statement or document describing the process and information relevant to the decision.

Other than the requirements that an EIA start early, be integrated in the planning and decision-making process, and consider alternatives, there are certain principal elements of the process that should be clearly demonstrated in an EIS. These are normally expressed in a logical order, although they may have evolved in the process as a result of a series of iterative steps relating to each other and feeding back as new information gives rise to an amended approach. Figure 1 is a demonstration of such a process (1). This process commences based on the objectives of a particular proponent or initiator. These objectives would provide initial focusing of the direction of the study. In the planning field the objectives can and should be very widely defined. The broad objectives give rise to the purpose of a particular study. The purpose could be stated

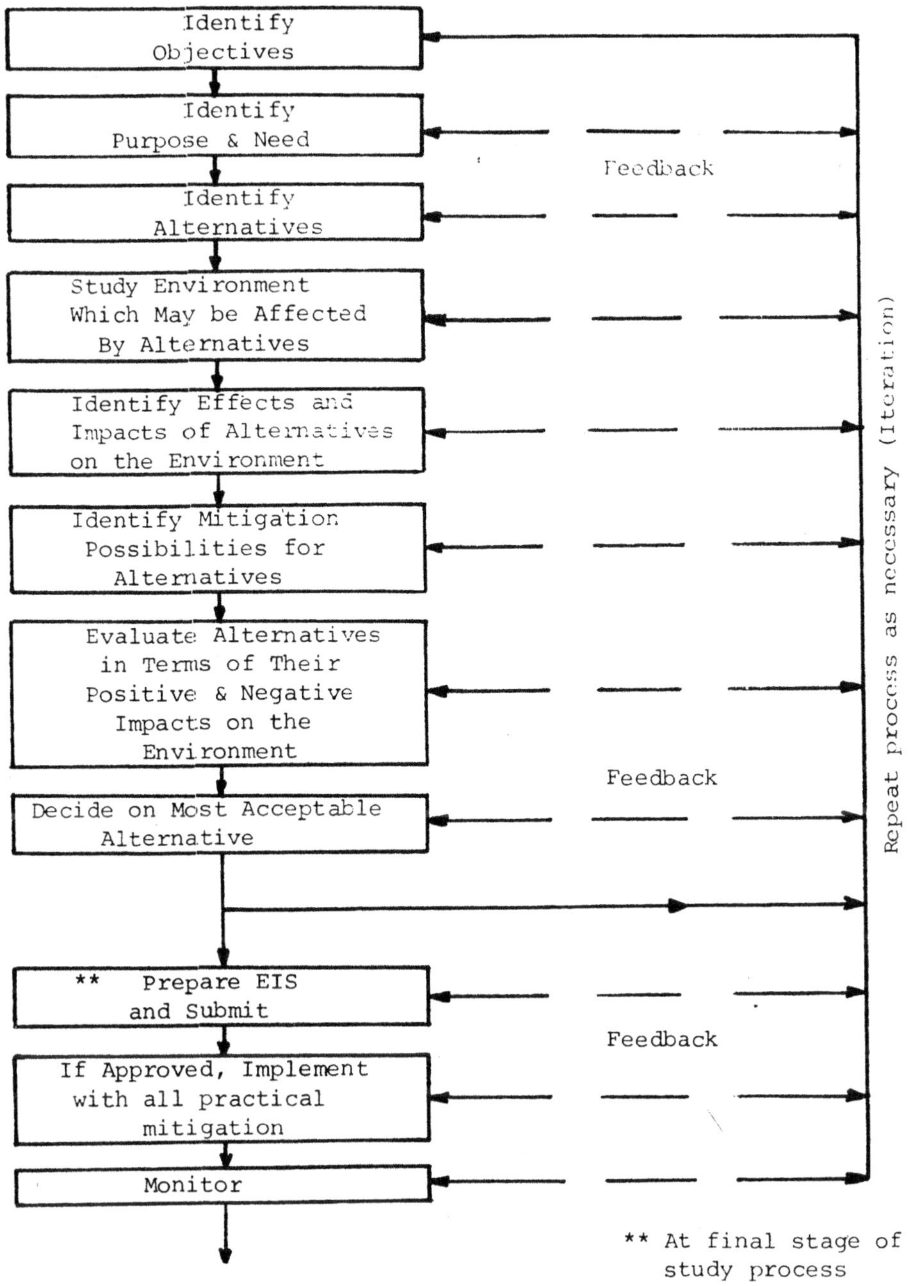

FIGURE 1: An Environmental Planning Process (Modified after Environmental Approvals Branch, Ontario Ministry of the Environment)

"... in terms of the intended effect on society, the particular problem which is to be solved or alleviated or the opportunity which is to be pursued..."(1). Related to the purpose is the need which would be fulfilled if the activity goes ahead. In effect, an answer to the question - why is it considered necessary? The alternatives that follow are the core of the EIS. Sometimes they are addressed as alternatives to the proposal, in which case the proposal is highlighted first. On different occasions no proposal is identified until later in the statement, so that the process of choosing can be demonstrated in a logical order - similar to that in which it actually occurred. On more rare occasions no proposal is put forward in the statement, only the alternatives on which the decision-makers must decide.

As Figure 1 shows, it is then incumbent on the researcher to identify the environment which could be affected. This is the descriptive part of an EIS based on fundamental base-line studies involving literature, previous EISs, data banks and field surveys. The activity (and alternatives) have the potential of producing effects at certain points in the environment caused by both locational aspects, emissions, inputs or outputs. These effects can be defined as the change in a particular environmental aspect such as air or water quality or change of features of the landscape or social environment. A distinction is made here as to what is an "effect" and what is an "impact". This is done to emphasize two separate stages of impact assessment. The first is the scientific or technical prediction of numerical values and the second is the interpretation of what the meaning of those numbers may be when related to the biotic, abiotic or social environment. Figure 2 is a schematic demonstration of this difference using an undefined activity. For simplicity the activity could be viewed as project rather then the more complex plan, although the concepts would be similar.

The list of emissions (or transformations when considering land occupation) in Figure 2 is not all encompassing; it merely offers some examples. Transmission, the second stage, is the movement of the emissions to any point(s) of concern or interest in the environment and can often be calculated with the aid of models. The interpretation of the emission or point values in relation to the environment by scientific, expert or other judgemental means determines the impact, although the further one proceeds into the judgement area, the greater the possibility of obtaining controversial results.

FIGURE 2: The Difference Between Effects and Impacts

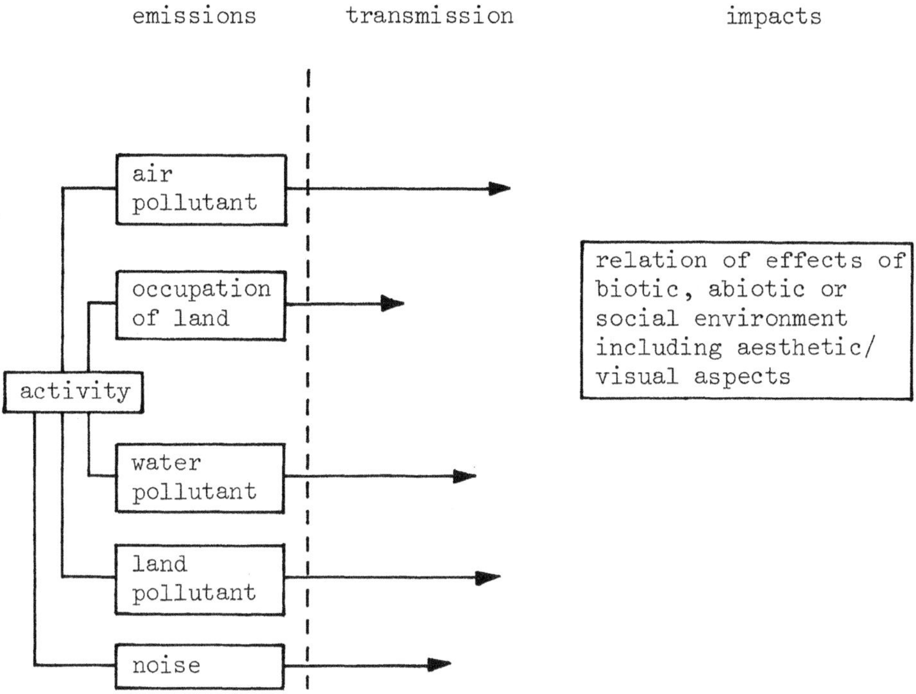

Returning to Figure 1, mitigation refers to the methods which can be used to minimize the effects or impacts and is a very important part of the EIS. It is also important to include descriptions of any techniques or methods used in the process to assist in prediction or evaluation. As the EIS is a decision-making and public document, care must be taken to make the analyses and descriptions clear.

These principal elements are common to project impact statements prepared in the past few years and great change should not be necessary because of application to the planning level. Because of the different scope and detail necessary at the planning level, certain problems can arise with the way in which the activity or its alternatives can be assessed, the environment described, and prediction and evaluation carried out. These fundamental questions are one reason why EIA has been slow to evolve at the planning level.

1.2 EIA at the Planning Level

The concept of levels of decision-making is important in grasping what is meant by "planning level". Because each country has a system of development tailored to its own needs, the types of actions taken prior to any project construction or emergence of a concrete activity can vary. However, common to many systems is the idea of orderly growth which must come about through some system of planning the type and location of development which should take place. Concern is therefore focused, in this paper, on the levels where basic goals, objectives, strategies and land uses are determined. In the Netherlands, for example, policy plans are prepared for a limited number of sectors which can effect the country's physical development as a whole (2). Regional plans begin to determine generalized land uses (3,4) and structure plans and allocation plans specific land uses (3). This series of four levels of planning will be explained in more detail in the section on "Physical Planning in the Netherlands" but the distinction between these four, referred to herein collectively as plans, and what is currently emphasized in the literature and in practice as planning for projects should be understood.

Consideration should also be given to the reasons for carrying out EIA on plans. The fact that there exists a wider variety of options at the planning level than at the project level cannot be overstressed. Recognition has been given in the Netherlands to the fact that the evaluation and weighing of alternatives must be based on an appropriate description of environmental effects (5). In addition, the point has been raised in an analysis of planning systems within the European Community that although the environment is supposed to receive recognition in plans, the degree of attention varies from place to place or province to province, and the goals are often expressed in "extremely general" terms and then interpreted too narrowly (6). Within countries, therefore, there is a general lack of consistency in the way the environment is addressed. An analysis in the US concluded that the environment has not been a factor in land use allocation (7). Instead decisions have been made on the basis of economic criteria within which the natural environment is viewed as an economic constraint. Essentially there is a failure to recognize that the complex factors making up the environment are interdependent and interrelated. As stated, "the need is to integrate a natural resource data base into the comprehensive planning and management process to be used in conjunction with economic and other criteria for determining optional land use allocations" (7).

Another problem is the division of responsibility for planning amongst a number of specialized agencies. This has led to a sectoral approach with a lack of effective co-ordination. This creates the predictable result that appropriate solutions in one area may exacerbate problems in another area. The existence of national agencies for planning (as in the Netherlands) can improve the system but only if a cross-sectoral interdisciplinary approach is taken, with appropriate attention being given to the environment.

Finally, it has been recognized that there exists a deficiency in methods appropriate to EIA at the planning level (8). This includes a lack of basic data, inadequacies in the methods of prediction and a lack of explanation of evaluation methodologies.

What are the factors which will cause planning level EIA to differ from project level? Some of the most important, summarized in Table 1, can be noted as follows:

a) The freedom to establish goals and objectives. The very broadest goals and objectives can be established with few (if any) previous constraints. The basic directions of the plan are open to discussion.

b) Availability of alternatives. Viable alternatives are more numerous at the policy and planning levels. Choices include not only technical but also institutional options and in government the ability to look across departmental lines. Willingness to include options may also be increased.

c) Long lead time in plan preparation and the period over which plans are to be effective. The former improves the ability to incorporate a) and b) above and to go through a true developmental program not constrained by the urgency surrounding project development. Care must be taken with the latter to remain flexible to changing circumstances outside the control of the plan. Alternatives can be maintained in the plan, to be pursued as the circumstances require.

d) Imprecise knowledge of the future, and therefore of the specific developments which will actually take place, can be a disadvantage when trying to produce a plan with firm commitments to choices.

e) Lack of a specific environment to study. At the highest level (policy planning) the "environment" can include a complete country or large region. Even physical (land use) plans can cover very large regions. In these areas specific field studies are difficult and costly. The problem is what to study and in what detail.

f) The question of what should be monitored. This presents a similar difficulty to the question of what environment

TABLE 1: Characteristics of Planning Versus Project EIA

ASPECTS OF PROCESS OR CONTENT	PROJECTS	PLANS
Goals and Objectives	Restricted Consideration	Potentialy broad - increasing with higher levels
Alternatives	Sites/Technologies	Technical/Institutional Cross departmental
Lead Time	Short	Long
Effective Period	Immediate Implementation	Medium to Long Term
Knowledge of Future	Reasonably Predictable	Imprecise
Environment to be Studied	Reasonably Specific	Not Specific
Environment to be Monitored	Specific	Limited

should be studied because of the lack of specifics and the very large areas affected.

An additional relevant thought when considering EIA and plan production is just how the two should relate to each other. One possibility is that during the course of normal planning procedures an EIA process is started in parallel. This would result in an EIS which refers to the plan, but does not incorporate all the details of the plan. This option could present some difficulties in demonstrating how the consideration of the environment influenced the choice or decisions in the plan.
or decisions in the plan.

The second possibility is to integrate the EIA process within the planning process. This would require that the EIA start at the same time as the work on the plan and that alternatives be generated and evaluated as a comprehensive process. All considerations would be given emphasis at the same time and a separate chapter or document could be produced as the descriptive vehicle showing the role of the environment in the plan decision-making.

A variation on this latter possibility can occur in jurisdictions not having an extensive planning process. In such cases the EIA can suffice for the process if the environment is broadly enough defined. The EIS represents the analysis of the alternatives with all factors in mind. The example of Ontario will demonstrate this alternative.

2 THE NETHERLANDS

2.1 EIA in the Netherlands

An extensive development program has taken place in the Netherlands to determine the scope and content of an EIA process, the form of legislation, and the techniques and methods most suitable (9). This program has resulted, in May 1981, in the presentation to Parliament of a draft bill on EIA which defined the framework for a system to be integrated within the existing decision-making structure. This bill will be supplemented by a variety of types of general, generic and specific guidelines and handbooks in a variety of forms, providing comprehensive technical advice on the process and content. To assist in the production of the bill and guidelines, a comprehensive series of research programs has been and is being undertaken (10). One of the programs was a series of trial EISs, including an example in the planning field, which provided specific procedural experience on how EIA might be done.

Great emphasis has been placed on the necessity of applying EIA to all levels of activities in the Netherlands which could have significant impacts on the environment from the planning to the implementation (project) level. Although the introduction of the field of application will be phased to allow for the development of procedures and experience, the planning level will receive a priority from the beginning.

To understand the application of EIA in the Netherlands to the planning level, consideration should be given to certain features of the overall program. As noted, a prime concern in its design was that it be integrated within existing decision-making. This was done so that no new decision points would be created, so that the overall process could be streamlined, and so that the decision-makers would be presented with all relevant considerations including those relating to the environment (11). It has been stated "... environmental impact assessment aims to give the interests of the environment a full place within the decision-making process. An important secondary purpose is the promotion of streamlining and acceleration of (licensing) procedures by systemization and harmonization of information on environmental consequences, and the use of such information" (11).

Integration has two sides when the proponent is a government ministry. Often this can mean that EIA implementation and decisions on projects are by the same people. Although this may be considered as a possible problem, certain participatory mechanisms have been built in to ensure all considerations are taken into account. These include the right of action groups, public, and advisors from other ministries and municipalities to participate at an early stage (scoping process, setting of guidelines) and to comment on the submitted EIS. The development of such procedures to focus and direct EIAs is considered of great importance and necessity in the Netherlands. These mechanisms also include the creation of a Commission of Independent Experts which will advise on the content and adequacy of every EIS. This Commission will have a permanent chairperson and a secretariat and a list of persons, with expertise in a variety of fields, a few of whom (8±) will be called upon to participate in any individual EIA which is undertaken.

A further feature is the definition of the environment. This is taken from the standpoint "The term 'physical environment' covers water, soil, air, man, animals, plants and inanimate objects. EIA will also have to take into consideration the consequences for the relationships between these objects ... and for aesthetic, material, and cultural-historical values (nature, landscape, and monuments). With regard to the consequences for man, both the physical and psychological consequences (livability) must be taken into consideration" (11). At the present time social, economic and other considerations have been left out of the definition. This has been done deliberately for two reasons. First, to raise the "physical environment" to a level of consideration equivalent with other factors already considered as routine, and second, to speed the introduction of EIA by keeping the definitions within the jurisdictional responsibility of the Ministries of Health and Environmental Protection, and Culture, Recreation and Social Welfare. These two ministries have co-operated in the research and development program and the preparation of the proposed law. Should other aspects be initially incorporated in the definition, a number of additional ministries would have to be involved, with greatly increased complications. However, the necessity to expand to a broader definition in the longer term has been recognized within the planning community (12) and thus a movement in this direction can be anticipated.

EIA is to be applied to all levels of activities. This will include policy plans, physical plans, programs and projects. The very comprehensive system of planning in the Netherlands flows from the broadest level of policy plan down to the physical and structure plans and finally the project level. EIA will be a part of each of these levels of development and thus a tiered system of assessment will evolve wherein decisions taken at higher levels can be incorporated into EIAs at subsequent levels (13). At the same time information gathered at the higher levels can also be used, as appropriate, in subsequent levels. In the Netherlands, the "incorporation by reference" idea presented in the Council on Environmental Quality (U.S.) regulations (14) is seen as a useful approach in this situation.

2.2 Physical Planning in the Netherlands

With the concepts behind EIA in mind, reference can now be made to the Dutch physical planning system. Planning is carried out at the national, provincial and local levels in a manner described as both "process planning" (5) and "active, dynamic planning" (2), (4), (15). The intent to be conveyed by these descriptions is that planning should not be fixed. It must be flexible and ongoing, allowing for changes which occur outside the structure or control of the plan.

As noted, planning flows from the policy to the plan (physical, structural, allocation) to the project level. The broadest plan is incorporated in the "Report on Physical Planning" which is a national policy document considering land use on a country-wide scale. It can be considered as a general framework for what is to take place, providing a reference on broad goals and objectives and a nation-wide physical planning structure.

The next level encompasses policy plans or the so-called structure schemes. These structure schemes address a variety of physically oriented subjects on which the government wishes to state a long-term policy. Examples would include traffic and transport, navigable waters, housing, electricity supply, military training areas, outdoor recreation, housing and so forth. A typical map taken from one of the structure schemes (electricity supply) is shown as Figure 3. It shows the national level of the planning with approximate ideas for the future.

The expected size and location of future developments within the scheme are indicated for 20 to 30 years and updating of the plans occurs about every five years. The steps involved in the creation of a structure scheme are:

a) analysis of the policy area concerned, especially with respect to the relationship with other areas of government care;

b) formulation of objectives and guidelines for the policy;

c) study of the possible future developments (alternatives) and formulation of the chosen policy in main outline;

d) elaboration of the chosen policy into measures and guidelines and into infrastructure projects in harmonious and functional compatibility with the physical planning structure;

e) ascertaining the physical and financial consequences of the measures and infrastructural projects mentioned under d) (evaluation);

f) description of the instruments available for realization of the chosen policy (2), 1).

The results of these steps, carried out in an iterative manner, are stated policy objectives which make it possible to proceed with the incorporation of the broadly defined physical structures in lower plans and/or commencement of design.

In the case of the structure scheme for traffic and transport, a six point goal was elaborated into 39 guidelines which had to be applied in the design of individual projects (2). In developing this structure scheme, a series of alternatives was studied in relation to the stated goal. To limit the number of possibilities, a series of four "fringe" alternative policies were investigated. These were realistic possibilities suggesting alternate future roles for the automobile. The "fringe" policies were checked against the goals and the guidelines and a choice was made for a policy

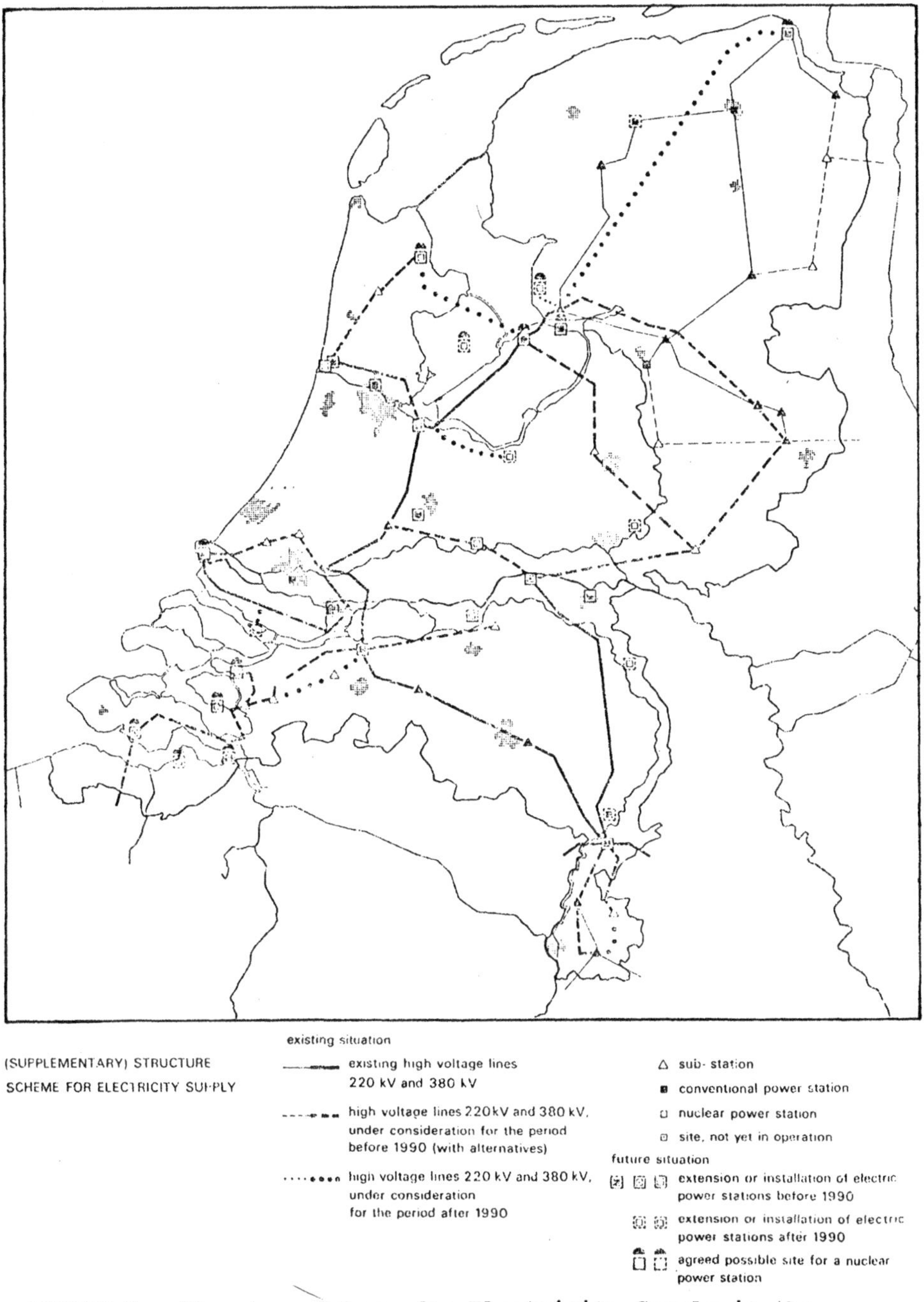

FIGURE 3: Structure Scheme for Electricity Supply in the Netherlands (Reproduced from *Planning and Development in the Netherlands* by Kind Permission of Van Gorcum Ltd.)

lying between the "fringes".

Another important level of planning occurs at the provincial level and is called the regional plan. Regional plans integrate the policies laid down by central government with the requirements of a region of the province. A regional plan sets out the most desirable development possibility for such land uses as residential, industrial, recreational, agricultural and so forth and indicates the associated infrastructure in existence or required. A typical example of a regional plan covers about 1200 square kilometres, and includes not only a region of a province, but also a number of municipalities. The regional plan is still a policy instrument and is meant to guide the establishment of more detailed structure and zoning plans which actually control the development of an area.

Structure plans describe the most desirable long-term development for a municipality or associated municipalities. Structure plans are not a common instrument (3). Most municipalities concentrate on allocation (or zoning) plans. These describe specific land use and are legally binding documents covering tens of hundreds of hectares. They would indicate types of development, densities, reserves for recreation, nature and so on. They also make allowances for the infrastructure (e.g. route locations) which has been more generally outlined at the national and regional level.

The preparation of formal EIAs on plans in the Netherlands has not yet started. However, in the development of the bill and the research program which preceded it, attention was paid to the unique requirements of EIA at the planning level. Two aspects of this consideration will be discussed as specific examples of EIA possibilities. The first of these is a trial EIA on a particular structure plan for an industrial park, carried out by the Ministry of Health and Environmental Protection, called the South East Brabant study. The second is an independent analysis of the ability of the current system of regional planning to meet EIA requirements carried out by the National Institute of Physical Planning.

2.3 South East Brabant

The South East Brabant study (16) was carried out as one of the "trial run" EISs prepared during the early research phase into the feasibility of introducing EIA. In addition to the Ministry of Health and Environmental Protection which funded the nine trial runs, the Commission of the European Communities also supported the study, demonstrating their interest in planning-level EIA. This particular EIA was concerned with the establishment of a 500 hectare industrial estate. The result of this was that the process was to be used as input into a structure plan and eventually into zoning plans prepared by the responsible local municipalities. The EIA was only one of six reports prepared as input to the structure plan. The structure plan was prepared according to the planning process set out in the Planning Act.

The EIA was undertaken outside the development of the plan but in parallel with certain crucial steps taking place in that process. It was considered a supporting document for the plan and took about eight months to produce. As a result of previous decisions, the intent to locate the industrial estate had already been stated. However, the no-go alternative common to EIA was still given some consideration, mainly as a reference state with which to compare the various development alternatives. If for any reason the estate was not developed, the region would retain its current agricultural character.

Accepting that the site had been determined, considerable flexibility was available on what might be placed on the site. It can be appreciated that this could then provide an almost infinite range of choices since the site is open to many industry types. Therefore, a decision was made to follow a scenario method for looking at particular possibilities and ultimately comparing their outputs to the existing norms and standards.

The four scenarios included (17):

a) maintenance of current status;

b) basic package - used national statistical data for industry types and adjusted these to the North Brabant situation;

c) most likely package - the second scenario was adjusted to meet with a policy to try and use the site for large scale industry; and

d) metal package - stressed the traditional types of industry already in the area, particularly the metal and building industries on the site.

Each of the last three scenarios provided sufficient information on industry types to be able to predict the quantities of the four principal emissions - air (combustion, process and traffic), water (sewers and surface), solid waste and noise.

The scenarios were then converted to actual zoning models to see how distribution of the industry types could take place and also to anticipate what land-oriented impacts would occur. To keep the layouts logical and comparable the site was divided into a grid. Criteria as listed in Table 2 were formulated to assist in the distribution of the industries on the site.

Weighting of the criteria by the consultants made various solutions possible, but to keep the number of alternatives to a minimum only four zoning models were developed from the latter three scenarios. As noted, the emissions from the industry types in each of the models could be calculated. This was done using statistical data and these were translated into point values (emission values) in the environment using transmission models. The results were compared to norms and standards to ensure they would meet with requirements or to demonstrate where changes in industry mix would be necessary. At the same time, the generalized layout plans gave indications of actual infrastructure requirements as well as pointing out the damage to physical features.

Special attention was paid to the landscape and the ways in which it could be protected. At this point, for instance, a decision was taken to adjust the layouts to protect a small valley and stream. The provisional EIS was then made public and served as an input into the planning process. No choice of a particular alternative was made.

TABLE 2: Criteria for Balancing Supply and Demand Zoning Models (Reproduced by Kind Permission of Grontmij, N.V. and Twijnstra Gudde, N.V.)

a) distance from built-up areas, in connection with environmental requirements (measured in distances varying from 100 to 2000 metres);

b) energy and cooling water requirements;

c) siting on waterways and/or railways;

d) extent of goods transport necessary;

e) passenger transport;

f) height of buildings in connection with their fitting into the landscape;

g) necessity for more robust foundations (e.g. piles).

How successful was this particular EIA process? First of all, its limitations should be considered. Since it was done as part of a series of experiments, costs, time and data gathering constraints were much in evidence. The actual decision was not to be taken on the EIA, but on the structure plan to which the EIA was one input. In addition, previous decisions greatly affecting this process were not well enough explained. Thus when the no-go alternative was included as one of the scenarios it was perceived by local farmers (the major interested public) as a real alternative instead of a reference point. This resulted in confusion on how to address the alternatives. The farmers also saw social problems as being a predominant issue, but due to the definition of physical environment used, these were not taken into account in the EIA.

On the benefit side, the EIS was seen as providing a better perception of what was to be expected, especially in the minds of the local and provincial authorities and the public (once the alternatives were made clear). Since there were still some alternatives available, debate could occur on what was to occupy the site. The scenario method provided a means of limiting alternatives to a reasonable number while at the same time giving a logical description of what could occur. Starting first with the scenarios, the focus was then moved to defining these on a fixed grid system, creation of a general zoning plan and ultimately a site layout plan. The cyclical decision-making process thus carried out by the proponents progressively narrowed the number of alternatives. This had benefits in terms of being able to restrict the choices and level of detail of the study, but gave the public little recourse at the end of the study when they became involved, to examine the options left for decision. A final benefit, as already noted, was that changes were made in the use of the site to protect the most important land form.

Certain conclusions were also drawn by the consultants as to how the process could be improved in the future (17). The recommendations included:

a) use of summary report to improve communication;

b) clear statement of the previous decisions relating to the current EIA;

c) incorporation of the EIA process into the processes of physical planning;

d) development of an "environmental management plan" to cover the implementation of the EIA.

2.4 Regional (Land Use) Plans

The second example from the Netherlands is based on an analysis of whether two types of plans (regional plan and structural outline plan) could meet with EIA requirements. This analysis is based on conditions thought to be necessary for a planning process to accommodate EIA (5) (6). These conditions are:

2.4.1 setting of goals and objectives;

2.4.2 surveying (existing environmental situation), prediction (of effects) and analysis (to determine impacts);

2.4.3 generation and evaluation of alternative plans;

2.4.4 taking of a decision followed by implementation and monitoring.

In addition, the procedures associated with these conditions should include publication in draft and final form, agency and public consultation and monitoring of plan implementation. The regional plan for East Gelderland was chosen, as it represented a good example of current, process-oriented, flexible planning. Taking each of the conditions in turn, the study found the following in respect of the East Gelderland plan:

2.4.1 Setting of goals and objectives

a) the main goal of the plan emphasized concern for the rural character of the area with an added qualification to allow for economic, social and cultural development.

b) subsidiary environmental goals included conservation of natural areas and areas having landscape value, and reduction of pollution of the environment.

2.4.2 Surveying, prediction and analysis

a) the surveys carried out were set forth in a series of environmental maps. These included maps of the landscape, ecological aspects, visual landscape assets and socio-historic assets.

b) surveys of social and physical developments showed a need to improve employment and an increasing demand for recreation. These were important inputs into the development of alternatives.

c) prediction of future conditions occurred as part of the description of alternatives.

d) analysis was carried out to the extent that conflicts were identified between certain land uses that existed or were predicted.

2.4.3 Generation and evaluation of alternative plans

a) the alternatives were established after fixing a number of starting points. These included the setting of certain limits (margins) to the study (i.e., the variables that would be considered) and the rules that would be applied. It is interesting to examine these three starting points in more detail as they provide an important mechanism for determining relevant or possible alternatives at the planning level. The limits included, first, those requirements set by any previously established goals and second, restrictions imposed by existing physical features (e.g. landscape or ecological elements and physical structures or infrastructure) or social features (e.g. population growth and housing). Within these physical and social limits alternatives were formulated through various combinations of the variables which included residential areas, industrial estates, outdoor recreation and traffic infrastructures. Superimposed on these was the agricultural/natural relationships which were seen to create the major conflict for future land use. This relationship itself provided three alternatives for the plan area, designated as

(i) the valuable elements in ecologically less valuable areas should be spared where possible;

(ii) attempts should be made to enrich areas which are less important from the landscape point of view by means of landscape structuring;

(iii) socio-historically valuable elements should be conserved as far as possible.

On matters related to health, the rules were:

(i) the location, nature and volume of activities

likely to disturb the physico-chemical environment must be carefully studied in relation to the function of the area concerned and those of adjoining areas;

(ii) a distinction must be made between noise susceptible uses and serious noise producing uses.

2.4.4 Taking of a decision, implementation and monitoring

a) on a regional plan, a decision is taken after a period of public discussion on the alternatives. The decision is made by the responsible authorities in the province. In the case of the East Gelderland study, a decision had not been taken at the time of the analysis. Consideration had, however, been given to a process of implementation which would include "a constant limited repetition of the previous phases of regional preparation ... (programming, provisional draft and establishment phases)" (5) with the results being placed before the provincial authorities every two to four years. This will maintain flexibility and allow adjustments to the plan as required. The constant evaluation will in effect allow the public and province to monitor the implementation of the provisions in the original plan and subsequent revisions.

Present planning procedures already allow for public participation in the setting of goals, alternatives, provisional draft and draft stages although they have not been without their criticisms (18). As an example of the participation permitted, the draft stage is open to inspection and comment for a few months (4). Consultation with other concerned agencies is also part of the normal planning process and, as mentioned above, monitoring can occur from both a public and agency perspective although the exact mechanisms are not spelt out.

It was therefore concluded, in the analysis, that the criteria laid down for successful integration of EIA (6) are fulfilled by the present physical planning process. Although physical planning is as concerned with economic and social aspects, the increased emphasis on the environment is perceived to be valuable, as it will stimulate research on environmental effects of physical activities, it will provide specific elaboration of environmental goals and will stress the role of alternatives.

Finally, the potential for expanding EIA beyond the current definition into social and economic aspects is seen as an important future possibility.

3 ONTARIO

3.1 The Ontario Situation

In Ontario, an environmental assessment system has been in place since the Environmental Assessment Act was proclaimed in October, 1976. Although the law applies to "... enterprises or activities or proposals, plans or programs in respect of enterprises or activities ..." by public bodies and to " ... major commercial or business enterprises or activities or proposals, plans or programs in respect of major commercial or business enterprises or activities ..." for the private sector, phasing of introduction has kept application substantially in the realm of public agencies and at the project level. Recently, however, statements of a more programmatic nature, including a few at the plan level, have appeared. Before examining an example, a number of the features of the Ontario system should be made clear.

The Act provides for a "front end" approval system wherein the first government approval given for any activity subject to the law must be that on the environmental assessment produced. This allows for consideration of a broad range of alternatives and saves on the commitment of resources by the proponent, for design and construction, until approval has been given for a particular course of action. The Act requires consideration of the no-go alternative, as well as alternatives <u>to</u> the proposal and alternative <u>methods</u> of carrying out the proposal. Alternative <u>methods</u> usually encompass the direct structural possibilities such as location, process, or construction methods. Alternatives <u>to</u> the proposal include institutional changes or new technologies and can include possibilities outside the mandate of the proponent.

The EA process is referred to as a planning process (1) which incorporates within itself the ability to include all factors relevant to the decision. This is due to the broad definition of environment which includes not only the physical environment but also the social, economic and technical environments. Although one ministry, the Ministry of the Environment, co-ordinates the program and is responsible for the legislation, all ministries can be involved in the setting of initial requirements and the review of an EA document. This unique review process results in a collectively written analysis called the government review, which is prepared over a variable period, after the submission of an EA document, depending on the complexity of the undertaking. The review is always made public.

The determination of what projects should be subject to EA is being done by a (loosely termed) "list approach". For the provincial sector (the most active, as it was the first sector to be included under the law) a "negative list" approach is used. This is because the Act stated all provincial government activities were liable to EA unless exempted. Thus, exemptions were authorized by the Minister of Environment upon application by governmental agencies. Exemptions were given to those activities which were already highly developed at the time of introduction, those which would not have a significant effect on the environment, or those which had to proceed out of urgency. Criticisms appeared as the number of exemptions continued beyond the introductory period and as certain controversial projects escaped the Act under the guise of commitment or urgency.

The municipal sector was covered by exemptions, designations and, in some cases, thresholds which specified certain limits on size or scope beyond which an EA would be required. The private sector, which the Act says must be designated (i.e. everything is out unless specifically defined as in) may follow something like a positive list route.

Decision-making is carried out in two stages - first on the acceptability of the EA document (i.e. is it adequate for a decision?) and second on the approval of the undertaking (the proposal). The decisions are to be made by the Minister of Environment unless he delegates the latter or both decisions to a public hearing body - The Environmental Assessment Board. This can be done either at the Minister's own initiative or if the Minister receives, and agrees with, a request from outside the Ministry.

As a result of the experience gained over the past few years two important areas of the procedure are being considered for amendment and one has already been amended. The first is in relation to the "front-end" of the EA planning process. Until now discussions or negotiations, on the content of an EA, have been carried out interministerially, intraministerially, and with the proponent on an informal basis. Serious attention is being focused on formalizing this preliminary consultation phase, so that there is consultation before the process really begins, the process is begun early, all affected groups or individuals are involved, and all important issues are identified. The second area is closer to resolution. This concerns the controversy over exemptions. The Premier of the Province recently announced the intent to establish a screening committee which can become involved in the determination of whether an activity should be subject to the law 2). This committee would probably be able to hear representations from proponents, ministries or the public and produce a public recommendation on the activity in question. This will add an unbiased voice to the discussions and should help in improving the credibility of the system.

Finally, of recent significance, is the passage of the Act to provide for the consolidation of hearings under certain Acts of the legislature of 1981 which will facilitate streamlining of hearings required for a particular activity. Instead of a number of major hearings as was possible under a number of different Acts, this Act enables the hearings to be combined and held at one time. A registrar designated under the Act, upon request, asks the chairman of the two principal bodies involved, the Environmental Assessment Board (for environmental assessments) and the Ontario Municipal Board (zoning, land use, etc.) to appoint members to sit together on the same activity at the same time. Joint procedures will be established and conclusions reached on matters to be heard or to be designated to other bodies. A decision at a hearing would be taken by the majority vote of the members. The potential thus exists for considerable saving of time and money.

3.2 Eastern Ontario Study

As an example of a planning level EA, and as a demonstration of the type of planning underway with EA in Ontario, a submission made by Ontario Hydro in July 1980 has been selected (19). This EA which was submitted as the first stage of a two-stage EA process had the stated objective (called a general purpose in the statement) of seeing how a reliable supply of electric power and energy could be provided for the people of Ontario at the lowest feasible cost, consistent with employee and public safety, taking into account the social, environmental and economic aspirations of the people of Ontario (19). The assessment covers a region of about 25,000 square kilometres. This plan-stage EA was prepared as a means of obtaining approval for both a type of alternative (transmission expansion) and an approach for locating and designing the facilities. This approach would lead to a route-stage approval via a separate or second stage EA which would be submitted once the plan-stage approval had been obtained.

To judge on the overall procedural approach taken in this EA, the same criteria will be employed as those which were used in the Regional (Land Use) Plans analysis in Part 2 of this paper. The overall adequacy of the EA will not be evaluated here, other than to consider a number of aspects important to the planning level. The review of the EA prepared by the Ministries and Agencies of the Provincial Government provides an in-depth evaluation (20).

3.2.1 Setting of goals and objectives

a) The broad goal of the study has been noted above.

b) The more specific purposes were identified in the EA as the supply of electrical energy to the year 2000, and the establishment of an interconnection with the neighbouring province of Quebec to enable energy transfers to take place for improving security and reliability.

c) The need for the former service was addressed in a previous inquiry into Electric Power Planning and for the latter in a joint study by Ontario Hydro with the neighbouring utility.

d) While the former need has been accepted in provincial policy, the Review gives a mixed opinion on the adequacy of the justification for the latter.

3.2.2 Generation and evaluation of alternative plans

Note: due to the organization of the EA and the early focusing on a particular type of alternative, this section is addressed ahead of "surveying, prediction and analysis".

a) A wide range of alternatives was examined to achieve the first purpose mentioned above. These include installation of additional sources of electrical energy generation (major and/or small units including co-generation), purchase of power from neighbouring utilities and installation of new bulk transmission facilities.

b) The "no-go" alternative (required in the Ontario EA system) was not discussed due to the previous acceptance by the government of the expected energy requirements in the region. The Review notes that the analysis of alternatives was based on a particular load forecast which may become outdated and thus change which alternative or combination of alternatives should have been chosen.

c) The Review also concludes that, with respect to additional sources, " ... the different alternatives appear to have been rejected on financial and technical grounds, with little or no reference on environmental factors, although the Act requires environmental factors to be taken into account in assessing and comparing alternatives" (20).

d) The Ontario Hydro analysis of alternatives (2 pages plus a 13 page appendix) leads them to the conclusion that new bulk transmission facilities were needed, and thus the preponderance of the report (over 100 pages) concentrates on the development of transmission facilities through the choosing and comparing of broad bands wherein transmission routes may be located.

3.2.3 Surveying, prediction and analysis

a) As a result of the alternatives evaluation-referenced above (Section 3.2.2), detailed environmental work concentrated only on locating transmission bands to fit with five alternative transmission plans. Transmission bands are geographical areas between the two points to be interconnected and are never less than 6 km in width, often much more. At a later date transmission lines will be sited within the bands.

b) Ontario Hydro has developed a system of constraint mapping for locating transmission bands.

c) The constraint mapping system involves a series of steps including:

(i) definition of study area

(ii) development of public involvement including establishment of working committee

(iii) inventory from reports, maps, aerial photos, field inspections, data computer-stored on a basis of 2 km x 2 km cells

(iv) identification of environmental factors (see Legend, Table 3) and objectives

(v) rank objectives within each factor

(vi) rank overall list of objectives (see Table 3)

(vii) prepare constraint maps

(viii) identify bands

(ix) evaluate and compare bands

(x) identify and approximate route study areas

d) In parallel with the environmental work, technical and economic studies were carried out to ensure the potential plans would meet with design criteria. The chosen alternative bands were based on technical, economic and environmental analysis.

e) The Review commends the constraint mapping approach as "an effective way of approaching the problem" (20) but expresses concern over certain technical aspects such as the

TABLE 3: Overall Ranking of Objectives (Reproduced by Kind Permission of Ontario Hydro)

1	UA	Human Settlement – "Urban and non-urban settlements with a population concentration of over 500".
2	HA	Terrestrial Communities – "Sensitive Biological Areas".
3	GA	Aquatic Communities – "Wetland (swamps, bogs and marshes)".
4	MA	Agricultural Production – "Areas of significant fruit and vegetable cultivation or very good arable land with a high occurrence of common field crops".
5	UB	Human Settlement – "Sensitive Human Settlement areas with populations of less than 500
6	AA	Appearance of the Landscape – "Visually and historically sensitive areas of the Rideau Canal System".
7	MB	Agricultural Production – "Very good arable land with a moderately high occurrence of common field crops or good arable land with a high occurrence of common field crops".
8	CA	Recreation – "Sensitive areas of major waterways (Rideau Canal, St. Lawrence and Ottawa Rivers)".
9	KA	Mineral Resources – "Existing and proposed surface and sub-surface extractions of metallic/non-metallic minerals and structural materials".
10	MC	Agricultural Production – "Very good arable land with a moderate occurrence of common field crops or good arable land with a moderately high occurrence of common field crops or arable soil with high occurrence of common field crops".
11	DA	Timber Production – "Forested land with the Ontario Land Inventory timber use capability of Classes 1, 2 or 3 with highest potential to produce timber and revenue".
12	SA	Wildlife Game Resources – "Wildlife Management Areas of productive waterfowl habitat (OLI Capability Classes 1-4, Degree of Effort A) including waterfowl nesting and staging areas".
13	CB	Recreation – "Wilderness and natural environment parks".
14	CC	Recreation – "Sensitive areas of regionally significant, designated linear systems, (canoe routes, foot trails, special interest roads)".
15	GB	Aquatic Communities – "Cold-water fish (trout) communities (lakes and streams)".
16	SB	Wildlife Game Resources – "Core deer yards which do not have core areas designated"; "unique moose concentration area".
17	AB	Appearance of the Landscape – "Landscapes that are visually dominated by water".
18	DB	Timber Production – "Forested land with the Ontario Land Inventory timber use capability of Classes 4 or 5, currently supporting mature or immature, most valuable species such as Hard Maple, American Beech, White or Red Pine, Yellow Birch, White or Red Oak".
19	UC	Human Settlement – "Existing institutional, commercial, industrial areas, including Department of National Defense Bases.
20	KB	Mineral Resources – "Potential aggregate from sand and gravel deposits within a critical distance of identified demand centres".
21	HB	Terrestrial Communities – "Undisturbed terrestrial communities characterized by late successional stages of forest ecosystem development"
22	MD	Agricultural Production – "Very good arable land with a low occurrence of common field crops or good arable land with a moderate occurrence of common field crops or arable land with a moderately high occurrence of common field crops".
23	UD	Human Settlement – "Existing airports of regional significance (excluding Ottawa International Airport)".
24	GC	Aquatic Communities – "Warm water fish (i.e. pickerel, bass, pike) communities (lakes and streams) in areas of erosion risk and spawning beds".
25	AC	Appearance of the Landscape – "St. Patrick Mountains".
26	HC	Terrestrial Communities – "Terrestrial Communities characterized by intermediate stages of forest ecosystem development in areas of erosion risk".
27	KC	Mineral Resources – "All potential aggregates from surficial sand and gravel deposits not including those identified within critical demand zones as described in Objective 20".
28	CD	Recreation – "Areas of concentration of cottages and resort development, vacation oriented and day-use parks".
29	SC	Wildlife Game Resources – "Peripheral deer yards".
30	GE	Recreation – "Canada Land Inventory Capability for outdoor recreation Class 1, 2, 3, and 4; provincial park reserves".
31	ME	Agricultural Production – "Good arable land with a low occurrence of common field crops or the remaining arable land".
32	DC	Timber Production – "Forested land with the Ontario Land Inventory timber use capability of Classes 4 or 5, currently supporting mature or immature species (less valuable than Objective 18) such as White Ash, White Cedar, Soft Maple, and Poplar".
33	UE	Human Settlement – "Proposed urban and non-urban settlements including proposed commercial, industrial and seasonal developments".
34	SD	Wildlife Game Resources – "Potentially productive waterfowl habitat requiring some form of management to reach optimum conditions (OLI Capability Classes 1-3, Degree of Effort B)".
35	AD	Appearance of the Landscape – "Remnant natural landscapes and natural river valleys".
36	UF	Human Settlement – "Communication tower fields".
37	GD	Aquatic Communities – "Warm-water fish (i.e. pickerel, bass, pike) communities (lakes and streams) in areas of low erosion risk".
38	AE	Appearance of the Landscape – "The Algonquin Highlands".
39	SE	Wildlife Game Resources – "Productive woodland wildlife habitat (OLI Capability Classes 1-4, Degree of Effort A)".
40	GE	Aquatic Communities – "Stream and lakes with no information on fish communities in areas of erosion risk".
41	HD	Terrestrial Communities – "Terrestrial Communities characterized by intermediate stages of forest ecosystem development".
42	DD	Timber Production – "Forested lands with the Ontario Land Inventory timber use capability of Classes 4 or 5, currently supporting mature or immature species (less valuable than Objective 32) such as Aspen and Jack Pine, unmerchantable species of all ages and cut-over or burned lands".
43	AF	Appearance of the Landscape – "Flat to very gently rolling landscapes that have little or no tree cover".
44	HE	Terrestrial Communities – "Terrestrial Communities characterized by early stages of forest ecosystem development".
45	AG	Appearance of the Landscape – "Landscapes in which man and nature are in harmony".
46	KD	Mineral Resources – "Potential aggregate and potential structural materials from sources such as bedrock · not including sand and gravel deposits as outlined in Objectives 20 and 27".

LEGEND

- U – Human Settlement Factor
- H – Terrestrial Communities
- G – Aquatic Communities
- M – Agricultural Production
- A – Appearance of the Landscape
- C – Recreation
- K – Mineral Resources
- D – Timber Production
- S – Wildlife Game Resources

grouping of constraints, the way maps are presented, the terminology and the methods used to establish and compare the band alternatives. Questions are also raised on the follow-through of the public participation program beyond the ranking of objectives stage. The fact that the consideration of existing rights of way was left out in the preparation of the constraint mapping is also being criticised.

f) However, the approach does follow a survey, prediction, analysis, framework. The choice of the constraint mapping system avoids too much prediction in relation to a particular environment. Instead this is accomplished in the discussion and weighing of the objectives.

3.2.4. Design implementation, monitoring

a) Two decisions are important at this stage - one already reached by Ontario Hydro for a particular alternative and the one to be taken by the government.

b) The first decision has been made through an examination of two levels of alternatives in accordance with the Act. That is, both "alternative <u>projects</u>" and "alternative <u>methods</u>" have been studied.

c) The second will be taken in accordance with the procedures described earlier. Ontario Hydro and members of the public have asked for a public hearing and thus the decisions on the acceptability of the document and the approval of the undertaking (proposal) will be made at that time by a hearing board. The hearing will be held under the Consolidation Hearings Act.

d) Implementation, if approved, will mean that Hydro will advance to the second stage EA - that concerned with actual route selection.

e) Monitoring can be accomplished by the various ministries and public as they participate in the second stage EA.

In the Ontario system, no official draft EA statement is legally required although proponents often do submit a draft document informally to the government for comment. Only a complete final document is required by the law. However, Ontario Hydro undertakes a system of public and governmental consultation allowing input throughout the process. Figure 4 shows the part of this participation process that related to the role of working committees in the study. The working committees, which are set up by Hydro, are composed of representatives of local groups and

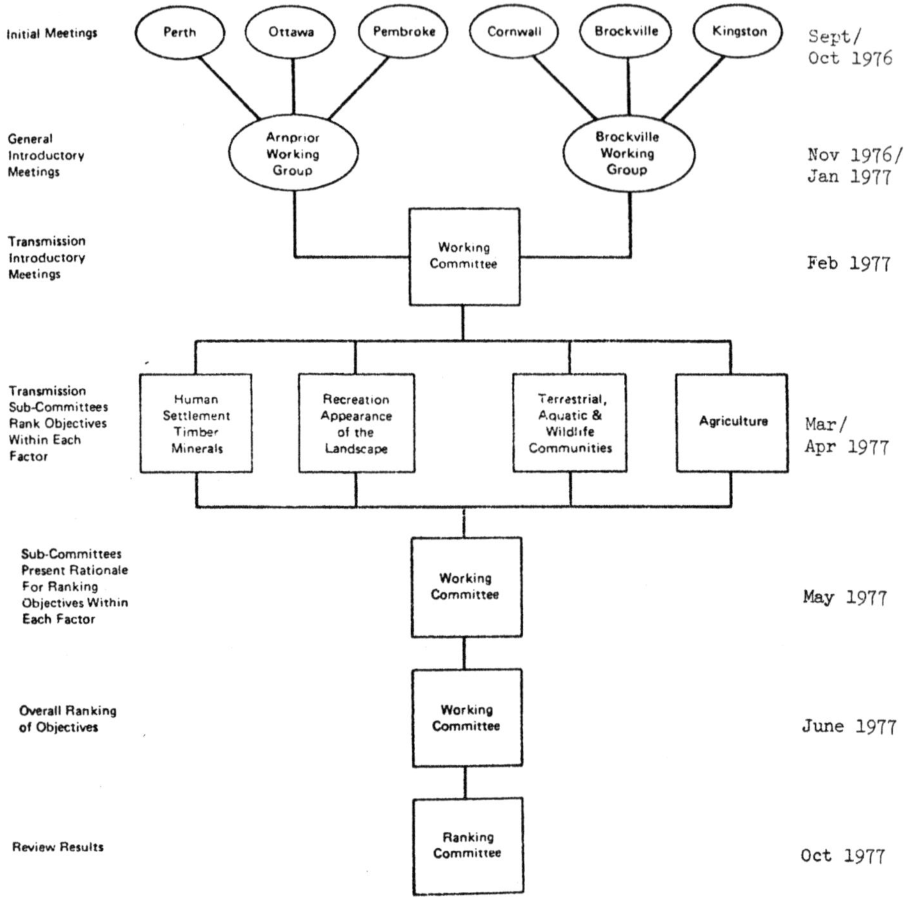

FIGURE 4: Portion of Public Participation Procedure Working Committee Structure (Reproduced by Kind Permission of Ontario Hydro)

ministries including members of the public.

In conclusion, the Ontario Hydro EA meets the content criteria for an EA process as laid down by Lee and Wood (8). This should be expected given that an EA procedure has been in existence in Ontario since 1976 and these criteria have been considered essential principles since that time. Of more importance is how well it has been done. The EA statement produced has interesting analytical processes including the method of constraint mapping which is applicable to large geographic areas. Although the discussion of very broad alternatives has been criticised for its lack of environmental content, the stepwise process in one EA statement from the wide range of alternatives to a particular alternative type and then the further development of this alternative type, shows a workable process for including alternatives in planning-level EA. The two-stage assessment process being followed in this case will also allow a general decision on broad alternatives to be taken before proceeding with actual route selections.

Although official planning exists for municipal and regional levels, the fact that this EA is not superimposed on any other planning process means that the EA itself can represent a complete planning procedure. This should be of interest to countries not having an intensive system of planning.

CONCLUSIONS

Two environmental impact assessment systems have been examined for their particular approach and relevance for application at the planning level. In the Netherlands a draft bill on environmental impact assessment has been presented to Parliament. Although an intensive system of planning already exists, advantages are seen for incorporating the provisions of the EIA program in policy, regional, structural and location plans.

This will commence from the earliest phase of introduction. In Ontario, which has the most highly defined EA system in Canada, emphasis has been placed on the project level but the first EAs on plans have also appeared.

The difficulty in applying EIA at the planning level is shown by the fact that the examples available tend to be at the lower end of the planning tier with certain generalized projects already in mind (an industrial estate and a transmission system). Further experience and continued emphasis is needed on associating EIA principles and procedures with higher level plans, but the trend can be seen moving slowly in a positive direction.

Two different analytical methods have been applied in two of the examples examined, a scenario approach from the Netherlands and a constraint mapping approach used by Ontario Hydro, while a third example was examined to determine the applicability of Regional Planning procedures/processes to EIA. Both the scenario and constraint mapping methods met with success particularly in being able to deal with large areas and a range of possible alternatives. Both appeared useful in being able to present information, although greater clarity in presenting the alternatives and describing their meaning is necessary. Difficulties are still evident in making sure the lay public can understand and participate in all stages of the process. Good potential is evident for the further development of such methods which will ease the prime difficulty of applying EIA to plans.

FOOTNOTES

1. The actual procedure is called the procedure of the Crucial Physical Planning Decisions - see Witsen, J. "Crucial Physical Planning Decisions". *Planning and Development in the Netherlands* 9 (1977) 99-114.

2. *The Globe and Mail,* Toronto, Ontario, July 28, 1981.

REFERENCES

1. Environmental Approvals Branch, Ontario Ministry of the Environment. *General Guidelines for the Preparation of Environmental Assessments* (Toronto, Ontario: Ontario Ministry of the Environment, 1978).

2. Meijer, P.G. "Structure Schemes for Infrastructure". *Planning and Development in the Netherlands* 9 (1977) 149-169.

3. Ministry of Housing and Physical Planning. *The Rules of Physical Planning in the Netherlands* (The Hague: Ministry of Housing and Physical Planning, 1977).

4. van Meel, W.A.S. "Regional Plans". *Planning and Development in the Netherlands* 9 (1977) 169-189.

5. in't Anker, M.C. and M. Burggraaff. "Environmental Impact Assessment in the Netherlands Physical Planning Process", pp. 143-154 in *Environmental Impact Assessment. Proceedings of a Seminar of the United Nations Economic Commission for Europe* (New York, New York: United Nations, 1979).

6. Lee, N. and C.M. Wood. *Environmental Impact Assessment of Physical Plans in the European Communities* (Brussels: Environment and Consumer Protection Division of the European Communities, unpublished).

7. Office of Policy Development and Research. *Integration of Environmental Considerations in the Comprehensive Planning and Management Process* (Washington, D.C.: U.S. Department of Housing and Urban Development, 1977).

8. Lee, N. and C.M. Wood. *Methods of Environmental Impact Assessment for Use in Project Appraisal and Physical Planning,* Occasional Paper No. 7 (Manchester, England: Department of Town and Country Planning, University of Manchester, 1980).

9. Jones, M.G. "Development of Environmental Impact Assessment in the Netherlands". *Environmental Impact Assessment Review* 1 (1980) 167-180.

10. Jones, M.G. and R.I. Seijffers. "Environmental Impact Assessment in the Netherlands". *UNEP Industry and Environment Newsletter* Special Issue 1 (1980) 6-8.

11. Ministry of Health and Environmental Protection. *Governmental Standpoint on Environmental Impact Assessment* (Leidschendam, The Netherlands: Ministry of Health and Environmental Protection, 1980).

12. Raad van advies voor de ruimtelijke ordening. *Summary Draft Recommendations on Environmental Impact Assessment* (s'Gravenhage, The Netherlands: Minister van Volkshuisvesting en Ruimtelijke Ordening en de Minister van Volksgezondheit en Milieuhygiene, 1979).

13. Jones, M.G. "Submission to the House of Lords" in House of Lords Select Committee on the European Communities. *Environmental Assessment of Projects* (London: Her Majesty's Stationery Office, 1979).

14. Council on Environmental Quality. "National Environmental Policy Act - Regulations. Implementation of Procedural Provisions". *Federal Register* 43 (230) (1978) 55978-56007.

15. Witsen, J. "Crucial Physical Planning Decisions". *Planning and Development in the Netherlands* 9 (1977) 99-114.

16. Grontmij, N.V. and N.V. Twijnstra Gudde. *Provisional Environmental Impact Statement South-East Brabant Industrial Estate* (Deventer/Amerstoort, The Netherlands: Grontmij, N.V. and Twijnstra Gudde, N.V., 1978).

17. de Kievit, A. *Environmental Impact Assessment of the South-East Brabant Industrial Estate,* Paper presented to seminar on Environmental Impact Assessment (Paris: Groupe de travail, Ministère de l'Environnement, et du Cadre de Vie, 1980).

18. Vonkeman, G.H. *Public Information and Participation,* Paper presented to United Nations Economic Commission for Europe Seminar on Environmental Impact Assessment, Villach, Austria, September 1979.

19. Ontario Hydro. *Eastern Ontario Plan Stage Environmental Assessment* (Toronto, Ontario: Ontario Hydro, 1980).

20. Ministries and Agencies of the Provincial Government. *Review of the Eastern Ontario Plan Stage Environmental Assessment* (Toronto, Ontario: Province of Ontario, 1981).

ROLE OF ENVIRONMENTAL IMPACT ASSESSMENT IN PLANS AND POLICIES

Xaver Monbailliu

Land Use Consultant, Paris

The role of regional and local planning is to ensure a harmonious, multi-purpose land use system by avoiding any future incompatibilities of development with its social and physical environment. Environmental impact assessment (EIA), for its part, deals with the short and long-term repercussions of a specific development project. EIA has examined, rarely, the consequences of a forward planning proposal on the regional or sub-regional scale. However, since one of the purposes of planning is to streamline harmonious development in order to avoid any environmental and socio-economic consequences, EIA should be introduced into the planning machinery. EIA is not a complete panacea for controlling the adverse effects of development, since at a project level it may fail to appreciate the cumulative and spatial aspects of impacts. Plans and policies provide the framework in which projects are undertaken and these may place constraints on available project options. Within this context, plans and policies ought also to be subjected to some form of EIA in order to foresee their consequences.

Present planning proposals, unfortunately, are often very vague and lack a profound insight into human and ecological interactions. They also raise questions concerning their robustness as indicated below:

(a) Why were nine French new towns designed for twice the number of inhabitants that they were likely to attract?
(b) Who could have foreseen in the late 1960s the economic consequences of energy crisis?
(c) How will planning proposals be affected by the new socialist government in France?

Politics, a civil war in an oil producing country, discovery of new natural resouces, and the advancement of science and technology, all make it virtually impossible for planners to produce valid guidelines beyond the horizon of the year 2000. Nevertheless, plan-making is needed to prevent piecemeal land use policies. Some of these land use proposals or planning schemes are being contested, however, by environmentalists or are the issue of controversial debate. Controversy may arise as the location of a planned industrial zone or a nature protection area may influence the price of land and thus the social stratification of a city or urban district. Other zoning proposals are sometimes ill-conceived and responsible for planning blight, or the deterioration of natural ecosystems. Land drainage and agricultural re-allotment schemes in Western Europe have destroyed many valuable wetlands and breeding territories of rare birds, such as the osprey.

Other interventions by man have led to the differentiation of ecosystems. For example, rice cultivation in the Camargue is partly responsible for the spectacular population increase of flamingoes. Areas once defined as industrial waste land have, thanks to natural re-vegetation processes, been re-classified as open air creation zones or conservation areas. All these examples serve to illustrate that an early awareness of the potential environmental impacts created by planning schemes, although difficult to achieve, is urgently required in order to safeguard the quality of our environment.

Physical Planning in France

Physical planning in France is organized on three different levels as indicated below:

(a) On the national scale, a goal-oriented 5 year national plan gives broad socio-economic guidelines for development.

(b) On the sub-regional level, the *Schéma Directeur d'Aménagement et d'Urbanisme* (SDAU) operates for most fast growing urban districts. A *Plan d'Aménagement Rural* (PAR) can be formulated for rural districts suffering from depopulation, urban overspill, tourist development, or for those likely to be affected by a large development project (for example a highway, airport or dam).

(c) On the local level, the *Plan d'Occupation des Sols* (POS) administers municipal land use (residential, industrial, agricultural and conservation). It also controls development (housing density for example).

It should be noted that the SDAU, the sub-regional structure plan, tries to streamline economic growth within three time horizons;

1990, 2000 and 2010.

Several other forms of planning schemes exist such as national park planning, development co-operation for new towns, rural funds for agriculture and forestry and planning schemes for mountainous areas or coastal waters *(Schema d'Aptitude et d' Utilization de la Mer)*. These planning schemes mainly deal with a specific goal, while a comprehensive multiple land use approach is absent. Although some regions utilize a large-scale development plan, for example, the Inter-Regional Development Plan for the Massif Central, a coherent physical planning structure is virtually non-existent at the departmental or regional level. Thus, if physical planning were ever to be subjected to EIA in France, only the micro-regional planning programmes would be directly affected (that is SDAU, PAR and POS).

EIA in France

EIA legislation in France dates from 10 July 1976 when the *Loi Relative à la Protection de la Nature* was passed. This law became operational on 1 January 1978. French EIAs, termed *études d'impact*, therefore have a legal framework and cover ecological, aesthetic and scientific topics. The same legislation stipulates that certain minor infrastructure proposals are excluded from the EIA system. Such exempted projects include:

(a) Most development projects costing less than 6 million francs.
(b) Transmission lines under 225 Kv.
(c) Areas with less than 200 units of caravans, although such camping grounds can easily occupy an area of 10 ha.
(d) Housing schemes less than 3000 m^2.
(e) All housing development zones to be constructed within the limits of local planning procedures (POS). It is to be noted that local plans do not evaluate the environmental repercussions of zoning and housing densities, and this loophole can have important consequences.

Existing legislation and the accompanying guidelines are rather vague and sometimes evasive as to the contents of an impact study. The diversity of landscapes and environmental conditions experienced in France inevitably make it difficult to produce concrete guidelines. Impact studies are frequently incomplete, since the interactions between a development proposal and it environment are poorly known. The legislation requires that an impact study should analyse the following items:

(a) Environmental analysis of the study area *(analyse de l'état initial)*.

(b) Study of environmental impacts.
(c) Analysis of alternative sites or solutions, with justification of the final choice.
(d) Exploration of measures to reduce or eliminate the predicted adverse environmental consequences.

For the more important public works, the EIA which is undertaken by the proponent has to accompany a special dossier requiring a building permit. This dossier is open to public inspection. For the less important projects, the impact study or impact notice, will be submitted for planning approval. The proponent must make public his intentions in a local or national newspaper and indicate where the impact study can be consulted. In more controversial or politically important projects, the Minister of the Environment can control the EIA study. This process eventually leads to an inter-ministerial meeting which can result in the refusal of the building permit.

The 1976 law contains an introductory statement declaring that every citizen has the moral right to protect the natural environment in his private or public life. Apart from this declaration, public participation in EIA procedures is limited to consultation of the impact study report when it is published. The public has no easy access to the law to affect the implementation of EIA studies. The EIA law does allow, however, judges of administrative courts to halt the progress of projects should a judge consider that an impact study should have been undertaken before a decision to proceed had been taken (1). To date, only about two dozen impact studies have been questioned by administrative courts. Proponents who finance an impact study often seek close ties with the local population during the preparation of the impact study. These contacts are sometimes at the political level with the *préfet* of the *région* or *département* or the village *maire*. Discussions at the political level often influence the final choice of the location of the proposed project. Occasionally, the proponent will hold a public hearing for local inhabitants or make use of opinion polls.

The current state-of-the-art of public participation was recently highlighted by the events surrounding two proposed nuclear power station sites at Plogoff (Britanny) and Golfech (NW of Toulouse). Local opposition has been so fierce that the new French government has recently abandoned both projects. It is perhaps interesting to know that the consultation process for the impact studies were held in so called 'annex town halls', which were in fact Citroën vans, as the local maire and citizens decided not to open their town hall for the public inquiry. A lawyer appointed by a citizens' group involved with the Plogoff station was not allowed to plead in court for some of the local protesters in custody.

Selection of Sites of Least Environmental Impact

Industrial installations and linear projects such as power lines or highways can often be located in one of a number of sites. Current practice in EIA does not allow for a thorough assessment of alternative sites because EIAs are usually implemented when commitments have been made in favour of a particular location. As a result, many EIAs often concentrate on the environmental impacts associated with a few sites and necessary mitigating measures rather than finding the most environmentally suitable site.

If the objectives of EIA were to be fulfilled, then it would be necessary to select sites of minimum environmental impact rather than the application of mitigation measures to sites of greater environmental consequence. It therefore becomes important to concentrate efforts on regional analysis and evaluation in order that areas of least impact can be defined in a more scientific and objective manner. To achieve this implies the need for forward planning. If a planning machinery were required to optimize land use so that environmental impacts were minimized, then EIA would be needed also for the most important planning decisions. Planning defines the geographical location of short- and long-term development and makes it possible to understand the complex relationship between multiple land use and the environment. It is within this planning framework that a balanced and harmonious land use system can be developed, especially if such planning entails the following:

(a) A comprehensive study of socio-economic and environmental priorities.
(b) An optimal multiple land use plan based on physical (e.g. soil quality) and ecological (e.g. conservation) interests) parameters.

Such comprehensive land use planning should facilitate therefore the search for a site of least impact for a development proposal, and enable a better understanding of the more intricate and complex repercussions that a particular development proposal might create. On the other hand, if planning concentrates a particular land use in a planning zone, for example individual housing schemes, then this may result in some specific environmental impacts occurring when the carrying capacity of that planning zone is exceeded. For example, the concentration of housing inevitably increases the impervious surface area with a consequent alteration in the hydrological régime of streams due to rapid water runoff.

The Assessment of Policies

Legislation and the drafting of goal-oriented policies (five year plans for example), can create unforeseen and sometimes far-reaching repercussions if no EIA were undertaken. Two years ago, an agricultural commission modified legislation regulating, among others the *Appelation Controlée* (AC) wines in some French regions. This legislation then allowed the Commune of Gigondas to enlarge its AC vineyards over all its municipal territory. This enlargement resulted in the bulldozing of the famous lace-like hills of "Dentelles de Montmirail", an area of outstanding natural beauty north-east of Avignon. Other laws or government decrees stimulating, for example,the planting of coniferous woodland, or agricultural re-allotment schemes can drastically alter terrestrial ecosystems. Regional funds such as the European Economic Commission, which may employ fiscal measures to favour a particular industrial sector or agricultural development, government directives and new technologies (for example deepwater sand extraction) may all contribute to the modification of landscapes and ecosystems.

Given the diversity of such policies it will be difficult to submit them all to an EIA. Nevertheless, it is important in our rapidly changing society to foresee the consequences they may create. A progressive application of EIA is required for those policies which may have significant effects over a wide area. This approach must inevitably lead to the application of EIA at the plan and policy levels.

Conclusion

EIA, to be totally effective, must become increasingly involved in the site selection process undertaken for development projects. In order to achieve this objective, the application of EIA at a plan level becomes important in order to fully comprehend the spatial and cumulative aspects of development. Equally, as plans are formulated within a policy framework, and as policies also may give rise to significant environmental consequences, it is important that EIA is also applied for certain types of policies. In conclusion, EIA must be applied at all levels of decision-making in order to reduce environmental degradation.

REFERENCE

1. Woltner, M.P. "France", pp. 33-38 in North Atlantic Treaty Organization Committee on the Challenge of Modern Society, ed. *The Role of Environmental Considerations in the Decision-Making Process*, CCMS Report No. 72 (Brussels: North Atlantic Treaty Organization, 1978).

EIA - THE RELATIONSHIP BETWEEN THE ENVIRONMENTAL SCIENTIST AND THE DECISION-MAKER: A BRITISH PERSPECTIVE

Robert G.H. Turnbull

Planning Consultant, Edinburgh

INTRODUCTION

One of the difficulties which arises in discussing EIA and the relationship between the scientist and decision-maker is that although 'environment' is an all-embracing word it is frequently used with a sectoral meaning. Reference is made to the physical environment, the natural environment, the built environment, the human environment, the social environment, all of which have differing but overlapping connotations. Speaking within one specific discipline it is easy simply to say 'the environment' and to forget the word may be used in a restricted sense.

One such restricted application of the word 'environment' can be seen in the concept of EIA which originated as a regular and integral component of the formal decision-making process preceding government actions in the US. The less formal approach to EIA in the UK has undoubtedly been effective in sharpening up the process of decision-making in relation to major development projects, but there is still considerable scope for widening the approach by combining organizational procedures and developing the contribution from scientists and others.

To date, EIAs which have been carried out in the UK have been primarily concerned with the 'physical' and 'natural' environments. This has come about because of the history and background of EIA and the fact that it has developed in the context of the town and country planning and pollution control system, and in particular, in relation to the granting of planning permission. While opinion varies on the type of assessment to be undertaken and on the categories of impact required in such an assessment there is a good case for bringing into the process a more explicit coverage of the likely scientific implications.

However desirable it may be to bring the question of scientific contributions within the framework of EIA and the decision-making process, a formidable effort of co-ordination and co-operation will be required. Environmental scientists, the medical profession, land use planners, landscape architects and other 'environmental managers' will be required to play a central role in collecting, analysing and presenting much information on a variety of inter-related topics. The major challenge is to combine and present an accurate and comprehensive picture of the direct and indirect, immediate and long-term impacts on man and his environment, where these are expected to be significant. Some of the difficulties are:

(a) The separate development of planning and health legislation and work in the field of the natural sciences, with which EIA is associated.
(b) Different professions and disciplines are involved in planning and scientific matters and there may be a lack of understanding of the possibilities of the two fields.
(c) Co-ordination and evaluation of scientific and other data relevant to a development decision will be a difficult task.
(d) There may be a need to look at policies as well as individual projects, and this much broader issue lies very much in the political arena.
(e) Putting values on and weighing the different environmental costs and benefits which are brought together in any EIA.

THE ROLE OF EIA IN DECISION-MAKING

In general terms, the role and function of EIA in the decision-making process is to generate and make available information on the environmental implications of a particular plan or development project. To be effective, an EIA must be sufficiently authoritative to command the respect and understanding of the decision-makers and their advisers. It must strike a balance between being too academic, or so lightweight that it is outweighed or dismissed by other supporting evidence or pressures. Above all it must give a clear indication of the environmental consequences of a plan or development proposal and, where possible, how these consequences might be mitigated.

DECISION-MAKING PROCESS

In the UK, there is a division of planning powers between local, regional and central government. A Minister is responsible to Parliament for the satisfactory operation of the system, but whenever possible the local authority is the executive authority for development control, with certain exceptions. Planning is a political act and the machinery must be part of government. The system favours strength at the local level, and at this level government knows its area and its problems, it is politically sensitive, and it knows the contribution individual, private, and voluntary enterprise can make to the process. It is less clear about the contribution that can be made by scientists.

Local government is required to initiate public participation in the planning process. There are statutory provisions requiring local authorities to make provision for publicity and public participation when preparing their development plans, so that the public can put their views in the formative, constructive stage of plan-making. In this way, the policies to be embodied in the plans may be better understood and accepted and the proposals in them implemented more successfully.

In the decision-making process, it must be accepted that there will always be cases when, after full consideration of the many factors involved, environmentally damaging projects will be authorized with or without conditions. Again, it must be appreciated that environmental considerations are seldom, if ever, the sole criteria governing decisions in a plan or project and may not even be the dominant ones, however important they may seem. A possible exception is where there is high risk and exposure to natural and man-made hazards.

Nonetheless, environmental considerations have come to play a greater role in the development planning and control process at national, regional and local levels. This growing interest in, and knowledge of, environmental matters has led to a strong public demand for environmentally sound decision-making. To meet this challenge, there must be a closer understanding and appreciation of the respective interests and contributions the planner and scientist can bring to bear on influencing the decision-making process.

Both planners and scientists are concerned with the protection of the environment, but their attitudes towards it may differ:-

(a) Planners desire to influence the environment to ease perceived problems for physical, social and economic reasons, whereas the scientists' desire is to acquire knowledge to understand it better. Both views deserve attention but only the planners try to effect direct changes in seeking to achieve the best use of land in response to development pressures.

(b) Planners tend to focus on a wide range of issues, some ill-defined, whereas scientists tend to focus on specific sets of variables and their relationships.

(c) Planners tend to make value judgements, particularly in relation to alternatives, whereas scientists may reject values as being unscientific.

TECHNICAL LANGUAGE

One of the side-effects of the growth of the environmental movement and the number of contributors to EIA has been the realisation of the need for a 'natural' or common language between the many contributors. Many of the subjects dealt with in EIA are highly technical, and some appreciation of the language used in the field concerned is virtually essential to understanding. Technical language is really, in my view, a form of shorthand, and without it some descriptions of impacts could become either imprecise or circumlocutory. Even technical words which are fairly familiar may be misunderstood. Few lay people understand what decibels measure, or how the logarithmic scale is used. It would be unreasonable to require every impact assessment to explain these basic concepts, and also, one suspects, difficult to compile an EIA without such terms.

INTERPRETATION

A research project carried out some years ago by the Institute of Operational Research (IOR) for the UK government, looked at the problems of incorporating specialized information (in some cases from the natural sciences) in the planning process. One of the problems identified was lack of understanding by planners of the terminology, and inability to gain access to relevant research findings. This is analogous to the situation of the interested layman being faced with an EIA. The solution suggested was that someone should act as an interpreter or intermediary. It may be that what is needed is a 'travelling circus' of skilled communicators who would be available to interpret specialist subjects to laymen concerned in impact analysis. The advantage of having an EIA and a skilled communicator before a public inquiry into particular development proposals could help focus discussion and avoid repetitious questioning and cross-examination of expert witnesses in the inquiry itself.

CONTRIBUTION BY SCIENTISTS TO EIA

Experience in dealing with the decision-making process at central government level in recent years suggests that much greater use and reliance is being placed on scientific input, particularly from such bodies as the Nature Conservancy Council, government research laboratories and universities. Similarly, local authorities are now more attuned to commissioning scientists to carry out specialist studies in relation to specific environmental interests.

A number of recent cases approved by the Secretary of State for Scotland have been accompanied with conditions in response to scientific evidence produced on matters such as noise, hazard from radio waves on chemical establishments, effects of ethylene and other gas emissions on vegetation and wildlife, protection of dune systems, radiation hazards, oil and other discharges into the marine environment. Unfortunately, some of the scientific contributions have led to controversy and disagreement between scientists themselves. In the case of a local public inquiry this left the Reporter to form his own judgement, and the decision-maker to commission more research with resulting delays and cost to the developer.

A classic example of a major success in protecting the environment and the ecological assets of the coastal dune system in the North-East of Scotland, was largely due to the initiative taken by the Zoology and Geography Departments of Aberdeen University in conjunction with environmental action groups. The setting up of the Environmental Liaison Group and its subsequent recognition by the local authorities as a source of scientific knowledge and advice also played an important role in the setting up of a Dune Management Group. This Group now monitors the effects of pipeline landfalls on the dune area and wildlife. It also marked the establishment of an 'environmental insurance' fund by a leading oil company, administered by the local authority on advice from the Group.

CONSULTATIONS IN THE DECISION-MAKING PROCESS

Most of the 'reconciliation' of economic development with environmental conservation is achieved within the scope of the physical system and its relevant legislation. But, as has already been indicated, a considerable amount of policy and vital action stems from related legislation on pollution, public health and the conservation of nature.

The extensive consultation necessary for the preparation of a development plan, for example, involves the planning authority in discussions with government departments and quasi-government agencies and public bodies, and also with local interest groups of all kinds, including both those concerned with development opportunities and those devoted to conservation and environmental protection. Planning authorities also take account of relevant proposals put forward in regional and sub-regional strategic plans and contingency plans, which over the last few years have increased in number.

Other public bodies that provide advice to planning authorities are concerned in particular with water and drainage, the countryside, with special reference to conservation of attractive landscapes and rural recreation, the preservation of flora and fauna and the ecological heritage in general (Nature Conservancy Council), and woodlands and forestry (Forestry Commission). Others are concerned with tourism and sport, while those that are responsible for major development programmes that often impinge on the natural environment, such as the electricity generating boards, British Gas Corporation and the National Coal Board likewise tender advice.

Thus a considerable amount of advice is made available to planning authorities, who themselves undertake surveys and make appreciations and evaluations of conditions in their areas. Frequently, consultant firms are engaged to investigate particular problems, and the results of studies made at research institutions and universities in the field of environmental conservation are normally made available to all interested bodies. Central government, acting mainly through various national research councils (the Natural Environmental Research Council is of particular importance) and the co-ordinated research programmes of the Scottish Office and the Department of the Environment, is responsible for sponsoring many major investigations into environmental problems, and government research establishments (Institute of Terrestrial Ecology) participate, together with academic bodies. Exchange of information at international level is playing an increasing role in putting the knowledge of scientists at the disposal of planners.

LIMITATIONS OF PLANNING

It is no doubt true that scientists sometimes disagree, but so do planners. It is also tempting to debunk planning, particularly long-term planning, because it is so easy to do. With all the information and advice available the planners and the decision-makers never seem to get it right. But planning has still to be attempted even if it is done in the full knowledge of what happens to the best laid schemes. Decision-makers and their planning advisers must still get on with it and try to plan in a way which will allow the necessary adjustments when events make fools of our forecasts.

SOME SUGGESTIONS

It is simple prudence, therefore, to avoid extravagant claims about what planning or EIAs can do or the degree of accuracy that can be attained. What decision-makers require is a clear and concise analysis setting out the key issues of a particular proposal or plan and as free as possible from any in-built biases. As a planner, let me venture to suggest how scientists could improve the effectiveness of their contribution to an EIA and its interpretability to decision-makers and the public.

(a) Keep advice simple, short and relevant to the decision to be taken. Make greater use of technical appendices.
(b) Avoid jargon and technical terms wherever possible.
(c) Avoid operating as an individual expert and get involved in the overall process at the earliest opportunity.
(d) Be clear about who the client is and what is required.
(e) Be clear about the time-scale involved and do not expect the decision-makers to commission long-term research.
(f) Remember that a public local inquiry may be held and contributions may have to be argued and defended.
(g) Be prepared to predict the consequences of the proposal, however difficult a task this is.
(h) Appreciate, as far as possible, other factors involved and frame contributions accordingly. Consult with other experts.
(i) Be prepared for controversy but avoid controversy about different scientific opinions. Be prepared to compromise.
(j) Do not expect too much understanding from other contributors but try to assess the value of their contributions in relation to other factors.

EIA - THE RELATIONSHIP BETWEEN THE ENVIRONMENTAL SCIENTIST AND THE DECISION-MAKER: A CANADIAN PERSPECTIVE

Robert R. Everitt

ESSA Environmental and Social Systems Analysts Ltd.
Vancouver, Canada

1 INTRODUCTION

Decision-makers have always had advisers, but only recently have environmental scientists been advising those in power. The stature of the environmental scientist has increased due to a growing imperative to integrate environmental concerns into public decision-making. Today few question the need to include environmental concerns at all stages of development, beginning with the initial planning of the project, although some may not go as far as Holling (1) in insisting that they should be integrated as equal partners with economic and social consideration so that design can benefit from, and even enhance, natural forces.

The earliest and most ubiquitous attempts at environmental science were biophysical inventories, ecological baseline studies, environmental impact statements, and monitoring programs. The basic premise behind these efforts seemed to be that more and better environmental information would lead to better decisions. The scientist in this case is seen as an omniscient generator of objective information.

Later efforts to integrate environmental dimensions into decision-making began to apply the techniques of systems ecology and system analysis. Relying on simulation of environmental systems, scientists have been able to explore alternative views of the world at the cost of admitting that no one view is true. Scientists became analysts and began to make subjective recommendations about alternative courses of action.

Recent efforts, under the guise of policy analysis, have

emphasized that the best approach is to evaluate, order, and structure incomplete knowledge to increase a decision-maker's understanding.

In a recent *Science* editorial on the subject of policy analysis, M. Granger Morgan (2) argues:

> "Good policy analysis recognizes that physical truth may be poor or incompletely known. Its objective is to evaluate, order, and structure incomplete knowledge so as to allow decisions to be made with as complete an understanding as possible of the current state of knowledge, its limitation, and its implication."

This paper first reviews criticism aroused by the approach that assumes that all is required is to provide good scientific information to the decision-maker for all to be well. It then discusses the experience of the author's group at the Institute of Animal Resource Ecology in applying systems analysis to a wide variety of resource management and environmental problems. Recent attempts by this group at ESSA to use policy analysis in integrating the environmental, socio-economic, safety, and economic consequences of large energy developments are then considered.

Out of this will emerge a set of precepts for the environmental professional, based on a set of propositions about the relationship of environmental professionals and environmental information to the public decision-making process.

2 THE RATIONAL DECISION PARADIGM: HAS IT LED US ASTRAY?

The rational decision paradigm (3) assumes:

> "Human beings make choices. If done properly, choices are made bv evaluating alternatives in terms of goals on the basis of information currently available. The alternative that is most attractive in terms of the goals is chosen. The process of making choices can be improved by using the technology of choice. Through the paraphernalia of modern techniques, we can improve the quality of the search for alternatives, the quality of information, and the quality of the analysis used to evaluate alternatives. Although actual choice may fall short of this ideal in many ways, it is an attractive model of how choices should be made by individuals, organizations, and social systems."

It is relatively easy to identify the role of the environmental scientist if one assumes that decisions are made within the context of the rational model. It essentially establishes the scientist as an information source on the environmental consequences of alternative courses of action.

2.1 Environmental Scientist as an Information Source

In this simple view of the world the decision-maker understands the problem and instructs the scientist to gather, analyze, and synthesize the existing data into information that is used to make the decision. Unfortunately the decision-maker very rarely knows what information needs to be gathered, and relies on the scientist to provide him with appropriate information. In spite of being unable to identify what information to feed into the decision-making process, scientists have little trouble in generating large amounts of data. Unfortunately most of the information is of little value to the decision-maker. Long species lists and volumes of descriptive information are of little value to a decision-maker who must decide whether a dam should be built or an energy project approved. The decision-maker needs to know what the impacts might be: in short, what the changes in the environment will be if the development goes ahead as planned or in some modified state.

Within the context of the rational decision paradigm it is the role of the scientist to provide this information to the decision-maker about what changes in the biophysical and social environment might take place as the result of a development.

2.2 More Information, Better Decisions?

One of the basic premises of the rational model is that more and better information is sufficient for better decisions. The next section discusses the invalidity of this proposition in relation to decisions on the environment by reviewing the criticism of environmental impact statements prepared under the U.S. National Environmental Policy Act (NEPA). This will show that the implicit belief in the rational decision paradigm has led us seriously astray in our attempt to integrate environmental concerns into decision-making.

3 LIMITATIONS OF ENVIRONMENTAL IMPACT ANALYSIS

The environmental scientist acts as an information source and as an adviser to the decision-maker. The importance of the environmental scientist has increased due to a growing imperative to integrate environmental concerns into public decision-making as evidenced by NEPA (1969) in the United States, by the Environmental Assessment Review Process (EARP) established in Canada in 1973, and by similar responses around the world. At the same time the integration of environmental concerns into corporate decision-making is, being nurtured (4). A major result of this effort towards integration has been the provision of more and better environmental information.

Initial efforts focussed on baseline studies that States (5), defined as a description of "the existing ecological condition and trends in the potentially affected region, providing a reference baseline from which environmental scientists can:

a) Predict the effects of the proposed action and recommend alternatives;
b) Define appropriate mitigation measures;
and c) Design future programs to monitor the accuracy of predictions and the effectiveness of mitigation."

While baseline studies have a place, it is clear that the decision-maker is more concerned with impacts, mitigations, and monitoring.

For the most part, decision-makers have been provided with information in the form of biophysical inventories, ecological baseline studies, environmental impact statements, and the results of monitoring. There is a growing concern that this information is not leading to decisions that take account of environmental factors (5).

> "As environmental scientists and resource managers look back over the last seven years of NEPA, we have to admit that much of the information gathered in the course of such studies (ecological baseline studies) was of little use to decision-makers, and even the useful information was often ineffectively used in the decision-making process."

States provides a historical perspective on why this is so, namely - more information does not mean better information and better information does not always lead to better decisions. Schindler (6) provides a different perspective that indicates that even the highest quality environmental information may fail to lead to environmentally sound decisions. He states:

> "Many politicians have been quick to grasp that the quickest way to silence critical 'ecofreaks' is to allocate a small proportion of funds for any engineering project for ecological studies".

Schindler goes on to critize the scientific rigour and credibility of impact statements. He questions not only the quality of the environmental information but its ultimate raison d'être.

Friesema (7), in discussing the relationship between environmental impact statements (EIS) and environmental management, outlines the problems with environmental impact statements as:

a) They are scientifically weak;
b) They are prone to the inclusion of excessive amounts of filler materials (species lists, and so on);
and c) They are terribly redundant, long and badly written.

Friesema's statement is a synopsis of the consensus of opinion about the poor quality of EISs. While he agrees that the EIS procedure needs to be reformed, he makes it clear that reformation is not sufficient for better decisions. He argues:

> "One can think of no reason to expect that policy-makers will read and directly use 'good' EISs more than they read and directly use inadequate ones ...".

Friesema puts the analysis of the weakness of environmental information in perspective when he presents an implicit assumption held by many environmental scientists:

> "The basic premise that underlies many of the analyses about weaknesses in the EIS process is that higher quality environment impact statements will lead to better policies and decisions. It presumes that if policy-makers are provided with succinct, accurate, and timely information based on the best scientific judgments, rigorously comparing and adequately describing the consequences of reasonable alternatives through the means of environmental impact statements, then those policy-makers will make decisions that are reasonable, rational and promote the protection and restoration of the environment."

In short, better information, better decisions. Before questioning the validity of the premise it will be instructive to look at the difficulties in generating "better information".

> "The following assessment addressed some of the impacts of the causal agents on the bowhead whale, but the majority of impacts cannot be analyzed because of the lack of quantitative or qualitative scientific information." (8)

> "In predicting the magnitude of impacts, one must consider the management techniques and their chances of success in limiting the impacts to those predicted. It becomes apparent that certain negative factors influence tremendously the ultimate impact but because these factors are hard to estimate, they are usually

neglected, with the hope that they won't be negative. These factors include: unpredictability of nature, cumulative and synergistic effects, secondary and tertiary effects, and human error and ignorance." (9)

Sentiments similar to the above quotations can be found in almost all environmental impact statements produced in North America Although there are many areas where the impacts can be adequately assessed, there are many others where they cannot be. Because of this it is not clear whether the "better information" required by the rational decision process will ever be available.

In order to discuss the validity of the premise that better environmental information will lead to better (environmentally sound) decisions, formal definitions of "better environmental information" and "environmentally sound decisions" must be given. "Better environmental information" is information that describes the existing ecological conditions and trends in the potentially affected area, providing a reference baseline from which environmental scientists can:

a) Predict the effects of the proposed action and recommend alternatives;
b) Define appropriate mitigation measures;
and c) Design future programs to monitor the accuracy of predictions and the effectiveness of mitigations.

An "environmentally sound decision" is a decision that promotes the conservation, preservation, protection, or restoration of the environment. It is now possible to state clearly three propositions

Proposition I: Better environmental information is sufficient for environmentally sound decisions.

Proposition II: Better environmental information is necessary for environmentally sound decisions.

Proposition III: Better environmental information is necessary and sufficient for environmentally sound decisions.

Proposition I has been under attack through most of this section and it is probably true that "better environmental information is not sufficient for environmentally sound decisions". Proposition II is the foundation of the entire infrastructure that has arisen in response to the need for environmental impact assessment. Should this proposition fail, industries will collapse and fortunes will be lost. The belief in proposition II and the belief that better environmental information exists allows for continued improvement in the techniques and methodologies for generation of environmental information.

The question now becomes, given that we accept proposition II, under what conditions will proposition III be true? The answer to this question lies in an understanding of the decision-making environment, and that is precisely where analysis must begin.

The analysis begins with an empirical investigation of experience in working with decision-makers, and environmental managers in the public and corporate sectors.

4 UNDERSTANDING THE INSTITUTIONAL MORASS

Painful experience has shown that descriptive studies have little impact on decision-makers; increasing understanding of the biophysical processes in an ecosystem, while more useful, has not had much success either. What is needed is to link one's understanding of environmental systems with an understanding of the institutions that are part of the decision-making process. Through two different but closely linked sets of experience an understanding of the institutional environment is emerging. One is the experience of applying the techniques of Adaptive Environmental Assessment and Management to resource management and environmental impact assessment in over sixty different cases; the other is experience of working as consultants to a large energy company on the environmental and socio-economic aspects of their projects.

4.1 Corporate Environmental Management

Almost all large corporations have personnel concerned with the environmental management of their projects. In Canada, the corporations, through these people, can and do have considerable influence over public decisions taken with regard to their projects. Petro Canada, Canada's national oil company, is perhaps one of the most aggressive and most effective in applying environmental management. Maurice Strong, former Executive Director of the United Nations Environment Program, who had the responsibility for launching Petro Canada, instituted a policy within Petro Canada that no major investment project would be approved without a full examination of environmental and social implications. These were to be given the same kinds of in-depth professional evaluation as the technical and economic aspects of the project.

It was easier said than done. Strong (10) recounts the resistance of geologists and engineers at all levels of the company hierarchy. Discussions with environmental managers from Petro Canada have revealed that the resistance is weakening, but some senior management still view the need to consider the environment as a necessary evil or, at best, an exercise in public relations. One of the major reasons is that environmentalists are regarded as speaking in moralistic terms; the language used and the presentations made are not well adapted to the style of decision-makers. This observation, made by environmental managers within Petro Canada, may have wider validity outside the company. In short, to affect decision-making, environmental professionals must learn the language of development. Mathews (11), in his paper "Moving Beyond the Environmental Rhetoric", offers a similar view.

In some ways the environmental managers may have a substantial impact on project planning and management. As noted earlier, Petro Canada project management is designed to include environmental aspects. Because of the emerging belief that it is best to spend environmental dollars at the start, in some cases the environmental aspects end up driving the project through its development phase. An environmental manager who has demonstrated his ability to deal with the environmental and social aspects of the project in the public arena can have tremendous influence on the direction of the project.

In essence,it is the environmental manager's ability to influence public decision-making that determines his effectiveness in influencing corporate decision-making.

The environmental manager serves two sets of decision-makers; corporate and public. I have discussed the relationship with corporate decision-makers; now I shall focus on that with public decision-makers. The simple view would suggest that the environmental manager acts solely as a provider of information, making sure that appropriate documents and impact statements are produced for the myriad of regulatory processes. But this view is wrong. While the EISs must be produced, this is only a small part of the influence a successful environmental manager has. In many cases the environmental manager must educate the regulators. This avoids the regulators being intimidated by the proponent's superior knowledge. This process is quite ubiquitous, and often involves pointing out the important areas of public concern. The product can be a positive atmosphere of co-operation.

To be effective in the public arena the environmental manager must establish a network of contacts and actively promote information exchange.

Ultimately the charge laid upon an environmental manager in a large corporation is to obtain environmental approval for the company projects. This places him in a high profile role in the public decision-making environment. His success in obtaining a favourable result depends on his abilities to take risks with a clear understanding of why they are necessary. Decisions, both public and corporate, are made throughout the development of a project from its inception to the finished product. Thus the environmental manager has considerable opportunity to influence both corporate planning and public perspectives of the project.

The experience of environmental managers with Petro Canada provides the environmental professional with a number of insights on how to affect decisions. These involve:

a) Being prepared to take risks;
b) Establishing a network of contacts within the institutions that participate in the decision process;
c) Being prepared to educate those who influence decisions;

and d) Learning the language of development and decision-makers.

On a recent project the group was asked to design and apply a methodology for siting a liquefied natural gas facility on the British Columbian coast. The site selection, instigated by environmental managers, included environmental, socio-economic, safety, and economic considerations. In a number of meetings with Petro Canada's project management the influence that the environmental manager had at the early stages was obvious.

Another interesting observation was made. The decision-making did not take the form of a single decision point based on the best available information; decisions were made and remade to adapt to changing and evolving conditions. The implications for the environmental professional are that he or she must adopt a flexible and responsive attitude to the decision process, making expertise and information pertinent to the shifting context of the decision environment.

4.2 Adaptive Environmental Assessment and Management

The discussion of corporate environmental management emphasized the development of an understanding of the institutional context and the importance of using that understanding to influence decision-making. There was little discussion of increasing our understanding of the environmental systems being affected. Adaptive Environmental Assessment and Management (AEAM) (1) emphasizes the need to increase our understanding of the environmental systems, thereby ensuring that one has something to say to the decision-maker. AEAM also recognizes the need to involve the decision-maker throughout the process in order to establish the necessary link between the environmental understanding and the decision-making institutions.

Four case histories have been selected from a possible 60 or more applications, not because they were the most successful but because they tell us most about the way environmental professionals interact with decision-makers. The first two deal with implementation of new policies and ideas; the second two deal with integrating research for impact assessment.

4.3 Case Histories

4.3.1 Forest and pest management. The success of the pulp and paper industry in New Brunswick, Canada has been maintained since the 1950s by an intensive insecticide spraying program. The target was the spruce budworm which periodically destroyed most of the native balsam of that province. The spraying program had reduced tree mortality but at a price; namely incipient outbreak conditions covering larger and larger areas. This produced escalating costs, greater dependence on continued spraying, public opposition, and no easily perceived alternatives. Some 20 government agencies have some say in the matter, and two key ones were at loggerheads - a federal agency for research and a provincial agency for management - with all the intertwined personalities, grievances, and territorial defence which that implies (12).

The scientific team of ecologists working at the University of British Columbia and with the International Institute for Applied Systems Analysis (IIASA) succeeded in building a comprehensive model that was consistent with the data available on budworm/forest dynamics. The use of this model with other policy design tools permitted discovery and evaluation of a range of possible policies for dealing with the management of the budworm/forest system (13).

All efforts to institute change, however, seemed to be frustrated at the eleventh hour. In despair, Baskerville wrote an explicit critique of federal, provincial, and the group activities for an IIASA policy seminar. There were three responses (12):

> "First, it is not true; second it is true but we cannot do things differently; third, Baskerville, how would you like a job as Assistant Deputy Minister?"

Gordon Baskerville, a senior scientist and academic, who became an assistant deputy minister in the New Brunswick government provides the best review of the increase in understanding of the institutions involved in resource management (13). His main points are:

a) Implementation of analysis must begin at the beginning of the project;
b) Implementation should focus on the people involved, rather than on the information or technology;

c) Implementation must be phased over time, requiring patience. It requires time for ideas to gestate, for inter-personal and inter-institutional adjustments to occur;

and d) Resistance to implementation is real and must be recognized as a formidable obstacle.

4.3.2 Salmonid management and enhancement. Salmon populations on the west coast of Canada are now at only 50% of their original levels, with the likelihood of collapses of major stocks only recently detected by public interests. Management of commercial and sport fishing faces the classic problems of mixed stocks, technology outstripping regulation, conflicting pressures from commercial, sports and environmental interests, and divisions between research and operational interests. A major investment into salmon enhancement facilities would produce more fish, with the potential of triggering the same sequence for spruce budworm management. Increased enhanced populations would lead to increased harvest of all stocks, so that the less productive natural stocks would be driven to collapse. The industry might be left precariously dependent on a few enhanced stocks that are vulnerable to collapse.

What could be done to help solve this dilemma? Carl Walters led a process beginning in 1974 that used modelling workshops to address a number of key problem areas,including stock relationships and fishing fleet dynamics. Much of the early work was done at the tactical operational level of the Canadian Department of Fisheries and Oceans. The salmon management program was part of an existing line department that had begun to initiate a change in the management approach when, for other reasons, the strategies and most of the tactical level staff involved were lost in an organizational change in 1978. However, a new group has begun to implement changes at both staff levels (14).

The experience of Al Wood is the best summary (14) of what was learned from this:

a) Implementation may elicit numerous bureaucratic survival responses. The key problems are people problems.
b) Support and commitment at the strategic level of the organization is very important. Trying to work upwards is a slow, costly, and frustrating experience.
c) New ideas have a definite gestation period for acceptance. Forcing ideas during this period may be counter-productive. It is better just to nurture them.
d) The larger the group affected, the greater the difference between present and proposed tactics

and the more difficult it will be to get
implementation.

4.3.3 Assessment of the environmental effects of Alberta oil sands development. The development of the Athabasca Deposit, one of several oil sands deposits in Alberta, has been the subject of intense interest for several decades. The Athabasca Deposit contains more than 600 x 10^9 barrels of bitumen reserves, and constitutes about 88% of known oil sands in Alberta. Consequently its potential to augment the oil supply of Canada has been a driving force in present development, and will continue to generate pressure for further development (15).

> "The Government of Alberta has an established policy of environmental legislation which allows for the orderly development of resources with a minimum of environmental damage. This policy resulted in particular attention being paid by regulatory agencies to the need for development of an environmental research program for the Athabasca oil sands region. Consequently, late in 1973, officials of Alberta Environment (Research Secretariat) and Environment Canada (Environmental Management Service) separately produced internal reports recommending a comprehensive environmental research program. Projections in each of the reports favoured an environmental research program lasting 10 years, with total costs estimated in the range of $30 million to $40 million."

The Alberta Oil Sands Environmental Research Program (AOSERP) began in April 1975. Early investigations were used primarily to establish large data bases. The data bases were intended to facilitate the construction of models to aid in predicting physical, chemical, biological, and social impacts of oil sands development.

In the fall of 1979, AOSERP sponsored an AEAM modelling project. By this time the program had been organized into four systems: Air, Land, Water, and Human. The objectives of the project were:

a) To construct a simulation model that would provide a mechanism of integration of the plethora of AOSERP data and information;
b) To delineate the interrelationships between the systems (Air, Land, Water, Human) that are basic to a general understanding;
c) To identify and evaluate data gaps and uncertainties about system function;

and d) To evaluate and recommend approaches to environmental management in the oil sands areas, including mechanisms for technology transfer from AOSERP to Alberta Environmental regulatory branches.

The participants in the project included the director of AOSERP, the heads of the Air, Water, Land and Human systems, the chairman of the Research Secretariat of Alberta Environment, as well as numerous government scientists and planners, and private consultants.

The AOSERP program had a troubled history even before the AEAM project began in 1979. The resultant project that went through a number of modelling workshops, refinement meetings, and policy sessions has had little impact on decision-making. The reason for this failure seems clear. The initial client, the Director of AOSERP, retired in the midst of the project, passing responsibility for the entire program to the Chairman of Research Secretariat who provided little direction until he resigned in early 1981. The hard lesson that we learned here is that without the key individual who becomes a critical partner with the environmental professional there is little hope of reaching the decision-maker. The presence of this key individual in the budworm and the salmon work contributed greatly to their success.

4.3.4 Assessment of the environmental effects of lakeshore development. The southern half of the Precambrian Shield in Ontario, Canada contains approximately 50,000 lakes 90 hectares or greater in size. Since the 1930s cottaging and associated water recreation activities have formed the economic base of the region. The increasing demand for cottages, coupled with an alarming decline in the health of may lake ecosystmes, indicated the need for a management strategy based on the understanding of impact of cottage development rather than one designed to take remedial measures. The need prompted the Ontario government to undertake a Lakeshore Capacity Study (LCS) which was designed to provide a way to forecast the probable effects of the development of subdivisions.

The ultimate goal of the LCS is to implement an integrated model that will aid the Ontario Ministry of Municipal Affairs and Housing (MMAH) in understanding the cottage subdivision approval process and improve the means for evaluating the impact of lakeshore developments on wildlife, fish, and water quality. The subdivision approval process currently requires that the subdivision plan be circulated by MMAH to a number of departments in the Ministries of Environment and Natural Resources for comment. The current process is cumbersome and time consuming, and a major objective of the LCS study is to streamline this process.

The group began working closely with the co-ordinator for the integration component of the LCS as well as the chairman of the Steering Committee, to plan the integration component, design and develop the integrated model, and bring about the preliminary stages of implementation.

In December 1980 the project began with a five day AEAM workshop attended by about 30 selected participants. These participants included managers of the component studies, members of the Steering Committee, planners from the Ministry of Municipal Affairs and Housing, scientists and managers from the Ministries of Environment and Natural Resources.

The results of the workshop were a preliminary model. A detailed report of the assumptions and functional relationships contained in the model was circulated and the model received considerable comment and criticism. In April 1981 a series of technical meetings with the scientists was held to make additions and refinements to the preliminary model. These technical meetings were a forerunner of an integration workshop held in May to bring the participants of the first (December) workshop back together to evaluate and game with the refinements of the model.

The integration workshop stimulated considerable enthusiasm to proceed with the model through an implementation phase. This involves making it user-oriented and will involve numerous demonstrations in the field.

The project, and in particular the workshops, were designed to involve decision-makers and policy-makers at most levels of the sub-division approval process. This has improved the project scientifically, and has increased awareness of the relationship between scientist and decision-maker. An interesting observation is the desire of the planners and managers to have a usable product. Another is the desire of the decision-makers to seek ways of simplifying the results. In summary, the decision-making process demands a simple, credible, and useful product from the environmental scientists. While some scientists showed reluctance to simplify, a general feeling is developing that the needs of the decision-maker must be heeded or the results of years of research may be wasted.

4.4 Institutional Lessons

In drawing on the experience of numerous environmental scientists, environmental managers, and decision-makers an attempt has been made to piece together an understanding of the institutions involved in environmental decision-making. Their experience provides us with a set of lessons about the behaviour of environmental institutions. In many ways these lessons can be used to yield a set of propositions that appear to be necessary conditions for effective integration of environmental interests into the decision-making process.

a) The environmental professional needs to be prepared to place himself at risk in order to establish credibility with his public.

b) The environmental professional needs to establish a network of contacts within the institutions that participate in the decision process.

c) The environmental professional should be prepared to educate those whose technical knowledge is inferior to his.

d) The environmental professional must learn the language of development and decision-making.

e) The environmental professional must begin to implement his analysis at the beginning of the project before the results are known.

f) The environmental professional must focus his attention on the people involved, rather than in the information being generated or the technology being used.

g) The environmental professional must realize that new ideas have a gestation period for acceptance. Forcing ideas during this period may be counter-productive.

h) The environmental professional must recognize that resistance to new ideas and information is real and is a formidable obstacle.

i) The environmental professional must obtain support and commitment from strategic levels of an organization.

j) The environmental professional must present his results in a simple, credible, and usable manner.

5 PRECEPTS FOR AN ENVIRONMENTAL PROFESSIONAL (RESEARCH/ANALYSTS)

In conclusion a set of precepts is offered for environmental professionals. They are intended to be a synthesis of the experiences outlined in this paper and are offered as a guide to those considering involvement in environmental policy-making.

5.1 Know Thy Client

Most scientists do not have a clear understanding of the decision-making process and often do not know or have little interaction with the decision-maker. To be effective the scientists must be good salesmen as well as good craftsmen (16). Any successful scientist knows the best way to ensure the success of a project is to talk to people who will be responsible for its results.

However, identifying the client is not a simple matter. For example Quade (17) in discussion of policy analysis quoted from the New York Times, November 16, 1970, page 25:

> "When a consultant works for government, who is the client? Is it the executive branch which engages him, or the legislative body which raises funds to pay him, or the citizen whose stake in his work may be far more vital than a stockholder's interest in a private corporation? The consultant properly employed by the city and the agencies which retain them believe beyond any question that the client is the executive branch of government. We conclude, however, that the client is not solely the executive branch, but the legislature as well and ultimately the city's citizens. It cannot be otherwise in a free society."

In discussions of some of the successes and failures of applying systems analysis to ecological management problems (18) we conclude that to be successful the analysis must be requested by the individuals who have the power to use its end results.

To be successful the environmental professional must have support and commitment from the strategic level of the organization that will use his information and analysis. But beyond this he must establish a network of contacts throughout those institutions responsible for decisions.

The key to success appears to be the ability to find a "wise man" who is an integral part of the institutional environment and to make him a critical partner in the work.

5.2 Learn to Communicate

Successful implementation of the results of research or analysis depends on the opportunity, ability, time and resources to communicate effectively. Hilborn (18) concludes that successful implementation of applied systems analysis of ecological problems requires large amounts of time devoted to communication. Effective communication is essential if environmental analysis is to have an impact on decision-making. Holling (1) devotes an entire chapter to communication techniques that have proven effective.

Every day of analysis can require six days of communication, mutual learning, trial-and-error, and interaction at all decision and staff levels. Moreover it requires as much creativity and professionalism as does analysis and requires considerably more wisdom and patience (12).

5.3 Have Something to Say

Knowing the client and knowing how to communicate with him are necessary but not sufficient to ensure that the environmental scientist will have an impact on the decision-makers. It is important that the scientist or analyst have something worthwhile to say.

In short, the information or advice being presented to the decision-maker must be good science and analysis. Hilborn (18) concluded that successful application of systems analysis to ecological management problems requires that the techniques used be defensible against technical criticism. Schindler (6) echoes the feelings of many in his criticism of the scientific rigour and credibility of impact statements.

5.4 Make the Message Clear

Most decision-makers do not have an immediate reaction to the results or recommendations of a scientific report. In most cases this is because it is not immediately apparent what the implications of the recommendations will be. However, in many instances this is because the results are incomprehensible or opaque.

The first rule in ensuring that the message is clear is to use techniques that are understandable and transparent. Hilborn (18) found this especially valid in applying systems analysis to ecological problems.

From the experience described in this paper making the methods, techniques, and results of the analysis transparent to the decision-maker is essential if any impact is to be made. One goes a long way towards this goal by avoiding the use of technical language and jargon.

The decision-makers require simple, credible, and useful advice. All environmental professionals would do well to pay heed to this need.

REFERENCES

1. Holling, C.S., ed. *Adaptive Environmental Assessment and Management* (Chichester, England: John Wiley, 1978).

2. Morgan, M.G. "Bad science and good policy analysis." *Science* 20(1978)971.

3. March, J.G. "The technology of foolishness", in J.G. March and J.P. Olsen, eds. *Ambiguity and Choice in Organizations* (Bergen, Norway: Universiteforlaget, 1976).

4. Bowie, D.B. *Incorporating environmental and social concerns into petroleum industry management: the Petro Canada experience* (Calgary, Alberta: Petro Canada, 1979).

5. States, J.B. "A practical approach to biological inventories for ecological baseline studies", in H.B. Lund *et al.*, technical co-ordinators. *Integrated Inventories of Renewable Natural Resources: Proceedings of the Workshop*, General Technical Report RM-55 (Fort Collins, Colorado: Rocky Mountain Forest and Range Experiment Station, U.S. Department of Agriculture, 1978).

6. Schindler, D.W. "The impact statement boondoggle." *Science* 192(4239)(1978)509.

7. Friesema, H.P. "Environmental impact statements and long-range environmental management", in R.K. Jain and B.L. Hutchings, eds. *Environmental Impact Analysis: Emerging Issues in Planning* (Urbana, Illinois: University of Illinois Press, 1977).

8. U.S. Department of the Interior, Bureau of Land Management. *Draft Environmental Impact Statement: Proposed Federal/State Oil and Gas Lease Sale, Beaufort Sea*(Anchorage, Alaska: U.S. Department of the Interior, 1979).

9. Alaska Highway Pipeline Panel. *Impact of the Dempster Corridor on the Mackenzie Delta*, Workshop held at Alaska Highway Pipeline Panel, Winnipeg, Manitoba, February 2-3, 1979.

10. Strong, M. *The role of engineers in resource development: servants or leaders?*, AMAX lecture delivered at Colorado School of Mines, Golden, Colorado, October 11, 1979.

11. Mathews, W.H. "Moving beyond the environmental rhetoric." *Mazingria* 4(2)(1980)6-14.

12. Holling, C.S. *Science for public policy: highlights of Adaptive Environmental Assessment and Management,* Research Report R-3 (Vancouver, British Columbia: Institute of Animal Resource

Ecology, University of British Columbia, 1980).

13. Baskerville, G. "Implementation of adaptive approach in provincial and federal forestry agencies", in International Institute for Applied Systems Analysis. *Adaptive Environmental Assessment and Management: Current Progress and Prospects for the Approach. Summary Report, First Policy Seminar,* CP-79-9 (Laxenburg, Austria: International Institute for Applied Systems Analysis, 1979).

14. Wood, F.E.A. "Experience in implementing adaptive management and assessment", in International Institute for Applied Systems Analysis. *Adaptive Environmental Assessment and Management: Current Progress and Prospects for the Approach. Summary Report, First Policy Seminar,* CP-79-9 (Laxenburg, Austria: International Institute for Applied Systems Analysis, 1979).

15. Smith, S.B., ed. *Alberta Oil Sands Environmental Research Program Interim Report,* covering period April 1975 to November 1978 (Edmonton, Alberta: Alberta Environment, 1979).

16. Wildavsky, A. *Speaking Truth to Power: The Art and Craft of Policy Analysis* (Toronto, Ontario: Little, Brown and Company, 1979).

17. Quade, E.S. *Analysis for Public Decisions* (New York, New York: Elsevier, 1979).

18. Hilborn, R. "Some failures and successes in applying systems analysis to ecological systems." *Journal of Applied Systems Analysis* 6(1979)25-31.

METHODS

INTRODUCTION TO METHODS FOR ENVIRONMENTAL IMPACT ASSESSMENT

Ronald Bisset

University of Aberdeen,
Aberdeen

INTRODUCTION

Many methods for environmental impact assessment (EIA) have been devised to aid identification, prediction and assessment of impacts and preparation of environmental impact statements (EISs). All these methods focus on impacts. Impact is a word much used in EIA, but little attention has been paid, in the academic and professional literature, to the nature of impacts. What is an impact? How do we know when an impact is likely to occur? These are questions which are addressed in this paper.

A discussion of the concept of impact opens the paper. Most attention will be paid to the socio-political character of impacts, but other aspects will be considered. Throughout the paper, reference will be made to EIA methods which are able to deal with some of the complex problems posed by the nature of impacts. A fuller, more descriptive discussion of many of these methods can be found in Bisset (1). This paper is essentially wide-ranging and discursive in attempting to deal with some basic issues related to impacts and EIA methods. It may provide a useful framework for other papers in this volume.

THE AIM OF EIA

EIA is a planning aid and is concerned with identifying, predicting and assessing impacts arising from proposed activities such as policies, programmes, plans and development projects which may affect the environment. The main aim of EIA is to improve decisions on development by increasing the quality and scope of information on likely impacts presented to decision-makers and members of the public. Although concerned with different levels of development activities (policies, plans and projects) EIAs are implemented mostly for major development projects such as highways, nuclear power stations and water-related projects. The importance of assessing policies and plans is increasingly recognized, but is proving difficult to implement because of institutional, political and technical problems. Consequently, most EIAs are undertaken for major projects in which a number of alternatives may be assessed.

In previous papers presented to this seminar, considerable attention has focused on the variety of institutional procedures which exist for the organization of national EIA systems. These procedures are generally administrative in nature and provide for the production of EISs to be used in decision-making. Usually, also, they attempt to ensure that EISs are produced to a uniform standard in accordance with established mechanisms. Procedures by themselves, however, cannot ensure that EISs contain structured information, produced in a "scientific" manner, likely to be of most use in decision-making. For this reason, methods to aid identification and assessment of impacts have been developed. Of course, the provision of information on impacts is no guarantee that it will be used by decision-makers, but that is a separate issue.

This paper deals simply with outlining EIA methods developed to ensure that the best possible information from EIA procedures is available to decision-makers and the public. All methods involve impacts. It is vital to consider the nature of impacts as they are the central concern of EIA methods.

THE POLITICAL DIMENSION OF IMPACTS

An important reason for considering the meaning and nature of the impact concept is that it may influence to some extent, the nature and operational characteristics of EIA methods. The use of these methods will subsequently have an effect on the content of EISs produced. Such EISs will form at least part of the information base upon which decision-makers and members of the public will come to conclusions concerning proposals. It is also useful to examine the impact concept to be aware of the complexity of the concept which forms most of the contents of EISs.

Reading the academic literature on EIA and a random selection of EISs leads one to the view that an impact is an event which results as a direct consequence of a prior event. However, not all such events would be included in EISs even if they were to arise from a development. A process of selection is constantly in operation to decide which events will be termed impacts and included in EISs and which will be excluded. For example, the use of pile-driving to prepare a development site gives rise to increased ambient noise levels. These increased noise levels are a direct result of the development activity. Are they an impact? As far as the air molecules are concerned circumstances have changed. Yet a description of the increased excitation rate of air molecules is unlikely to figure in many EISs. For decision-makers and the public it is the further indirect or second-order "impact" of these increased noise levels which are of concern. Will these levels disrupt the sleep of local residents at night? Will they annoy them during the day by making normal conversation more difficult? If these events were likely then they would in most cases be included in EISs as impacts.

Those likely to experience these events would consider them to be impacts as would most EIS writers and other people not likely to experience the noise levels. However, if we continue with this noise example, such unanimity may break down. For example, interrupted sleep patterns may affect work performance and increase family tensions. Thus, a second-order event may result in further events, some of which may not be perceived by those experiencing them. Individuals may blame increased noise levels for disrupted sleep, but not realize that increased family tension is due to this factor. For the operators of a development the impacts of increased noise levels may be thought of only in terms of the number of complaints received. This example shows some of the complexities involved in the concept of impact.

This example of the series of changes which can result from an initial increase in ambient noise levels indicates a major point about EIA. Not all environmental changes will find their way into EISs nor will all individuals or social groups agree on which changes should be classed as impacts for inclusion in EISs. It matters little what terminology is used - whether one distinguishes between effects and impacts, or changes in physical and environmental parameters and impacts - the fact remains that not all changes induced by a development will be classed as impacts. Trying to devise a suitable nomenclature, although laudable and perhaps necessary, cannot disguise the nature of the social and political choices involved in selecting and giving meaning to those likely changes resulting from a proposed project which will be termed impacts and included in EISs.

The social or political choices involved in EIA work are not made by society as a whole. It would be much easier if they were, as much disagreement in project planning would vanish. However, society is not a monolithic structure exhibiting an unitary structure. Societies are made up of groups and individuals with differing interests and values. Therefore, the allocation of significance to environmental changes depends on who is aware of the changes involved. For example, a particular change in the composition of a wildlife habitat may be seen as an ecologically significant local impact by local conservation groups. Other interest groups, for example, the unemployed looking for new work opportunities from a project, might consider the likely habitat change to be of no significance or not even recognise it as an impact at all.

Not only can non-expert social groups and individuals assign different values to environmental changes, but those preparing EISs, also, may act in this manner. For example, the scope of an EIS prepared by the regional office of the US Environmental Protection Agency (EPA) for a proposed wastewater treatment plant in Wyoming, reflected the values of EPA staff (2). A number of alternative sites were examined and, of these, one site was favoured by EPA whereas a different site was favoured by the council of the local community requiring the treatment plant.

The EIS was, therefore, a component of a political process revolving round the proposed plant. It was subject to the value choices of EPA staff, who ensured that the EIS considered the impacts of induced urbanization, thought to be harmful by EPA staff in many different ways. The impact of the expected urbanization was divided into a number of separate impacts; for example, loss of agricultural productivity, induced development cost, growth inducement and adherence to existing planning policies. These separate impacts were all different forms of expressing the same general impact. A scoring system was used to compare these impacts for the alternative sites. Each individual form of the general impact scored worst for the site favoured by the local council. Thus the format of impact identification was used in the EIS to further political ends. This EIS, like many others, one suspects, was not a purely technical, scientifically "objective" document.

Another characteristic of this social aspect of "impacts" is the time element. As techniques develop and more is understood of the relationships between environmental, social and health factors and their combined influence on individual well-being, "new" impacts may be recognised. Therefore, improved knowledge of the causal relations between development activity and environmental/social change leads to public recognition of impacts which were previously unsuspected.

This attempt to show the socio-political dimension of EIA is not new (3). Many working in EIA will consider the argument unoriginal both theoretically and practically. However, the significance of this aspect of EIA is often ignored or not considered by those producing EISs and developing EIA methods. For example, many EISs describe, in considerable technical detail, the likely concentrations of differing air or water pollutants at various locations relative to the sources of emission, but fail to relate these environmental changes, which are, in themselves, neither good or bad, to the interests and concerns of those groups which may use (economically, aesthetically, symbolically or by recreation) the resource being affected. Environmental changes are not converted into impacts and described in terms of their consequences for people. This cannot be done for all changes. Some, for example ecological changes, may affect only specialized social groups such as scientists and may only be considered impacts by such groups. An attempt should be made to trace all changes to social interests and how they will be affected.

Many environmental assessment methods leave out this social dimension. They are concerned solely with measuring environmental changes and ascribing a significance to them. For example, a change in dissolve O_2 in a river may be predicted and some concept of the importance of this change applied to the prediction. However, we have seen that such direct changes lead on, in many cases, to indirect impacts on social and individual interests. Methods which compartmentalize the environment into discrete units such as BOD levels, SO_2 concentrations and noise contours do not show the indirect effects of changes in these parameters. They stop short of attempting to assess the effect of the changes on people. Data on changes in these discrete units are frequently manipulated arithmetically and aggregated, often with weighting,to produce composite impact scores. Such scores are also of limited relevance because they do not express the ways in which social groups will be affected differentially by a proposal.

The importance of always tracing environmental changes to people is reinforced when it is realized that impacts (beneficial or harmful) from projects are not evenly and homogeneously distributed among social groups or individuals. Everyone does not benefit or suffer equally. Some people may only benefit, suffering none of the deleterious impacts borne by others. In addition, the mix of beneficial and harmful impacts is likely to be differentially distributed. It is important for decision-makers and members of the public to appreciate that certain social groups may be subject to more harmful and fewer beneficial impacts, while other groups may be in a more favourable position. This type of information is essential for public debate on the future of a proposed project.

To summarize, EIA is not a purely technical exercise. Value judgements are made through the entire process. At this point it should be stated that EIA is also an activity in which "objective", scientific description and prediction is a crucial component. The prediction of water pollution impacts, for example, requires use of dispersion models which provide data on concentrations at varying locations. Scientific method and a balanced viewpoint are very important, but the role of value judgements throughout the EIA process has tended to be ignored. Politics, as well as science, is part of EIA. This duality needs to be appreciated by those involved in EIA procedures.

CHARACTERISTICS OF IMPACTS

Apart from a demonstration of the "social" character of impacts, it is relevant to consider some other aspects of the environmental and social changes which are often termed impacts and included in EISs. One of the most potent views of "impact", and one which seems to have been incorporated into a large number of EIA methods, consists of the conception of an impact as being solely limited to the change in a single environmental or social parameter, caused by a particular activity associated with development. Methods which demonstrate this rather simplistic view of impacts include Leopold-type matrices (4) and both simple and complex checklists, for example the Environmental Evaluation System (EES) (5). Figure 1 shows a section of the Leopold matrix and Figure 2 shows the list of parameters to be considered when predicting impacts for a proposed project using EES. In both these cases, impacts are treated in a static manner and are considered as entities which can be easily separated, compartmentalized and removed outside time and space. All that is important is the individual changes in parameters which can be treated in isolation from each other.

For example, if the EES checklist were being used, impacts would be identified for each of the parameters listed likely to be affected by a project. In the case of "water quality" it is possible for all of the listed parameters, from basin hydrologic loss to turbidity, to undergo change. Each change or impact would be considered in isolation. No guidance is given on how, if at all, changes in one parameter might affect the others. Furthermore, nothing is said about the possibility that impacts may alter with time.

Since the introduction of formal EIA procedures in the USA there has been a slowly growing realization that a more complex notion of the nature and behaviour of impacts is needed for EIA. It was recognized in the early years of EIA that a change in one parameter might influence other related parameters and not just remain in its initial "box". The realization of the chain of impacts which might result from an initial impact led to the development of methods such as networks (6,7) and systems diagrams (8,9). Examples of each of these methods are shown in Figures 3 and 4. Basically, linked parameters likely to be affected are shown, and expected impacts can be traced by following the representation of the linked components in the system under study.

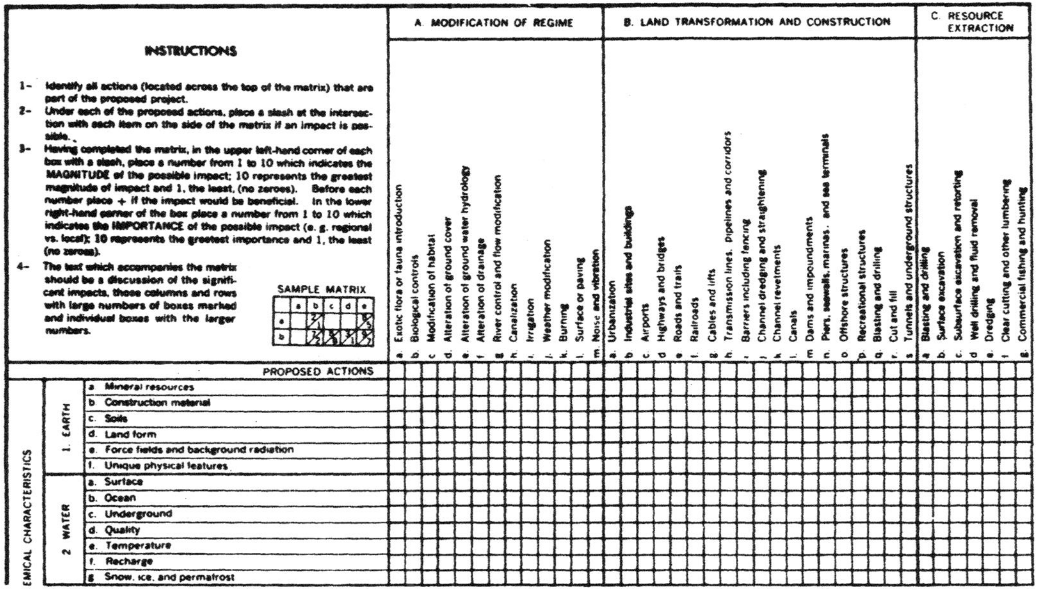

Figure 1: Section of the Leopold matrix (Reproduced by kind permission of the U.S. Geological Survey)

ECOLOGY

Terrestrial species and populations
Browsers and grazers (14)
Crops (14)
Natural vegetation (14)
Pest species (14)
Upland game birds (14)

Aquatic species and populations
Commercial fisheries (14)
Natural vegetation (14)
Pest species (14)
Sport fish (14)
Waterfowl (14)

Terrestrial habitats and communities
Food web index (12)
Land use (12)
Rare and endangered species
Species diversity (14)

Aquatic habitats and communities
Food web index (12)
Rare and endangered species (12)
River characteristics (12)
Species diversity (14)

Ecosystems
Descriptive only

AESTHETICS

Land
Geologic surface material (6)
Relief and topographic character (16)
Width and alignment (10)

Air
Odour and visual (3)
Sounds (2)

Water
Appearance of water (10)
Land and water interface (16)
Odour and floating material (6)
Water surface area (10)
Wooded and geologic shoreline (10)

Biota
Animals—domestic (5)
Animals—wild (5)
Diversity of vegetation types (9)
Variety within vegetation types (5)

Man-made objects
Man-made objects (10)

Composition
Composite effects (15)
Unique composition (15)

PHYSICAL/CHEMICAL

Water quality
Basin hydrologic loss (20)
Biochemical oxygen demand (25)
Dissolved oxygen (31)
Faecal coliforms (18)
Inorganic carbon (22)
Inorganic nitrogen (25)
Inorganic phosphate (28)
Pesticides (16)
pH (18)
Streamflow variation (28)
Temperature (28)
Total dissolved solids (25)
Toxic substances (14)
Turbidity (20)

Air quality
Carbon monoxide (5)
Hydrocarbons (5)
Nitrogen oxides (10)
Particulate matter (12)
Photochemical oxidants (5)
Sulphur oxides (10)
Other (5)

Land pollution
Land use (14)
Soil erosion (14)

Noise pollution
Noise (4)

HUMAN INTEREST/SOCIAL

Education/scientific
Archeological (13)
Ecological (13)
Geological (11)
Hydrological (11)

Historical
Architecture and styles (11)
Events (11)
Persons (11)
Religions and cultures (11)
" Western Frontier " (11)

Cultures
Indians (14)
Other ethnic groups (7)
Religious groups (7)

Mood/atmosphere
Awe inspiration (11)
Isolation/solitude (11)
Mystery (4)
" Oneness " with nature (11)

Life patterns
Employment opportunities (13)
Housing (13)
Social interactions (11)

Figure 2: List of environmental parameters in the Environmental Evaluation System (Reproduced by kind permission of the *Journal of Environmental Management*).

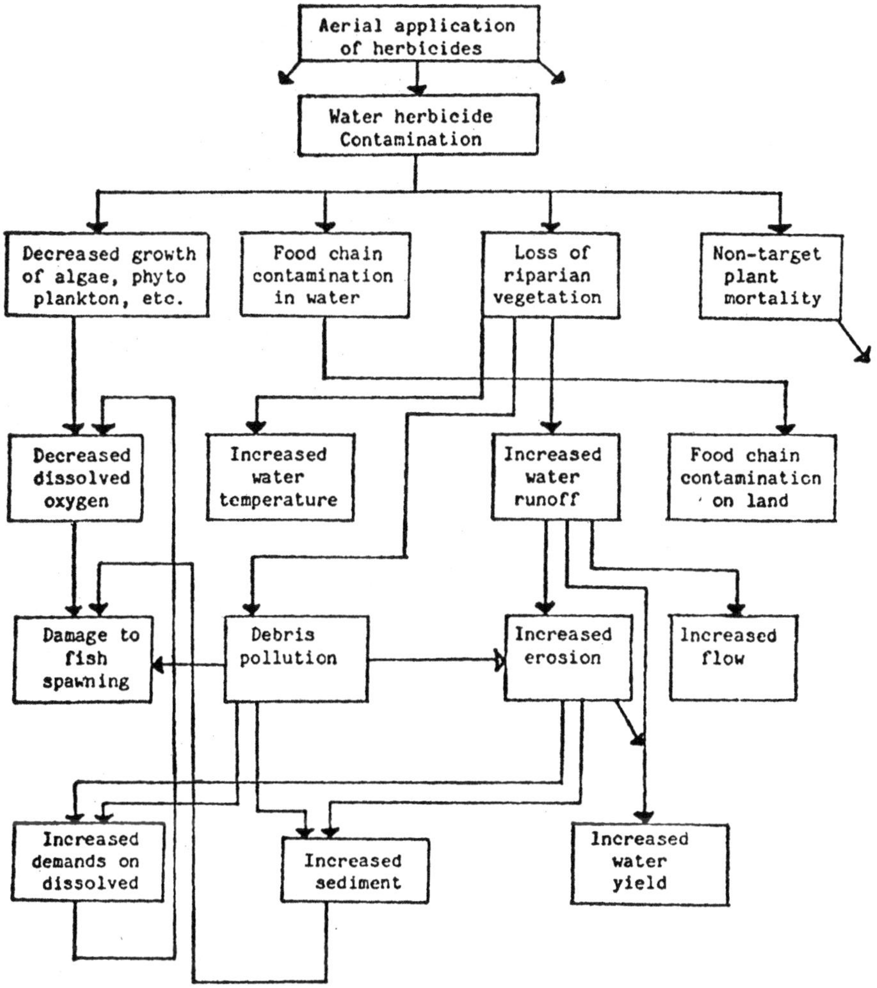

Figure 3: Section of the IMPACT network
(Reproduced by kind permission of Dr E. Thor)

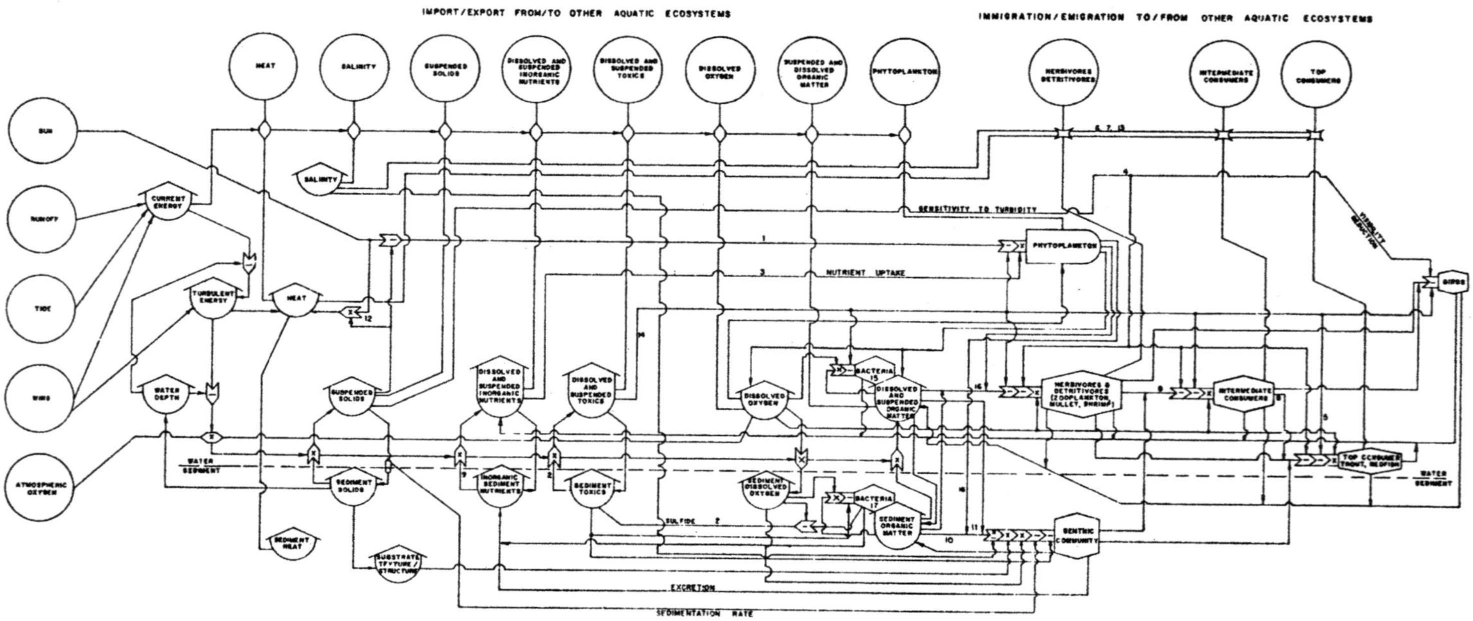

Figure 4: Ecological system diagram for medium salinity bay (Reproduced by kind permission of Texas General Land Office)

Knowledge of links makes it easier to understand how changes in one factor can affect the other factors.

As well as a realization of the importance of linked indirect impacts, there has been an understanding of the spatial attributes of impacts. Impacts can occur in the vicinity of a project, for example, fluorosis caused by an aluminium smelter. Alternatively, they can be observed at considerable distances. For instance, the impacts of acid rain in Scandinavian countries are thought to be caused, in part, by hydrocarbon combustion in the UK and northern, central Europe. The use of computerized overlays has attempted to deal with the spatial aspects of impacts. Such methods have been applied, usually, to the assessment of linear developments such as highways and pipelines. This method originates in the work of Ian McHarg (10) although he did not restrict its use to such projects. Krauskopf and Bunde (11) have developed a computerized overlay method for addressing alternative highway corridors. In this method, entitled the Regional Environmental Allocation Process, data on a large number of environmental and cultural variables are collected on a 1 km square cell basis and stored in a computer. These variables are subjectively weighted and the combination of weights and variables results in the creation of symbolic maps of the study area. Cells with the highest value are the most restrictive to highway location and are represented by the most dense print character. Using the computer to display sensitive areas allows severity and spatial distribution of the impacts of various highway alternatives to be determined. Also, new routes can be generated by using a programme to identify routes of least impact.

A more recent EIA method for considering, explicitly, the spatial dimensions of impacts has been developed by Voogd (12). Voogd argues that EIA should be "region" oriented and not "project" oriented. The method involves the use of a number of matrices and is quantitatively complex. Use of the method is exemplified by assessing future development of water-based recreation in part of the Waddensee.

Another aspect of the nature of impacts concerns the relationships between impact-causing factors. For example, chemicals emitted to the environment (such as pollutants in aqueous effluents discharged into an estuary) can each result in an impact. Many chemicals, however, can interact synergistically - where the combined effect of two toxic chemicals is greater than if their individual toxicities were additive. Other types of interactions between chemicals also occur, making the cumulative assessment of impacts an important part of EIA. This type of analysis is very difficult to accomplish, but this does not mean that it should be ignored. The Voogd method not only attempts to deal with the spatial aspects of impacts, but also with the cumulative aspect of impact behaviour.

In addition to the spatial and cumulative aspects of impacts, there is the very difficult problem of probability. Due to lack of knowledge of the responses of ecosystems to externally induced changes, it is often impossible to be precise when identifying, predicting and assessing impacts for EIS preparation. In many instances the likelihood of an impact occurring is uncertain. The difficulties of prediction lead to many impact predictions in EISs being couched in ambiguous language, or being qualified by statements which reduce any expression of certainty in the statements describing the impact. The probabilistic nature of many impacts makes decision-making difficult and consequently, some attention (although not very much) has been paid in the development of EIA methods to incorporating probability.

Two EIA methods which attempt to deal with the probability dimension are the Optimum Pathway Matrix, and a method based on multi-attribute utility functions. The former method was developed by the Institute of Ecology of the University of Georgia to evaluate alternative highway routes (13). 56 environmental parameters likely to be affected were identified and impacts on each parameter from the alternative routes were normalized, weighted and aggregated to produce composite impact scores so that alternatives could be compared directly. Once this had been done an assessment was made of the amount of error that might be present in the prediction of impacts. It was realized that predictions of the likelihood of impacts occurring might be inaccurate. Those who developed and applied the method argued that this approach allowed the probability of different alternatives having greater or lesser environmental impact to be considered. A sensitivity analysis showed that an estimated error of 50% did not affect the result of the initial aggregate score comparison.

The method based on multi-attribute utility functions was used to assess alternative nuclear power station sites on the Pacific north-west coast of the US (14). In this example only two ecological impacts were assessed, impacts on "salmonids" and "biologically important areas". Measures of the probability of impacts of certain magnitudes occurring were incorporated. For example, the probability of the extent of salmonid losses in a particular river was assessed. In one case, it was decided that the most probable impact would be a 15% loss of the annual average number returning to a particular stream to spawn. Probability estimates were included in the method to give overall scores for impacts to enable useful comparisons to be made.

Another characteristic of impacts concerns their reversibility. Some impacts are irreversible, that is, once they have occurred, reinstatement of original conditions is impossible. For example, the vegetation cleared from a building site cannot subsequently be replaced unless buildings are removed. However, other impacts are reversible. Noise levels during construction might rise by 10dB(A) above ambient levels. Once the construction activity causing the increase ceased, noise levels might return to previous ambient levels (if no other noise-producing activities were introduced). In between these two extremes there are gradations of reversibility. In some cases, it is possible to approximate to a previous situation. For example, derelict sites can be reinstated artificially to a vegetation cover which resembles the situation prior to disturbance. Reversion to a previous condition can be achieved in two ways. First, it can occur naturally through ecological processes such as colonization and succession. Secondly, it can be achieved artificially through direct human intervention.

The Optimum Pathway Matrix method is one of the few EIA methods which takes account of reversibility. Long-term irreversible impacts are given a greater weighting than short-term reversible impacts. Thus the reversibility character of each impact can be incorporated in the overall aggregate impact score for each alternative highway route assessed using the method.

The final impact characteristic to be considered in this paper relates to the temporal and dynamic nature of impacts. Not only do different impacts arise at the various stages in the development of a project (for example, the construction phase can be characterized by impacts which do not recur in other phases such as the operational period), but also individual impacts can exhibit different attributes over time. An impact may manifest itself immediately a causal factor is present. For example, noise levels from construction work may cause irritation to nearby people once work begins. Alternatively, an impact may only appear after a long period of time has elapsed and critical assimilative thresholds have been exceeded. The lethal effects of the bioaccumulation of heavy metals in aquatic organisms is an example of such a mechanism at work.

A variation on the Leopold matrix produced by Parker and Howard (15) has attempted explicitly to incorporate the time dimension by applying significance values (on a scale of 0-4 of increasing significance) to impacts for time periods from the day in which a development action is undertaken to a period of 15 years hence. In the example provided in Figure 5 it can be seen that in the case of the impact of drainage alteration on soils, immediate effects on the first day (given a score of 1) are not serious. However, the impact changes as time passes. The last number, 4, represents a severe, long-term impact. Conversely, the impacts of drilling fluid discharges on soils decrease in severity over time. The impact is given a score of 3 for the immediate time period after the discharge and declines over time until there is no impact after 15 years.

	MODIFICATION OF HABITAT	MODIFICATION OF SURFACE OR PAVING	ALTERATION OF DRAINAGE	WELL DRILLING AND FLUID REMOVAL	EMPLACEMENT OF TAILINGS AND OVERBURDEN	DRILLING FLUID DISCHARGE
SOILS	11111	11111	12344	11000	33333	32110
SURFACE, INCLUDING STREAMS	11111	11111	11111	33211	11110	33211
STABILITY (SLIDES & SLUMPS)	12321	11111	11110	-----	11000	11000
MICROORGANISMS	44432	21111	33211	11111	33211	44432
BENTHIC ORGANISMS	-----	-----	-----	-----	-----	-----

Figure 5: Section of matrix used to assess alternative drilling sites in Antarctica (modified after Parker and Howard).

The Parker and Howard matrix deals only with a rather simple view of the dynamic nature of impacts; its concept of time is limited. Impacts are projected onto a static baseline condition. Only the change of impacts over time is considered. The behaviour of the baseline upon which the impacts are being projected is not considered. It is now realized that environmental and social systems are changing constantly. There are various dynamic trends which determine the structure and functioning of habitats or ecosystems at any one time. To properly assess impacts it is necessary to project the baseline conditions onto the same future timeframe as the impact predictions. The concept of "moving baselines" is crucial for an adequate appreciation of likely impacts. The dynamic interaction between projected impacts and trends in the baseline will create a new situation. The "new" baseline will be acted upon by the impact-causing factors and additional changing impacts may characterize the altered systems. These systems may become very sensitive to any one of the impact-inducing factors, to which they were previously indifferent. Conversely, a system may react to nullify an impact and,if critical thresholds are not exceeded, an impact may decline and eventually disappear. A method which can cope with system dynamics is the simulation model. Holling and his colleagues have been at the forefront of attempts to apply holistic models to prediction of impacts (16). The paper by Everitt, later in this section, describes the principles and utility of this approach.

CONCLUSIONS

This paper has deliberately over-stated the case for recognition of the political aspects of impact identification and assessment. The discussion has been aimed at stimulating awareness of some of the implicit assumptions and concepts which underly many EIA methods and the decisions made by those implementing EIA studies. The discussion of other aspects is offered in the hope that it may be of some utility in reminding those responsible for preparing or reviewing EISs of those aspects which should be considered in EIA. It is realized that there is no solution, at present, to many of the complexities involved in presenting an account of impacts which is as full as possible. However, it is argued that a discussion of the various characteristics of impacts would improve the quality of EISs and would lead to a more informed, though not necessarily a more conclusive, public debate. Attention to these details in EISs may be unlikely to improve decision-making, but would provide more "honest" EISs. This would be a small step forward in the science/art of EIA.

REFERENCES

1. Bisset, R. "Methods for environmental impact analysis: recent trends and future prospects". *Journal of Environmental Management* 11(1980)27-43.

2. Elliott, M.L. "Pulling the pieces together: amalgamation in environmental impact assessment". *Environmental Impact Assessment Review* 2(1)(1981)11-38.

3. Matthews, W.H. "Objective and subjective judgements in environmental impact analysis". *Environmental Conservation* 2(2)(1975)121-131.

4. Leopold, L.B. *et al. A Procedure for Evaluating Environmental Impact,* U.S. Geological Survey Circular 645 (Washington, D.C.: U.S. Geological Survey, 1971).

5. Dee, N. *et al.* "An environmental evaluation system for water resources planning". *Water Resources Research* 9(3)(1973)523-535.

6. Thor, E.C. *et al.* "Forest environmental impact analysis - a new approach". *Journal of Forestry* 76(1978)723-725.

7. Thor, E.C. *IMPACT - A New Approach to Environmental Impact Assessment,* Paper presented to the Society of American Foresters National Workshop, Lexington, Kentucky, 10 August, 1978.

8. Gilliland, M.W. and P.G. Risser. "The use of systems diagrams for environmental impact assessment: procedure and an application". *Ecological Modelling* 3(3)(1977)188-209.

9. Texas General Land Office. *Ecological Systems Diagrams· Theory and Construction,* Technical Paper 12 of the Activity Assessment Routine (Austin, Texas: Texas General Land Office, 1978).

10. McHarg, I. *A Comprehensive Highway Route-selection Method,* Highway Research Record No. 246 (Washington, D.C.: Highway Research Board, 1968).

11. Krauskopf, T.M. and D.C. Bunde, "Evaluation of Environmental Impact Through a Computer Modelling Process", pp. 107-125 in R. Ditton and T. Goodale, eds. *Environmental Impact Analysis: Philosophy and Methods* (Madison, Wisconsin: University of Wisconsin Sea Grant Program, 1972).

12. Voogd, H. *Monitoring Environmental Qualities in Regional Planning,* Paper presented to the Annual Conference of the British Section of the Regional Science Association, London, September, 1980.

13. Institute of Ecology. *Optimum Pathway Matrix Analysis Approach to the Environmental Decision-Making Process* (Athens, Georgia: University of Georgia Institute of Ecology, 1971).

14. Keeney, R.C. and G.A. Robilliard. "Assessing and evaluating environmental impacts at proposed nuclear power plant sites". *Journal of Environmental Economics and Management* 4(2)(1977)153-166.

15. Parker, B.C. and R.V. Howard. "The first environmental monitoring and assessment in Antarctica: the Dry Valley Drilling Project". *Biological Conservation* 12(2)(1977)163-177.

16. Holling, C.S. ed. *Adaptive Environmental Assessment and Management* (Chichester, England: John Wiley, 1978).

EIA MANUALS: GENERAL OBJECTIVES AND THE PADC MANUAL

Brian D. Clark

University of Aberdeen,
Aberdeen.

INTRODUCTION

As EIA has grown in importance there has been a need to provide advice and assistance to those involved in the process of impact assessment. This has ranged from the development of specific methods to implement particular requirements of the EIA process, to broader advice and guidance. One such feature has been the development of EIA manuals. This chapter will consider the general objectives of EIA manuals and then describe in some detail one such manual that has been produced in the UK for use by central and local government, developers, consultants and the general public who may be affected by development proposals.

EIA MANUALS

The basic objective of manuals is to give advice as to how impact assessments could be undertaken for specific development types. Many of the manuals that have been published, and it should be noted that the majority have been produced in the United States, provide advice about identifying, predicting and displaying potential impacts. Many, however, do far more than this and include advice on the structuring of impact assessment, screening and scoping, responsibility for undertaking assessment and review procedures, personnel, cost, implementation and monitoring.

As has already been noted, most manuals have been produced in the United States. Often these have been commissioned by federal agencies because they are useful to agency staff in translating the generalized requirals of NEPA, CEQ guidelines and agency guidelines into detailed advice as to how specific development types should be assessed. This is important because, for example, information requirements for an environmental impact statement (EIS) on a research and development proposal differ from those needed for a highway project. Some manuals have been prepared for other organisations, such as the manuals to help local governments assess proposed projects (1,2). These manuals are relevant to the needs of planners and environmental scientists involved in impact assessment both within the United States and elsewhere.

The U.S. Department of Transportation and the U.S. Army Corps of Engineers have prepared more impact statements than other agencies and several manuals for highway proposals and civil and military engineering projects have been produced (3). The U.S. Army Corps of Engineers has responsibility not only for military engineering projects, but also for a wide range of civil engineering activities. Manuals have been commissioned to cover different types of civil engineering projects (4,5,6). The U.S. Army Construction Engineering Research Laboratory has developed a computerized system which can be used in the preparation of impact statements for differing activities (7). The system generates information on potential impacts when provided with data on the nature of the proposal and the surrounding area.

Manuals have been produced for other types of development including waste-water treatment plant and management plans (8), naval development (9), point pollution sources (10), coal-related energy development research projects (11), nuclear power stations (12,13), coal-fired power stations (14) and airport developments (15).

Whilst the greatest number of EIA manuals have been produced in the United States, the concept of manuals is now being adopted in many other countries. Several European countries that have adopted either a mandatory or informal EIA system have now produced guidance manuals of either a generic type or relating to specific development types. In the UK for example a manual was produced for the assessment of major industrial developments (16) and for major developments (17).

Another noticeable development has been the introduction of manuals in those developing countries where an EIA system has been introduced. It is perhaps unfortunate that in many cases the developing countries have adopted EIA manuals from the United States and it is slowly being realized that these may be inappropriate for the local environmental conditions and the distinctive decision-making frameworks which exist.

One further area where manuals are now being produced relates to the activities of a number of international agencies. The United

Nations Environment Programme, for example, has produced a document called *Guidelines for Assessing Industrial Environmental Impact and Environmental Criteria for the Siting of Industry* (18), which is a manual that gives advice as to how an EIA system could operate. It includes a wide range of advice on many aspects of EIA and can be thought of as a flexible approach to encourage the use of EIA. The World Bank has also produced an advice manual which indicates the kind of environmental factors that should be evaluated when international funding agencies are considering investment projects in developing countries (19).

THE PADC MANUAL

The manual (17) was formulated for use in the UK by central and local government, developers and the public. It was produced in an attempt to resolve two recurrent problems encountered by planners assessing proposals, namely, the difficulty of obtaining sufficient detailed information from prospective developers, and the lack of a systematic assessment procedure for the appraisal of proposals. Components of the assessment process are grouped into three sets of related activities (Figure 1). Although the approach is shown as a sequence of events, many of the activities may be performed concurrently. The method allows for public involvement throughout the process, both formally and informally. It should be noted that in the UK, the assessment of a proposal is undertaken by the government and not by the project proponent. The manual was also formulated so that EIA could be incorporated into both existing land use planning legislation and decision-making structures.

ACQUISITION OF INFORMATION ON A PROPOSAL

Initial Discussion with the Developer

Often a developer may only have preliminary details available concerning the project. It may then be necessary to invite the developer to prepare a short prospectus about the proposals, providing:

a) A list of key siting criteria of the proposed development, and information on sites considered and reasons for selection of the favoured site.

b) Brief details of the proposed development and the processes it will use when operational.

The depth of analysis, at this stage, will be limited by the information available, but some assessment of the proposal in relation to existing plans and policies may be possible. Informal negot-

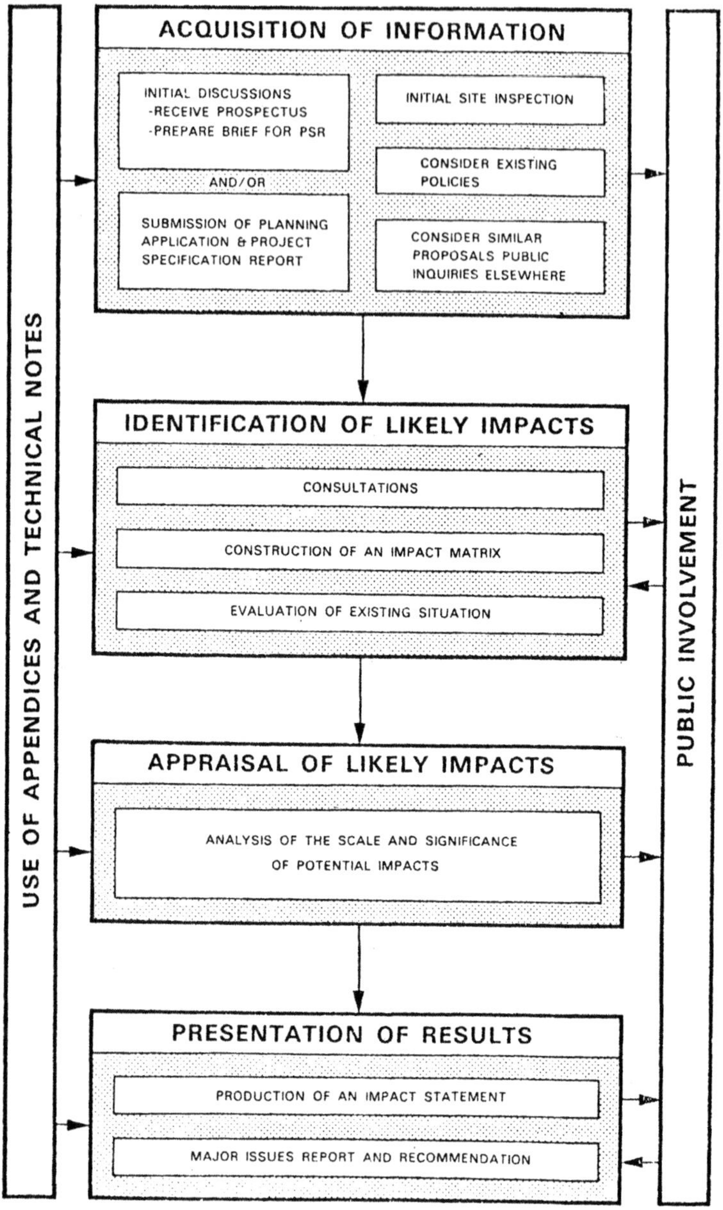

Figure 1: Linked activities in the appraisal method (Reproduced by kind permission of Her Majesty's Stationery Office)

iations with the developer could include a discussion of alternative sites, and if a formal application appears likely, a planning authority could begin the collection of relevant information, such as previous experience with similar developments and public inquiry reports where available.

Brief for a Project Specification Report (PSR)

Much of the information necessary for an assessment could be obtained by inviting the developer to complete a PSR, to be submitted close to the time that an application for planning consent is made. The PSR ought to provide the following details:

a) Physical characteristics of application site
- Land requirements
- Site layout (detailed plans at varying scales)
- Marine access characteristics (where appropriate)

b) Employment characteristics
- During construction phase
- When development is operational

c) Financial data
- Wage and salary levels
- Expenditure on locally produced inputs

d) Infrastructure requirements
- Raw material demand
- Transport requirements
- Water demand
- Electricity demand
- Gas demand
- Housing demand

e) Factors of environmental significance
- Noise levels
- Vibration levels
- Gaseous emissions
- Particulate emissions
- Odours
- Dust
- Discharge of aqueous effluents
- Solid wastes

f) Emergency services
- Fire and medical services
- Control of pollution at marine facilities

g) Hazard

A planning authority may wish to omit or supplement the above where appropriate.

The PSR will form the basis for a comprehensive appraisal of potential impacts. Applicants may benefit from the preparation of a PSR since the submission of detailed information at an early stage should promote more rapid and efficient processing of the application. If the development is approved, then this information may also be useful in the formulation of planning conditions 1). Developers may find it difficult to provide all the detailed information at this stage, although it is normal for some fairly detailed investigations concerning the development to have been prepared for major investments. Some information may be approximate, being based on past experience, but this can if necessary be improved on at a later date.

The information in the PSR can be checked by government departments or other suitable qualified consultants in order to bring to light unrealistic expectations which could increase or make the development impractical.

If information is withheld on the grounds of confidentiality then the authority ought to proceed with analysis on 'worst case' assumptions.

Initial Site Inspection

There is no need to stress the importance of site inspection for obtaining information about the proposed site and its surroundings, checking the accuracy of the submitted plans and assessing whether there are matters requiring special consideration.

Consideration of Existing Planning Policies

The planning authority can consider the proposal in relation to existing policies, previous planning decisions in the vicinity and the policies of statutory bodies.

Consideration of Similar Proposals

Information concerning previous experience with similar proposals ought to be of use in the assessment process, however it is important to determine the applicability of such experience to the present proposal.

THE IDENTIFICATION OF LIKELY IMPACTS

Consultations

Many consultations are statutory, although others can be undertaken on an advisory basis. Such consultations may be assisted by the publication of the PSR. A list of possible consultations can be prepared. In order to minimise delay, initial consultations may be requested within a period of one month, although further detailed advice from some consultees can be obtained later.

Construction of Impact Matrix

Once the information relating to a proposal, the site and its environs has been obtained, it is possible to identify the potential impacts. One of the most simple and effective ways of achieving this is by using a generalized impact matrix. An example of such is shown in Figure 2. In this matrix, characteristics of a development proposal are listed on one axis and site characteristics on the other. When it is thought that the interaction of components on the two axes is likely to result in an impact, the appropriate intersection is marked. Depending on the proposal to be appraised and the level of detail required, expanded matrices dealing with specific impacts may be constructed (Figure 3).

In both matrices it is important to consider separately the construction and operational phases of development, as some types of impact may occur only in one phase. On occasion it may be necessary to add an additional, for example, a restoration phase for mineral developments.

Evaluation of the Existing Situation

It is necessary to establish the nature of the existing environmental, social and economic conditions in the area surrounding a proposed development so that likely impacts can be identified and their implications assessed. Some of this information may be obtained from existing structure plans, regional reports, local plans, etc. Such plans will have considered many of the topics listed below and will have defined a clear context within which individual applications can be judged. Information on ambient levels of various pollutants may also be obtained from the relevant authorities. Information may be required on:

Physical characteristics of the site and its surroundings

a) Land
b) Water
c) Climate
d) Land use and landscape character

CONSTRUCTION PHASE

Characteristics of the Existing Situation	Immigration	Severance	Transport of raw materials	Transport of employees	Site preparation	Dust and particulates	Employment	Local expenditure	Water demand	Vibration	Noise	Gaseous emissions	Odours	Aqueous discharges	Solid waste disposal	Hazard
Climate																
Land uses		×														
Water quality					×	×								×		
Landscape quality					×									×	×	
Ecological characteristics																
Population density																
Tourism																
Employment structure							×									
Unemployment							×									
Local economy								×								
Traffic			×	×												
Water supply									×							
Sewerage																
Finance	×															
Education	×										×					
Health service facilities	×															
Housing	×									×	×					
Emergency services	×															×
Community structure	×															
Culture	×															

OPERATIONAL PHASE

Characteristics of the Existing Situation	Immigration	Structures	Severance	Water demand	Local expenditure	Employment	Transport of raw materials	Transport of employees	Transport of products	Noise	Vibration	Gaseous emissions	Odours	Dust and particulates	Aqueous discharges	Solid waste disposal	Hazard
Climate																	
Land uses			×									×					×
Water quality															×	×	
Landscape quality		×										×			×	×	
Ecological characteristics																×	
Population density	×																
Tourism																	
Employment structure						×											
Unemployment						×											
Local economy					×								×				
Traffic							×	×	×								
Water supply				×													
Sewerage															×		
Finance	×																
Education	×																
Health service facilities	×									×							
Housing	×									×	×	×					
Emergency services	×																×
Community structure	×																
Culture	×																

Figure 2: Generalized impact matrix (Reproduced by kind permission of Her Majesty's Stationery Office)

	CHARACTERISTICS OF THE PROPOSED DEVELOPMENT			
	CONSTRUCTION PHASE		OPERATIONAL PHASE	
Characteristics of the Existing Situation	NOISE	VIBRATION	NOISE	VIBRATION
Housing Schemes				
Ancient Monuments				
Educational Establishments				
Isolated Houses				
Hospitals and Old People's Homes				
Important Habitats				

Figure 3: Example of an expanded matrix (Reproduced by kind permission of Her Majesty's Stationery Office)

Ecological characteristics of the site and its surroundings
e) Habitats, communities and species

Human activity patterns in the area
f) Demographic aspects
g) Employment structure
h) Transport

Infrastructure services
i) Electricity
j) Gas
k) Water
l) Sewerage
m) Solid waste disposal
n) Finance
o) Education
p) Housing
q) Telecommunications

Social and community services
r) Health service facilities
s) Emergency services - fire and ambulance

Existing levels of environmental pollution
t) Air pollution
u) Water pollution
v) Noise and vibration
w) Radioactivity

APPRAISAL OF LIKELY IMPACTS

Although there is no method of systematically identifying and describing every potential impact of a proposed development, the approach outlined aids a comprehensive assessment of those which are significant. It is suggested that the appraisal should be based on the PSR, information on the existing situation and the impact matrices.

Where a likely impact was identified in a matrix, it is useful to undertake an analysis of the scale and significance of the potential change. This will help to establish not only the likely impacts on the local environment and community, but also the extent to which these impose constraints during both construction and operational phases. Technical notes are appended to the manual in order to assist the assessment of impacts.

Analysis of the Scale and Significance of Potential Impacts

It will usually be useful to determine whether a potential

impact is:

a) Beneficial or adverse. Certain impacts which are benficial to some social groups may have detrimental effects on others.
b) Short-term and/or long-term. Some impacts resulting from construction activities are short-lived, whereas others may last beyond the operating life of an installation.
c) Reversible or irreversible. Some impacts may be reversible in the sense that naturally occurring ecological processes may repair damage caused by adverse impacts. In other cases this cannot occur and the impacts are irreversible.
d) Direct and/or indirect. Impacts such as noise nuisance can arise directly from the characteristics of a development. Others are the result of combinations of impacts: loss of tourist-based employment may be caused by the depletion of fish stocks in a river resulting from a discharge of aqueous pollutants.
e) Local and/or strategic. Some impacts may be of only local significance; however, the hazard implication of developments such as natural gas liquids pipelines may be of strategic importance for a local authority area.

PRESENTATION OF RESULTS

Production of an Environmental Impact Statement (EIS)

The Environmental Impact Statement provides the format for the assembly of the conclusions regarding the potential impacts, in which each impact investigated may be considered in turn and its implications clearly stated. Details of how to prepare an EIS are provided in an appendix to the manual; however the EIS ought to include the following items:

a) A brief description of the proposed development.
b) A brief description of the local area.
c) Potential impacts.
d) Mitigating actions.
e) Examination of effects on the area if existing trends were to continue.
f) Consultations and objections.

A consideration of the likely situation without the particular development is also useful and ought to include:

a) Alternative employment projects in the area.
b) Dynamic environmental factors which may alter the 'value' of a site.
c) Changes due to planning policies, such as a change in the

industrial base.

This document could be available for public consultation and could be presented to a planning committee as part of the background material.

Major Issues Report and Recommendations

A Major Issues Report should be prepared, interpreting and summarising the findings of the EIS, and presenting them in the form of a simplified summary of options. Advice on how this may be structured is provided in an appendix to the manual. The report would serve two important functions:

a) It could be published, to indicate both the likely impacts of a proposed development and the implications of refusing an application.
b) It could be presented to committee as factual support for recommendations made by officers.

Summary sheets may provide a suitable means of drawing together technical information from various reports produced as part of the assessment procedure. Summary sheets (see Figure 4) are intended to be objective statements of the facts as accurately as they can be determined, although a written interpretation may be considered necessary.

In making recommendations it will be necessary to be aware of the priorities that exist within the area regarding such factors as amenity, employment and rate income. A list of priorities would allow those involved to examine, and possibly question, these priorities.

Impacts can be either ranked or weighted to indicate their relative importance. However this approach may create an illusion of objectivity which conceals the subjective nature of such judgements. A discussion of ranking and weighting is considered to be outside the scope of this paper.

PUBLIC INVOLVEMENT

A major development will inevitably generate considerable public interest. This constitutes a two-way process in which it is considered desirable to seek to involve the public by providing information and an opportunity for comment. Public involvement has an important dual role in assessment:

a) Initial public reactions to a proposal will indicate those potential impacts which most concern the public.

Potential impact of proposed development	Classification of impact	Description of potential impact
Loss of *Zostera* feeding areas of brent geese	Indicate where appropriate B = Beneficial **A** = Adverse St = Short term **Lt** = Long term R = Reversible **I** = Irreversible **D** = Direct In = Indirect L = Local **Sg** = Strategic	The reclamation programme of 56 hectares associated with the development proposals would cause disruption of siltation patterns in the estuary. Inundation of *Zostera* beds in the north bay by water heavily laden with fine silts would lead to loss of approximately 60% of this important habitat. It can be expected that there would be short term increases in the number of waders in this particular area; once the construction phase is complete and these silts are not replenished the number will return to its former level. A thin layer of anoxic mud will cover the beds. These would probably not be reinvaded by *Zostera*, so that 60% of this area would be permanently lost. The over-wintering brent geese, which constitute 20% of the world population, feed exclusively on *Zostera* and the reduced area of grazing would not support the present number of geese.

Figure 4: Example of a summary sheet (Reproduced by kind permission of Her Majesty's Stationery Office)

b) Once an EIS and major issues report have been produced, public comment on their content will play a role in the decision on an application.

Once a decision has been reached, it is equally important to provide respondents with a summary of all comments received, the criteria by which such comments have been evaluated, and how in conjunction with technical evidence they have contributed to the decision making process.

CONCLUSION

Most accounts of EIA methods contain discussions of the main types, and these are usually sub-divided into checklists (simple or scaling-weighting), matrices, networks, overlays and models (20,21). This focus has been criticized as being too restrictive: "... the standard definition of the range of assessment methodologies is too narrow" (22). Most EIA methods have been devised to aid some of the tasks involved in EIA (such as identification of impacts and presentation of results) or to help in the assessment of particular categories of impacts, for example, ecological impacts. Only a few methods can be used for all EIA activities and such methods (Environmental Evaluation System, etc.) have not met unanimous approval.

Manuals contain comprehensive guidance on how to assess particular projects and on the type of data which should be included in EISs. Often, the format and means of presenting information in EISs is suggested. In addition, manuals may contain a particular type of method, such as a simple interaction matrix, to help the implementation of assessment. However, the scope of manuals is broader than most methods. They are concerned with more than the ability of particular methods to carry out some EIA activities, such as the ability of networks to identify impacts. Unlike methods they may contain techniques for prediction of impacts. Manuals have received little attention in the EIA literature because there are large numbers in existence and many are project-specific. In contrast matrices, networks and other methods can be applied to a wide variety of projects in differing environmental settings.

FOOTNOTE

1. In the U.K. planning authorities may grant planning permission but impose constraints, termed "planning conditions", on the development.

REFERENCES

1. Schaenman, P.S. *Using an Impact Measurement System to Evaluate Land Development* (Washington, D.C.: The Urban Institute, 1976).

2. Research and Planning Consultants. *Activity Assessment Routine, Ecological Systems Component: Users Manual* (Austin, Texas: Texas General Land Office, 1978).

3. Wolsko, T.D. *et al. Environmental Impact Handbook for Highway Systems* (Illinois: Argonne National Laboratory, 1974).

4. Ortolano, L. *Impact Assessment in Water Resources Planning*, Paper to Short Course on Impact Assessment in Water Resource Planning, Ann Arbor, Michigan, 4 June, 1973.

5. Warner, M.L. *et al. An Assessment Methodology for the Environmental Impact of Water Resources Projects* (Columbus, Ohio: Battelle Columbus Laboratories, 1974).

6. Bradley, M.D. *Environmental Impact Statements in Water Resources Planning and Decision Making* (Tucson, Arizona: Arizona University, 1976).

7. Thomas, S.E. *et al. Computer-Aided Environmental Impact Analysis for Industrial, Procurement, Research, Development, Test and Evaluation Activities: User Manual* (Champaign Illinois: U.S. Army Construction Engineering Laboratory, 1978).

8. U.S. Environmental Protection Agency. *Manual for Preparation of Environmental Impact Statements for Wastewater Treatment Works. Facilities, Plans and 208 Areawide Waste Treatment Management Plans* (Washington, D.C.: U.S. Environmental Protection Agency, 1973).

9. Quimette, J.R. *Environmental Impact Model Development of Naval Operations* (China Lake, California: Naval Weapons Centre, 1975).

10. U.S. Environmental Protection Agency. *Environmental Impact Assessment Guidelines for Selected New Source Industries* (Washington, D.C.: U.S. Environmental Protection Agency, 1975).

11. Warner, M.L. *et al. Energy from Coal, Guidelines for the Preparation of Environmental Impact Statements* (Columbus, Ohio: Battelle Columbus Laboratories, 1975).

12. Young, J.R. *et al. Information Requirements for Controlled Thermonuclear Reactor Environmental Impact Statements* (Richland, Washington: Battelle Pacific Northwest Laboratories, 1975).

13. U.S. Nuclear Regulatory Commission. *Preparation of Environmental Reports for Nuclear Power Stations* (Washington, D.C.: U.S. Nuclear Regulatory Commission, 1976).

14. United Engineers and Construction Inc. *Guidelines for the Preparation of an Environmental Impact Report for a Fossil-Fueled Steam Electric Generation Station* (Philadelphia, Pennsylvania: United Engineers and Construction Inc., 1976).

15. Greiner Environmental Sciences Inc. *Environmental Assessment of Airport Development Actions* (Baltimore, Maryland: Greiner Environmental Sciences Inc., 1977).

16. Clark, B.D. *et al. The Assessment of Major Industrial Applications: A Manual,* Research Report No. 13 (London: Department of the Environment, 1976).

17. Clark, B.D. *et al. A Manual for the Assessment of Major Development Proposals* (London: Her Majesty's Stationery Office, 1981).

18. United Nations Environment Programme. *Guidelines for Assessing Industrial Environmental Impact and Environmental Criteria for the Siting of Industry,* UNEP Industry and Environment Guideline Series, Vol. 1 (Paris: UNEP Industry and Environment Office, 1980).

19. World Bank. *Environmental, Health and Human Ecological Considerations in Economic Development Projects* (Washington, D.C.: World Bank, 1974).

20. Skutsch, M.McC. and R.T.N. Flowerdew. "Measurement techniques in environmental impact assessment". *Environmental Conservation* 3(3)(1976)209-217.

21. Clark, B.D. *et al. U.S. Environmental Impact Assessment: A Critical Review,* Research Report 26 (London: Departments of Environment and Transport, 1978).

22. Coleman, D.J. "Environmental impact assessment methodologies: a critical review", pp. 35-59 in M. Plewes and J.B.R. Whitney, eds. *Environmental Impact Assessment in Canada: Processes and Approaches* (Toronto, Ontario: Institute for Environmental Studies, University of Toronto, 1977).

METHODS FOR ENVIRONMENTAL IMPACT ASSESSMENT: THEORY AND APPLICATION (EMPHASIS ON WEIGHTING-SCALING CHECKLISTS AND NETWORKS)

L. W. Canter

Professor of Civil Engineering and Environmental Science and Co-Director, National Center for Ground Water Research, University of Oklahoma, Norman, Oklahoma, United States of America

Several activities are required in the conducting of an environmental impact study, including impact identification, preparation of a description of the existing environment, impact prediction and assessment, and selection of the proposed action from a set of alternatives being evaluated to meet identified needs. The objectives of the various activities differ as do the usable methodologies for accomplishing the activities. The term methodology as used herein refers to structured approaches for achieving one or more of the basic activities. Since the passage of the National Environmental Policy Act (NEPA) in the United States, which became law on 1 January 1970, numerous methodologies have been developed to aid in achieving the various activities in the environmental impact assessment process. The purpose of this paper is to describe checklists and networks as types of methodologies developed for usage in environmental impact studies. Background comments are provided on the purposes for methodologies as well as a classification scheme. Brief information is included on matrices, with primary attention devoted to four types of checklists and, more specifically, to checklists which include importance-weighting of environmental factors, and the scaling or ranking of the impacts of alternatives on each factor. Information is also included on the use of networks and energy system diagrams and several other emerging types of methodologies. The desirable characteristics of methodologies are identified along with general research needs in this topical area.

BACKGROUND COMMENTS

Environmental impact methodologies can be categorized into interaction matrices and checklists, with networks and energy system diagrams representing variations of interaction matrices. Interaction matrices range from simple considerations of project activities and their impacts on environmental factors to stepped approaches which display interrelationships between impacted factors. Checklists range from simple listings of environmental factors to descriptive approaches which include information on measurement, prediction and interpretation of changes for identified environmental factors. Checklists may also involve the scaling or ranking of impacts of alternatives on each of the environmental factors under consideration. Scaling techniques include the use of numerical scores, letter assignments, or linear proportioning. Alternatives can be ranked from best to worst in terms of potential impacts on each environmental factor. The most sophisticated checklists are those involving the assignment of importance weights to environmental factors and the scaling or ranking of the impacts for each alternative on each factor. Resultant comparisons can be made through the development of a product matrix which consists of multiplying importance weights by the scale or rank for each alternative.

Methodologies can be useful throughout the impact assessment process, with certain ones being of greater value for specific activities. Table 1 identifies four activities and useful methodologies for achieving the requirements of the activities. Matrices and networks are particularly useful for impact identification, while weighting-scaling checklists find greatest application in the final evaluation of alternatives and the selection of a proposed action. It is not necessary to use a methodology in entirety in an environmental impact study; it may be instructive to use portions of methodologies for certain requisite activities.

While numerous methodologies have been developed since the passage of NEPA, and still additional methodologies are being tested, there is no "universal" methodology which can be applied to all project types in all environmental settings. It is unlikely that an all-purpose methodology will be developed due to lack of technical information as well as the need for exercising subjective judgment about predicted impacts in the environmental setting wherein the potential project may occur. Accordingly, the most appropriate perspective is to consider methodologies as "tools" which can be used to aid the impact assessment process. In that sense, every methodology should be project and location specific, with the basic concepts derivable from existing methodologies. Methodologies do not provide complete answers to all questions related to the impacts of a potential project or set of alternatives. Methodologies are not "cookbooks" in which a successful

Table 1: Applications of Methodologies in Impact Assessment Process

Process Activity	Useful Methodologies
Impact Identification	Matrices - simple stepped Networks Energy system diagrams
Describing Existing Environment	Matrices - simple stepped Networks Energy system diagrams Checklists - simple descriptive
Impact Prediction and Assessment	Energy system diagrams Checklists - descriptive scaling
Selection of Proposed Action	Matrices - simple stepped Checklists - descriptive scaling weighting-scaling

study is achieved by meeting the requirements of the methodologies. Methodologies must be selected based on appropriate evaluation and professional judgment, and they must be used with the continuous application of judgment relative to data inputs as well as analysis and interpretation of results.

One of the purposes for using methodologies is to insure that pertinent environmental factors that need to be considered are included in the study. Most methodologies contain lists of environmental factors ranging from about 50 to 1,000 items, with the majority having between 50 and 100 items. Another purpose for using methodologies is to aid in planning baseline studies in locations where relevant environmental data are lacking. If information were not available on the factors identified in appropriate methodologies, this would indicate that field studies would be necessary (1).

One of the most important reasons for using methodologies is that they provide a means for evaluation of alternatives on a common basis. Within the United States the analysis of alternatives in environmental impact studies has often been less than desirable. In many cases, alternatives have been eliminated from detailed consideration based only on economic comparisons. Usage of structured methodologies can provide the basis for evaluation of alternatives using a common framework of decision factors. Methodologies can also be useful in evaluating the cost-effectiveness of proposed impact mitigation measures. Evaluation of a proposed project with and without mitigation will enable a clearer delineation of the effectiveness of potential measures.

An important element in environmental impact studies is the communication of resultant information to other practitioners, regulatory agencies, and the general public. Some methodologies have features which are particularly useful in communicating impact information in summary form. Finally, in the U.S., NEPA requires that agencies shall utilize methods and procedures which will ensure that unquantified environmental amenities and values may be given appropriate consideration in decision-making along with more traditional economic and technical considerations.

While numerous impact methodologies exist, and more are being developed, they are not always used in environmental impact studies. In 1975, Dee, *et al.*(2), identified three reasons for the general non-use of methodologies: a) different perspectives on the role of environmental impact statements (EIS's) held by methodology developers and practitioners; b) limited direction provided by administrative guidelines related to EIS implementation; and c) political processes surrounding major development proposals. Additional reasons for limited usage of methodologies include: a) information gap due to lack of knowledge of available methodologies on the part of many practitioners, b) non-existence of current technology during the early years following the passage of NEPA, and c) general reluctance by practitioners to use approaches perceived to be time- and cost-consuming.

Despite the general perception that methodologies have not been extensively used, there is evidence of considerable usage if EISs are systematically examined. For example, Canter (3) reported on a comparative analysis of 28 EISs dealing with municipal wastewater programs. The EISs in the sample group represented approximately one year of EIS preparation work by the U.S. Environmental Protection Agency in their construction grants program. Table 2 summarizes the methodologies used in 20 EISs, with the totals consisting of four descriptive checklists, four ranking checklists, six scaling checklists, one weighting-ranking checklist, and five weighting-scaling checklists. For the eight EISs in which no selection methodology was identified, it is possible that one or more methodologies, or parts thereof, were used but simply not documented.

Usage of impact assessment methodologies is expected to increase in the U.S. due to the Regulations concerning impact assessment, issued by the Council on Environmental Quality (CEQ), which became operational in 1979, and to the growing emphasis on public participation. The CEQ Regulations indicated that the analysis of alternatives represented the heart of the process, and that agencies should insure the professional integrity, including scientific integrity, of the discussions and analyses by extensive use of appropriate methodologies (4). Active public participation programs encourage practitioners to utilize the most appropriate methodologies in environmental impact studies. When decisions related to alternatives have to be presented and defended in the public arena, usage of systematic methodologies for comparisons of alternatives is anticipated to increase. Finally, another reason for the anticipated increase in the use of methodologies is the increasing knowledge base on the part of practitioners for the conduction of environmental impact studies.

INTERACTION MATRIX METHODOLOGIES

Interaction matrices were one of the earliest types of methodologies utilized in the U.S. The simple matrix refers to a display of project actions or activities along one axis, with appropriate environmental factors listed along the other axis of the matrix. When a given action or activity is anticipated to cause a change in an environmental factor, this is noted at the intersection point in the matrix and further described in terms of magnitude and importance considerations. Many variations of the simple interaction matrix have been utilized in environmental impact studies. It is beyond the scope of this paper to present

Table 2: Methodologies Used in Environmental Impact Statements on Municipal Wastewater Programs (Reproduced by kind permission of Information Resources Press, Inc.)

Location and Description of Methodology
Scarborough, Maine -- Ranking checklist: Alternatives compared in terms of costs, primary and secondary environmental effects, public health risk, administrative feasibility, and political opposition. Each alternative ranked from most to least acceptable for each factor.
Cranston, Rhode Island -- Weighting/ranking checklist: Each factor assigned a Weighting Factor Multiplier (weight); each alternative given a ranking relative to each factor from 1 to 5, with 1 denoting the most favorable rating and 5 the least favorable rating. A summation of products of weights and ranks yielded a composite score for each alternative.
New Milford, Connecticut -- Scaling checklist: The factors were water quality and quantity, air, biology, state land-use policy, town land-use policy, aesthetics, and historical/archaeological sites. Six scales of impact were used: significant beneficial (+3), minor beneficial (+2), insignificant beneficial (+1), insignificant adverse (-1), minor adverse (-2), and significant adverse (-3). Scale values totalled for a given alternative.
Henrico County, Virginia -- Descriptive checklist: The impacts of the alternatives were qualitatively and quantitatively (where possible) displayed in a summary table. The factors considered include historic/archaeological sites, terrestrial impacts, social impacts, other impacts, and engineering and economic comparisons.
Grand Strand Region, South Carolina -- Weighting/scaling checklist: Each factor assigned an importance weight; each alternative scaled from 0-10 relative to each factor (0-4 denoted detrimental, 5 denoted neutral or no change, and 6-10 denoted beneficial). A summation of the products of the weights and scales yielded a composite score for each alternative.
City of Greensboro, Guilford County, North Carolina -- Weighting/scaling checklist: Each factor assigned a relative importance by an advisory committee, the adverse impacts of each alternative relative to each factor assigned scaled values by a multidisciplinary professional group; and a figure of merit calculated for each alternative by summing the products of the relative importance

Table 2 (continued)

Location and Description of Methodology
weights and the scaled values.
Calumet Tunnel System, Chicago, Illinois -- Ranking checklist: The alternatives were ranked from best to worst for each of a series of factors.
Lower Des Plaines Tunnel System, Chicago, Illinois -- Ranking checklist: The alternatives were ranked from best to worst for each of a series of factors.
Denver, Colorado -- Scaling checklist: Each alternative compared in terms of environmental impacts (beneficial and adverse), engineering factors (feasibility and reliability), institutional ability to implement, and costs. An overall scale value for each alternative was assigned, and five scales were used: unacceptable, marginal, acceptable, favorable, and most favorable.
Jackson, Wyoming -- Weighting/scaling checklist: Relative importance weights from 1 (low) to 3 (high) were assigned to each of the 22 factors utilized. Impact scaling was accomplished by assigning values ranging from +5 (long-term beneficial) to -5 (long-term detrimental) for both primary and secondary impacts. Summation of the products of the weights and scales (algebraic addition of primary and secondary scale values) yielded a composite score for each alternative.
Boulder, Colorado -- Scaling checklist: Each alternative was screened relative to treatment effectiveness, costs, and compatibility with other plans. Three alternatives were eliminated following preliminary screening, and the remaining five were evaluated by a scaling checklist for short-term impacts and a qualitative discussion of secondary and long-term impacts.
Steamboat Springs, Colorado -- Descriptive checklist: Alternatives compared in terms of costs and impacts on water quality, geology, soils,biology, air quality, and land use.
East Bay Dischargers Authority, California -- Weighting/scaling checklist: The factors included short-term and long-term primary impacts, as well as secondary impacts and monetary costs. Each of these factors assigned qualitative relative-importance weights (maximum or moderate or minimum), and each alternative scaled relative to each factor in terms of being excellent, good, fair, or poor.

Table 2 (continued)

Location and Description of Methodology
East Central Contra Costa County, California -- <u>Weighting/scaling checklist</u>: Relative-importance weights assigned to individual factors. Each alternative scaled relative to each factor, and final evaluation consisted of a rating for each alternative.
Orange County, California -- <u>Descriptive checklist</u>: Summary tables used to display relative comparisons of alternatives.
Los Angeles County, California -- <u>Descriptive checklist</u>: Each alternative compared in terms of monetary cost, energy consumption and resource conservation, water quality and aquatic biology impacts, air pollutant emissions, impacts on people and neighborhoods, impacts on rare or endangered species, and impacts on areas of special significance (biological, historical, archaeological, or paleontological).
North Monterey County, California -- <u>Scaling checklist</u>: Factors grouped into physical/biological impacts, economic impacts, and social impacts. Each alternative scaled relative to each factor in the physical/biological and social impacts groups -- scales used were beneficial, adverse, problematical (unknown or open to question), and none; and each alternative was ranked from 1 (most acceptable) to 6 (least acceptable) for each economic impact factor.
Jerome, Idaho -- <u>Ranking checklist</u>: Each alternative ranked from 1 (most desirable) to 8 (least desirable) for monetary costs, environmental impacts, and legal compliance, and rankings for each category for each alternative were summed.
Jackson County, Oregon -- <u>Scaling checklist</u>: Each alternative scaled for adverse impacts only, and scaling was major adverse (+), moderate adverse (O), minor adverse (-), and no adverse impact (N).
Jacksonville, Oregon -- <u>Scaling checklist</u>: Each alternative scaled relative to short-term impacts, ability to meet project and environmental objectives, and long-term impacts. Monetary costs also considered.

detailed information; however, Table 3 provides a listing of example matrices.

CHECKLIST METHODOLOGIES

Checklist methodologies range from listings of environmental factors to highly structured approaches involving importance weightings for factors and the application of scaling techniques for the impacts of each alternative on each factor. Simple checklists represent lists of environmental factors which should be addressed; however, no information is provided on specific data needs, methods for measurement, or impact prediction and assessment. Simple checklists were extensively used in the initial years following the passage of NEPA, and they represent a valid approach for providing systemization to an environmental impact study. Table 4 contains a listing of the environmental factors in a simple checklist utilized for gas pipeline projects in the United States (21).

Descriptive checklists refer to methodologies that include lists of environmental factors along with information on measurement and impact prediction and assessment. These checklists are widely used in environmental impact studies. Carstea, *et al.* (22) developed a descriptive checklist approach for projects in coastal areas. The methodology addressed the following actions/projects: riprap placement; bulkheads; groins and jetties; piers, dolphins, mooring piles, and ramp construction; dredging (new and maintenance); outfalls, submerged lines and pipes; and aerial crossings. For each of the actions/projects, environmental impact information was provided on potential changes in erosion, sedimentation, and deposition; flood heights and drift; water quality; ecology; air quality; noise; safety/navigation; recreation; aesthetics; and socio-economics.

Several descriptive checklists have been developed for water resources projects. Figure 1 contains a list of 62 environmental factors related to the environmental quality account used for project evaluation in the United States (23). For each factor information is included on its definition and measurement, prediction of impacts, and functional curves for data interpretation (where curves available or easily developed). Descriptive checklist methodologies for reservoir projects were developed in 1973 by the Environmental Impact Center (EIC) (24), and in 1974 by Battelle-Columbus Laboratories (BCL) (25). The EIC method focuses on variables for the biological and physical/chemical environment. Information is also included on socio-economic factors and their interrelationships with natural and human systems. For certain biological and physical/chemical factors, information is provided on impact prediction. The BCL methodology includes information on the technical aspects of impact prediction for water quality and

Table 3: Example References on Matrix Methodologies

Reference and Comments
(5) -- Computerized matrix for identifying impacts to be addressed for U.S. Army military activities.
(6) -- Priority-trade-off-scanning using 3 types of matrices.
(7) -- Use of series of three matrices in defining environmental setting, identifying impacts and summarizing the features of alternatives.
(8) -- A cross-impact simulation technique for impact identification and prediction is described.
(9) -- Use of interaction matrix to identify and visually display potential impacts.
(10) -- Matrix displaying interrelated clusters of high-valued ratings.
(11) -- Stepped or linked matrices to display interrelationships between impacted assessment variables.
(12) -- Matrix analysis and narrative description for wastewater management systems.
(13) -- Concepts of cross-impact matrix and disruption matrix.
(14) -- Compares several matrix approaches for impact assessment.
(15) -- Cross-impact matrix procedure for dealing with long-range impacts.
(16) -- Matrix used to evaluate impacts of alternative land uses in terms of cost-benefit approaches.
(17) -- Matrix for evaluation of trade-offs between economic and environmental objectives in water resources planning.
(18) -- Matrix for nuclear power plant site selection.
(19) -- Use of matrix or stepped matrix approaches in conjunction with linear vector or non-linear evaluation systems.

Table 3 (continued)

Reference and Comments
(20) -- Matrix for summarizing impacts of water resources projects on stream characteristics.

Table 4: Simple Checklist Methodology for Gas Pipeline Projects (Reproduced by kind permission of U.S. Federal Power Commission)

Category	Comments
Land Features and Uses	Identify present uses and describe the characteristics of the land area. Land uses -- Describe the extent of present uses, such as, in agriculture, business, industry, recreation, residence, wildlife, and other uses, including the potential for development; locate major nearby transportation corridors, including roads, highways, ship channels, and air traffic patterns; locate transmission facilities and their placement (underground, surface, or overhead); identify water resources. Topography, physiography, and geology -- Provide a detailed description of the topographical, physiographical, and geological features within the area of the proposed action. Include U.S. Geological Survey Topographic Maps, aerial photographs (if available), and other such graphic material. Soils -- Describe the physical characteristics and chemical composition of the soils, including the relationship of these factors to land slope.
Species and Ecosystems	Identify those species and ecosystems that will be affected by the proposed action. Species -- List in general categories, by common and scientific names, the plant and wildlife species found in the area of the proposed action and indicate those having commercial and recreational importance. Communities and associations -- Describe the dominant plant and wildlife communities and associations located within the area of the proposed action. Provide an estimate of the population densities of major species. If

Table 4 (continued)

Category	Comments
	data are not available for the immediate area of the proposed action, data from comparable areas may be used. Unique and other biotic resources -- Describe unique ecosystems or rare or endangered species and other biotic resources that may have special importance in the area of the proposed action.
Socio-economic Considerations	If the proposed action could have a significant socio-economic effect on the locl area, discuss the socio-economic future of the area without the implementation of the proposed action; describe the economic development in the vicinity of the proposed action, particularly the local tax base and per capita income; and identify trends in economic development and/or land use of the area, from both an historical and a prospective viewpoint. Describe the population densities of both the immediate and generalized area. Include distances from the site of the proposed action to nearby residences, cities, and urban areas and list the populations of these areas. Indicate the number and type of residences, businesses, and industries that will be directly affected and those requiring relocation if the proposed action occurs.
Air and Water Environment	Describe the prevailing climate and the quality and quantity of the air and water resources of the area. Climate -- Describe the climatic conditions that have prevailed in the vicinity of the proposed action: extremes and means of monthly temperatures, precipitation, and wind speed and direction. In addition, indicate the frequency of temperature inversions, fog, smog, and destructive storms such as hurricanes and tornadoes. Hydrology and hydrography -- Describe surface

Table 4 (continued)

Category	Comments
	waters, fresh, brackish, or saline, in the vicinity of the proposed action and discuss drainage basins, physical and chemical characteristics, water use, water supplies, and circulation. Describe the ground water situation, water uses and sources, aquifer systems, and flow characteristics.
	Air, noise, and water quality -- Provide data on the existing quality of the air and water (indicate the distance(s) from the proposed action site to monitoring stations) and the mean and maximum noise levels at the site boundaries.
Unique Features	Identify unique or unusual features of the area, including historical, archeological, and scenic sites and values.

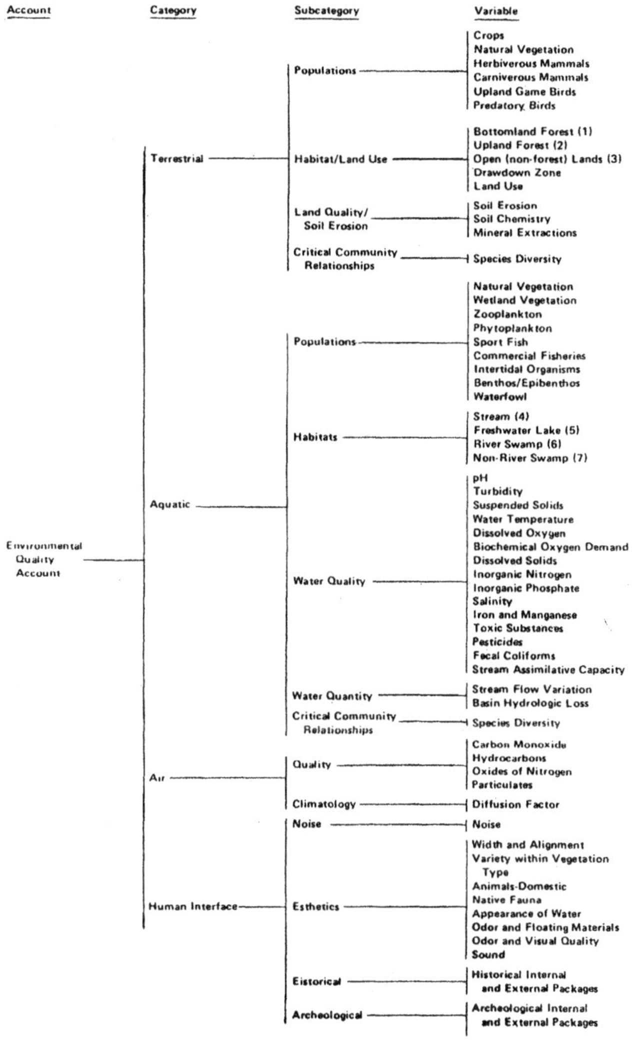

Figure 1: Structure of the Environmental Quality Account (Reproduced by kind permission of Ann Arbor Science Publishers, Inc.)

ecological impacts (26). A descriptive checklist for dredging projects is also available, with the methodology including 84 environmental factors along with information on impact prediction for many of them.

Descriptive checklists have also been used for impact studies associated with military activities (27). Examples of actions evaluated include: construction; operation, maintenance and repair; training; mission change; real estate; procurement; industrial activities; research, development, test and evaluation; and administration and support. The methodology consists of using an interaction matrix to identify potential impacts and thus generate a descriptive checklist. Forty-six environmental factors included in the methodology are listed in Table 5. A similar approach has been developed for the U.S. Air Force (28).

Descriptive checklists are also used for transportation projects and land development projects. The transportation methodology addresses social, economic, and physical impacts of highway construction and operation (29). Social impacts include those related to community cohesion, accessibility of facilities and services, and displacement of people. Economic impacts relate to those on employment, income and business activity, residential activity, property taxes, regional and community plans and growth, and resources. Physical impacts address changes on aesthetics and historic values, terrestrial and aquatic ecosystems, air quality, noise and vibration. For each of the identified environmental factors, workable state-of-the-art methods and techniques for impact identification, data collection, analysis and evaluation are included. A descriptive checklist methodology containing 47 factors for housing and other land development projects is summarized in Table 6 (30). The "bases for estimates" column presents a simplified, brief listing of key data and models (if any) needed for the factor. The methodology has been field-tested in Phoenix, Arizona; Indianapolis, Indiana; and Montgomery County, Maryland.

Another major group of checklists can be classified as scaling or ranking checklists. Scaling refers to the assignment of algebraic- or letter-scales to the impact of each alternative being evaluated on each identified environmental factor. Ranking checklists refer to those approaches in which alternatives are ranked from best to worst in terms of their potential impacts on identified environmental factors. These types of checklists are useful for comparative evaluations of alternatives, thus they provide a basis for selection of the preferred alternative. One example of a scaling checklist is the methodology developed for the social well-being account for water resources projects (31). This methodology contains 389 factors descriptive of the socio-economic environment, with assessment of baseline conditions involving comparisons with appropriate regional or national data. Impact

Table 5: Environmental Factors in Descriptive Checklist for U.S. Army (Reproduced by kind permission of U.S. Department of the Army)

Category	No.	Factor	No.	Factor
Air	1	Diffusion factor	6	Carbon monoxide
	2	Particulates	7	Photochemical oxidants
	3	Sulphur oxides	8	Hazardous toxicants
	4	Hydrocarbons	9	Odor
	5	Nitrogen oxide		
Water	10	Aquifer safe yield	17	Biochemical oxygen demand
	11	Flow variations	18	Dissolved oxygen (DO)
	12	Oil	19	Dissolved solids
	13	Radioactivity	20	Nutrients
	14	Suspended solids	21	Toxic compounds
	15	Thermal pollution	22	Aquatic life
	16	Acid and alkali	23	Fecal coliform
Land	24	Erosion	26	Land use patterns
	25	Natural hazard		
Ecology	27	Large animals (wild and domestic)	31	Field crops
			32	Threatened species
	28	Predatory birds	33	Natural land vegetation
	29	Small game	34	Aquatic plants
	30	Fish, shell fish, and water fowl		
Sound	35	Physiological effects	38	Performance effects
	36	Psychological effects	39	Social behavior effects
	37	Communication effects		
Human	40	Life styles	42	Physiological systems
	41	Psychological needs	43	Community needs
Economic	44	Regional economic stability	46	Per capita consumption
	45	Public sector revenue		

Table 6: Descriptive Checklist for Land Development Projects (Reproduced by kind permission of the Urban Institute)

Factor	Bases for Estimates
I. Local Economy	
Public Fiscal Balance	
1. Net change in government fiscal flow (revenues less expenditures).	Public revenues: expected household incomes by residential housing type; added property values. Public expenditures: analysis of new service demand, current costs; available capacities by service.
Employment	
2. Change in numbers and percent employed, unemployed, underemployed, by skill level.	Direct from new business; or estimated from floor space, local residential patterns, expected immigration, current unemployment profiles.
Wealth	
3. Change in land values.	Supply and demand of similarly zoned land, environmental changes near property.
II. Natural Environment	
Air Quality	
Health	
4. Change in air pollution concentrations by frequency of occurrence and number of people at risk.	Current ambient concentrations, current and expected emissions, dispersion models, population maps.
Nuisance	
5. Change in occurrence of visual (smoke, haze) or olfactory (odor) air quality nuisances, and number of people affected.	Baseline citizen survey, expected industrial processes, traffic volumes.
Water Quality	
6. Changes in permissable or tolerable water uses and number of people affected - for each relevant body of water.	Current and expected effluents, current ambient concentrations, water quality model.
Noise	
7. Change in noise levels and frequency of occurrence, and number of people bothered.	Changes in nearby traffic or other noise sources, and in noise barriers; noise propagation model or nomographs relating noise levels to traffic, barriers, etc.; baseline citizen survey of current satisfaction with noise levels.
Wildlife and Vegetation	
8. Change in diversity and population size (abundance) of wildlife and vegetation (including trees) of common species.	Wildlife and vegetation inventory; expected removal of cover or changes to habitats.
9. Change in numbers of rare or endangered species.	Changes in amount and quality of (a) habitat by animal type; (b) green space, or (c) number of mature trees.

Table 6 (continued)

Natural Disasters	
10. Change in number of people and value of property endangered by: flooding, earthquakes, landslides, mudslides, and other natural disasters, by frequency of occurrence.	Flood plain and other hazard maps; changes in local topography and sewering; change in percent permeable cover; stream flow and hydraulic models.
III. Aesthetics and Cultural Values	
Attractiveness	
11. Change in number and percent of citizens who are satisfied with neighborhood appearance.	Baseline citizen survey of ratings of current attractiveness and identification of problems and assets; visual simulation of proposed development using retouched photos, drawings or 3-D models for assessing future preferences using a sample of citizens.
View Opportunities	
12. Change in number or percent of citizens satisfied with views from their homes (or businesses).	Baseline citizen survey; geometric analysis of structures to identify view opportunities before and after development.
Landmarks	
13. Number and perceived importance of cultural, historic, or scientific landmarks to be lost, made less accessible, or made more accessible.	Inventory and importance ranking of landmarks; survey of citizens and scholars regarding importance.
IV. Public and Private Services	
Drinking Water	
Availability	
14. Change in frequency duration and severity of water shortage incidents, and number of people affected.	Current usage, expected new demand; projected supplies.
Quality	
15. Changes in salinity and other indices of drinking water quality and safety, and number of people affected.	Expected effluents from new development; purification process used; current and expected usage; profile of underground water systems.
Hospital Care	
Emergency Care Availability	
16. Change in number of citizens	Maps of population distribution and emergency facilities; number of emergency vehicles (if any), expected calls, and dispatch policy.
Availability/Crowdedness	
17. Change in potential bed need versus bed supply of area hospitals, by type of clinical service (medical, surgical, pediatric, obstetrical).	Current patient hospital bed days per 1000 population by sex-age group and medical service; available bed capacities; expected population by sex-age group.

Table 6 (continued)

Crime Control	
Crime Rate	
18. Change in rate of crimes in existing community.	Current crime rates and case histories of similar neighborhood changes; changes in community lighting, sightlines, hiding places, people mix.
Feeling of Security	
19. Change in percent of people feeling a lack of security from crime.	Baseline citizen survey plus the data above.
Fire Protection	
20. Change in fire incidence, property loss, and casualty rates.	Incident rates by occupancy types; people mix; available water supply; available fire suppression equipment and manning; likely building materials; site plan if available.
Recreation-Public Facilities	
Overall Satisfaction	
21. Change in number and percent of households satisfied with public recreation opportunities.	Baseline citizen surveys, and expected changes in facilities and environment (noise, air quality, dangers).
22. Change in number or percent of households using facilities (viewed relative to nominal capacity), by facility.	Citizen survey.
Accessibility	
23. Change in number and percent of households with access to various types of recreation facilities within x minutes travel, by type of facility and mode of travel.	Maps of facilities and distribution of population; citizen survey of travel mode.
Recreation-Informal Settings	
Overall Satisfaction	
24. Change in number or percent of households satisfied with recreation in informal outdoor spaces in neighborhood.	Baseline citizen survey and observation of current usage patterns; physical environment changes expected.
Availability	
25. Change in availability of informal physical settings for recreation and number of people affected.	Changes in open space and physical environment expected.
Education	
Accessibility/Convenience	
26. Change in number and percent of households satisfied with accessibility of schools.	Citizen survey; changes in available path, nearby traffic conditions en route to schools.
27. Change in number and percent of students within x minutes, by type of school and travel mode.	Map of school and population distribution; busing records.

Table 6 (continued)

28. Number and percent of students having to switch schools or busing status.	Relation of capacity to expected demands, and school board policy.
Crowdedness	
29. Change in school crowdedness indicators; e.g., student-teacher ratios, number of shifts.	Citizen survey; expected change in noise, traffic hazards, air quality, other hazards.
Transportation-Mass Transit	
Satisfaction	
30. Change in number and percent of households satisfied with mass transit service.	Citizen survey, expected service changes, expected change in factors affecting satisfaction.
	Usage levels, from fares and surveys.
Accessibility	
31. Change in number and percent of citizens residing (or working) within x feet of public transit stop.	
Transportation-Pedestrians	
Satisfaction/Accessibility	
32. Change in number and percent of households satisfied with walking conditions and walking opportunities in their neighborhood.	Baseline citizen survey, estimated changes in physical walking conditions; additions or removals of desired destinations.
Safety	
See measures 33 and 34 below.	N/A
Transportation-Private Vehicles	
Safety	
33. Change in number and percent of households satisfied with traffic safety (vehicle and pedestrian).	Baseline citizen survey, changes in traffic and traffic controls; circulation patterns.
34. Change in number and severity of accidents per 1,000 persons by pedestrians and riders.	Accident frequency and causation data; changes in traffic and traffic controls, circulation patterns, expected traffic volumes.
Travel Time	
35. Change in vehicular travel times between selected origins and destinations, by time of day and day of week.	Current traffic volumes; changes in street layout, width and traffic controls; estimated net new vehicular trips.

Table 6 (continued)

Parking Availability	
36. Change in average time needed to find acceptable parking space within x feet of residence (or desired destinations) in neighborhood of development, by time of day and day of week.	Current spaces available; new demand and supply; math model for estimating parking times.
37. Percent of drivers finding neighborhood parking satisfactory.	Baseline citizen survey; expected changes in supply and demand for spaces.
Shopping	
38. Change in number and percent of households satisfied with shopping opportunities.	Baseline citizen survey; change in physical conditions around shopping areas.
39. Change in number and percent of households within x minutes travel time to shopping, by type store and mode of travel.	Map showing location of stores and population, before and after development.
Energy Services	
40. Change in the frequency and duration of energy shortages, and the number of people affected, by fuel type.	Current and expected usage and supply in community; design and construction of buildings, type of manufacturing activity expected.
Housing	
41. Change in number and percent of housing units that are substandard and the number of people living in them.	Current housing stock conditions number to be removed or improved.
42. Change in number and percent of housing units relative to need, by type of housing (price, owner/rental, number of bedrooms, style, etc.)	Current profile of housing units added or destroyed; past housing chain effects in distribution of population by income level, indicators of latent demand for housing.
V. Other Social Impacts (in addition to those included above)	
People Displacement	
43. Number of residents (or workers) displaced by development, and whether satisfied with move.	Number of persons living in building to be destroyed; special survey of them.
Special Hazards	
44. Number of children physically at risk from "special" hazards created by development (e.g., machinery, junk, unguarded deep water).	Physical outdoor changes expected.

Table 6 (continued)

Sociability/Friendliness	
45. Change in social interaction patterns (e.g., frequency of neighboring, community activities).	Baseline survey of current neighboring and community activity patterns; changes in availability of community and small group meeting places; changes in physical barriers (e.g., highways, fences, heavy traffic, buildings which hinder access from one area of a neighborhood to another or footbridges or removal of barriers linking the areas); changes in people mix.
Privacy	
46. Change in number and percent of households satisfied with privacy in outdoor areas around home.	Citizen survey; geometric analysis of sightlines; changes in sight and sound barriers.
Overall Contentment with Neighborhood	
47. Change in number and percent of citizens satisfied with their residential (or work) neighborhood.	Citizen survey using data from other measures.

scaling is achieved through the assignment of one of the following five values: ++, very positive; +, positive; 0, neutral; and --, very negative. These scale assignments are not additive for the environmental factors; however, they do provide an important and direct way of comparing socio-economic impacts for a given factor across several alternative plans.

A methodology for evaluation of transportation route alternatives involves scaling of the impacts of alternatives on a relative basis from -5 to +5 (32). The -5 designation denotes the most detrimental impact of a set of alternatives on a given environmental factor, while a +5 would denote the most beneficial impact. An assignment of 0 would indicate that the alternative does not cause any change in the factor. This methodology further indicates that the point assignments are additive for a given alternative, and considers the algebraic sum as well as number of plus and minus ratings and the ratio of plus ratings to the total. This approach is useful for summarizing impacts and displaying trade-offs between alternatives. It also minimizes the need for specific impact calculations, and it is particularly useful for route locations for corridor-type projects, including highways, pipelines, transmission lines, and waterway navigation.

Letter-scaling of impacts is included in a methodology developed for the U.S. Department of Housing and Urban Development 1975 (33). The methodology includes 80 environmental factors and specific information on data needs and associated calculations for determining anticipated changes in the factors. The scaling system consists of the assignment of a letter grade ranging from A+ to C- for the impact of each alternative on each factor. To serve as an illustration, water quality is one of the 80 factors in the methodology. Human activity associated with housing projects may affect surface and ground water by several mechanisms. Contaminated stormwater, runoff, sediment, and the uncontrolled discharge of sanitary wastes are prime causes of surface water degradation; while septic tank seepage, leaky sewer lines, the improper disposal of cleaning agents, and contamination of recharge zones are some of the causes of ground water pollution. In turn, water of poor quality, whether degraded by human or natural causes, affects human activities by limiting the range of uses of a given water body. Some key water quality impacts related to housing development projects can be expressed in terms of the following questions:

a) Does the drinking water supply meet federal or state standards?

b) Will sewage be adequately accommodated and treated?

c) Will receiving waters meet federal, state or local

standards?

d) Will ground water suffer contamination by surface seepage, intrusion of salt or polluted water from adjacent water bodies, or from other contaminated aquifers?

This methodology provides suggestions for the assignment of impact scales, with Table 7 containing a summary of the recommended assignments for water quality impacts. Specific information of this type is in the methodology for each of the 80 environmental factors.

Another example of a scaling checklist methodology was developed in 1977 for water resources projects (34). The critical element is the establishment of evaluation guidelines for each of the identified factors. The evaluation guideline is defined as the smallest change in the highest existing quality in the region that would be considered significant. The guidelines are specific to the project setting, and their establishment will enable the user to determine the magnitude and direction of each environmental effect. The highest existing quality is defined as the best environmental quality for that factor measurement that currently exists in the region. For example, assume that the highest existing quality for dissolved oxygen in the region is 8.0mg/1. If a reduction of 1.5 mg/1 in that stream would be considered significant, then the evaluation guideline is 1.5 mg/1. It is suggested that more than one individual be involved in the development process for evaluation guidelines, and that the following three steps be used:

a) Identify the highest existing quality in the region.

b) Determine the change in the highest existing quality that would be considered significant. This change should be based on professional judgment or the use of prediction techniques combined with professional judgment.

c) Document the rationale for the selection of the highest existing quality and the selection of the guideline.

Once the evaluation guidelines have been identified, impact prediction calculations are required for each of the identified factors, with their interpretation being made in the context of the guidelines. The interpretation concept is shown in Figure 2. If the anticipated change in the environmental factor is within the evaluation guideline, this is identified as no significant effect. If the change is greater than the evaluation guideline

Table 7: Assignment of Scales for Water Quality Impacts in Housing and Urban Development Methodology (Reproduced by kind permission of A. M. Voorhees and Associates)

Scale	Comments
A+	Clearly beneficial effects are likely to occur.
A	Water quality standards are met for water uses intended by project. Wastewater will be discharged into wastewater treatment system with available capacity.
B	Water quality standards are met for water uses intended by project. Wastewater may not receive "best available" treatment. Ground water may be polluted. Ameliorative actions are possible.
C	Existing water quality is at or below official standards. Project may cause pollution of ground water.
C-	Project will cause surface or ground water quality standards to be exceeded.

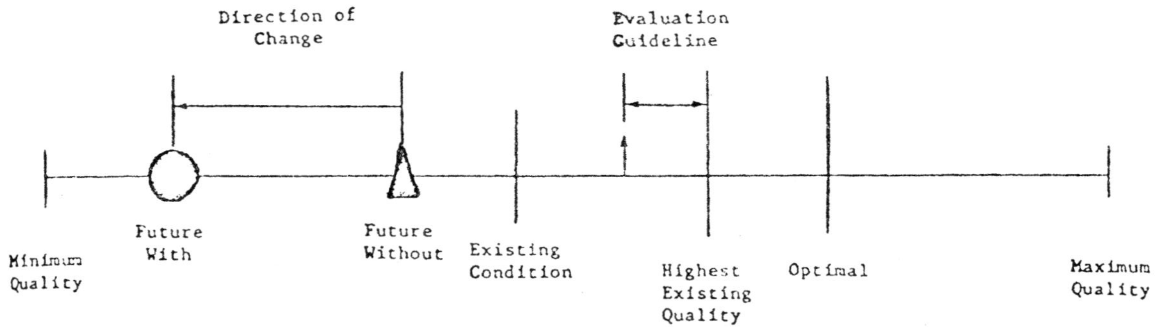

Figure 2: Usage of Evaluation Guideline (Reproduced by kind permission of Battelle-Columbus Laboratories)

and moves the factor from the baseline condition toward its optimal value, this is defined as a significant beneficial effect. If the change is greater than the guideline and away from the optimal condition, this is defined as a significant adverse effect. Figure 2 displays a significant adverse effect.

Table 8 displays the application of a scaling approach for evaluation of 15 treatment alternatives considered for meeting the wastewater management needs for the resort complex of Twining, commonly known as Taos Ski Valley, in northern New Mexico in the U.S. (35). This usage involved the assignment of numerical scales from 1 to 3 for each of eight key issues. As noted earlier, this type of presentation is useful for comparison and evaluation of alternatives on a common basis.

Weighting-scaling checklists refer to methodologies that embody the assignment of relative importance weights to environmental factors and impact scales for each alternative relative to each factor. Weighting-ranking checklists involve importance weight assignments and the relative ranking of the alternatives from best to worst in terms of their impacts on each environmental factor. Numerous weighting-scaling and weighting-ranking checklists have been developed for environmental impact studies (36). These types of methodologies represent adaptations of routinely used multiple-criteria or multiple-attribute decision-making techniques. These techniques have been used for more than a decade for various types of decisions not necessarily related to environmental impact studies.

An early weighting-scaling checklist was developed for water resources projects (37). This methodology, called the Battelle Environmental Evaluation System (EES), was developed for the U.S. Bureau of Reclamation and it contained 78 environmental factors. The specific factors and their associated 17 components and 4 categories are shown in Figure 3. Each of the categories, components, and factors were assigned importance weights by an interdisciplinary team based on usage of the ranked pair-wise comparison technique; and resultant importance weight points (parameter importance units or PIU's) are shown in Figure 3 by the numbers adjacent to the four environmental categories, in the right-hand corner of the boxes representing the intermediate components, and in the parentheses in front of each environmental factor. The higher the number, the greater the relative importance.

Impact scaling in the Battelle EES is accomplished through the use of functional relationships for each of the 78 factors. Functional relationships refer to graphical means for transforming factor measurements (baseline or predicted values) into subjective evaluations. Figure 4 illustrates the functional relationship for turbidity in water. Objective measurements are transformed into a

Table 8: Evaluation of Treatment Alternatives for Twining, New Mexico (Reproduced by kind permission of U.S. Environmental Protection Agency)

ISSUES are identified by letter, as follows. A = Cost. B = Reliability.
C = Water quality. D = Water supply. E = Construction impacts.
F = Land use (growth). G = Community conflict. H = Benefits to region.

Name of alternative	Brief description of alternative	Ratings on ISSUES a/							
		A	B	C	D	E	F	G	H
1. Conventional new plant	Build new plant in Twining using advanced technology which needs energy, chemicals	1	2	2	2	1	2	1	3
2. Aquaculture	Build new plant in Twining using new technology which grows algae in ponds inside a greenhouse	1	2	2	2	1	2	1	2
3. Septic tanks	Abandon the sewer system; everyone uses septic tanks	1	1	3	2	1	1	1	3
4. Greywater/ blackwater	Some sewage goes to existing (renovated) plant; toilet wastes to composting toilets	2	3	2	2	1	2	1	3
5. Storage and discharge	Store the winter sewage and discharge it in spring when river can better assimilate nutrients	1	2	3	1	1	2	1	3
6. Total detention	Pipe sewage to ponds near Arroyo Seco and evaporate it	3	1	1	3	3	2	2	2
7. Land application (farming)	Pipe sewage to ponds near Arroyo Seco and let farmers use it for irrigation	2	2	2	1	3	2	3	2
8. Pipeline to Rio Grande	Renovate existing Twining plant; build pipeline to Rio Grande (west of Arroyo Hondo) and discharge sewage there	2	2	1	3	2	2	2	3
9. Holding tanks	Store sewage in ponds in Twining and use trucks to haul it to the Town of Taos sewage plant	1	2	1	3	1	2	2	3
10. Regional plant in Hondo Valley	Pipe sewage to a new regional sewage plant near Arroyo Hondo	2	3	1	3	3	3	3	1
11. Regional plant in Taos	Pipe sewage to Town of Taos sewage plant	2	3	1	3	2	3	3	1

Table 8 (continued)

12. Rehabil- itation	Renovate existing plant and add on advanced processes for nutrient removal	1	3	2	2	1	2	1	3
13. Total treatment	Build very complex plant in Twining which would turn sewage into drinking water for use by Twining	3	3	1	2	1	2	1	3
14. Greywater evaporation	Combine alternatives 4 and 6: compost toilets, greywater evaporation	3	3	1	3	3	2	2	2
15. Aquaculture + snowmaking	Combine alternatives 2 and 5	2	2	3	1	1	2	1	2

a. 1 = less impact (best rating); 2 = intermediate impact; 3 = more impact.

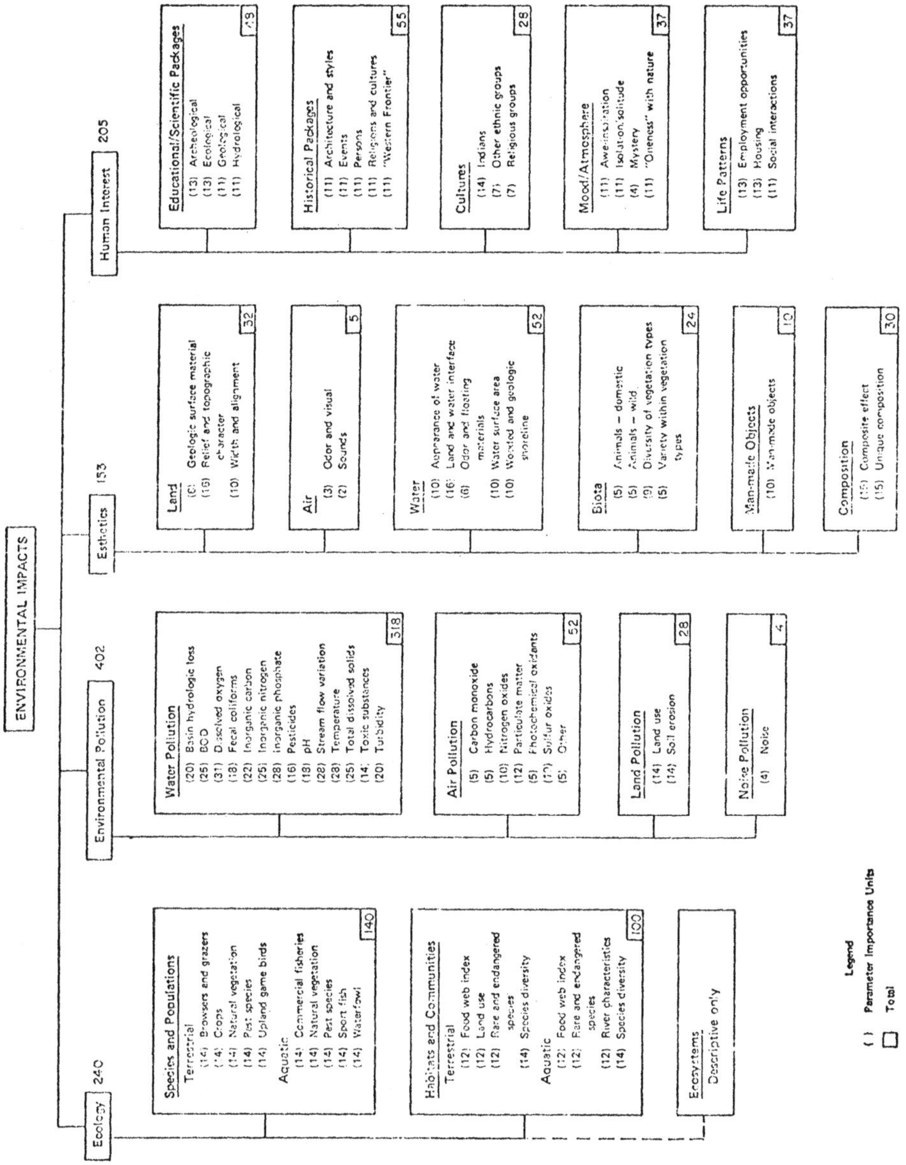

Figure 3: Battelle Environmental Evaluation System (Reproduced by kind permission of Battelle-Columbus Laboratories)

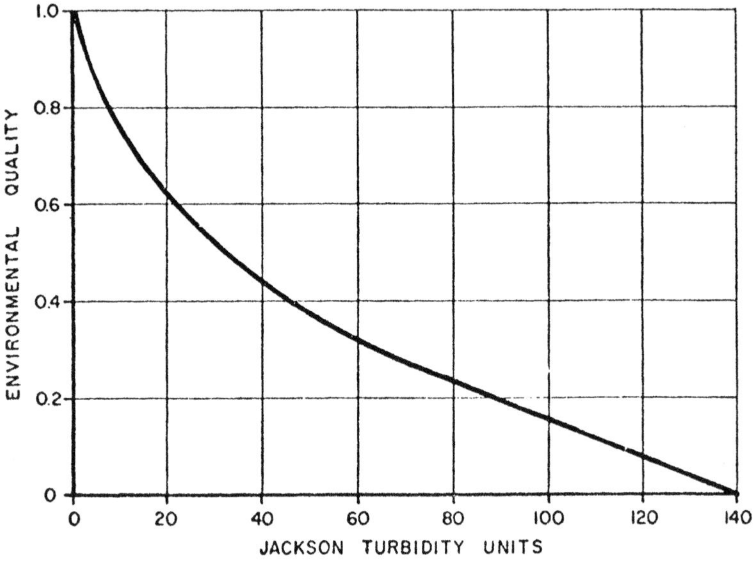

Figure 4: Functional Relationship for Water Turbidity (Reproduced by kind permission of Battelle-Columbus Laboratories)

subjective interpretation of environmental quality (EQ) based on a scale of 1.0 for good quality and 0.0 for poor quality. Usage of functional relationships is not limited to the Battelle EES (38-41). Summaries of functional curves utilized for a variety of environmental impact methodologies are available (23). In addition, environmental indices for water and air quality and aquatic and terrestrial habitats utilize functional relationships (42-44). Detailed steps for the development of functional relationships are described elsewhere (45) with the general approach as follows:

a) Decide on the units of measurement of the environmental factor being evaluated. Estimate the range of the units of measurement, with this constituting the range associated with the x-axis in the functional curve.

b) Decide on the good, or more desirable, and bad, or less desirable, portions of the range of measurements. Desirable conditions include non-limitations in usage of the resource as well as factor measurements in compliance with established environmental standards. Undesirable portions of the range of measurements relate to limitations in resource usage and non-compliance with standards. These decisions should be based on the collective judgment of an interdisciplinary team. The Delphi procedure has been used in the development of functional curves, with this constituting a valid approach for decision-making.

The basic concept of the Battelle EES is that an index expressed in environmental impact units (EIU's) can be developed for each alternative and baseline environmental conditions. Mathematical formulation of this index is as follows:

$$EIU_j = \sum_{i=1}^{n} (EQ)_{ij} \, (PIU)_i$$

where EIU = environmental impact units for jth alternative,

EQ = environmental quality scale value for ith factor and jth alternative, and

PIU = parameter importance units for ith factor.

Usage of the Battelle EES consists of obtaining baseline data on the 78 environmental factors and, through use of their functional relationships, conversion of the data into EQ scale values. These scale values would then be multiplied by the appropriate PIUs and aggregated to obtain a composite EIU score for the baseline setting. For each alternative being evaluated it would be necessary to predict the anticipated changes in the 78 factors. The predicted factor measurements would be converted into EQ scale values using the appropriate functional relationships. The EQ scale values would then be multiplied by the PIUs and aggregated to arrive at a composite EIU score for each alternative. This numerical system provides an opportunity for displaying trade-offs between the alternatives in terms of specific environmental factors, intermediate components, and categories. Professional judgment must be exercised in the interpretation of the numerical results, with the focus being on comparative analyses rather than specific numerical values.

As noted earlier, the Battelle EES was developed for the environmental impact considerations relating to water resources projects. The general approach can be applied to other project types through selection of pertinent environmental factors, assignment of importance weights, and the development of appropriate functional relationships for the factors. The approach used in the Battelle EES has been applied to a rapid transit system (40), a waterway navigation project (39), and to highway projects, pipeline projects, channel improvement projects, and wastewater treatment plants (38).

Another example of a weighting-scaling checklist for water resources projects is the Water Resources Assessment Methodology (WRAM) developed by the U.S. Army Corps of Engineers (41). The concept of the methodology is displayed in Figure 5, with the details listed in Table 9. Key elements include the selection of an interdisciplinary team; selection and inventory of assessment variables (environmental factors); impact prediction, assessment, and evaluation; and documentation of the results. The element including weighting and scaling is impact prediction, assessment, and evaluation. The weighted rankings technique (a paired-comparison technique) is used to determine the relative importance coefficient (RIC) for each assessment variable. Mechanics of the technique are summarized in Table 10. Importance weight assignments are required for each study and should reflect the importance of the variables in the given geographical location.

Impact scaling in the WRAM can be accomplished through the use of functional graphs (relationships), linear proportioning, or the development of alternative choice coefficients (ACCs). Linear scaling is associated with considering the largest impact and calculating the relative proportions of each change based on assigning the largest impact a scale value of 1.0. It should be noted that changes could be either beneficial or detrimental, and linear scaling should delineate between them. Development of ACC's involves the comparative evaluation of the impacts of each alter-

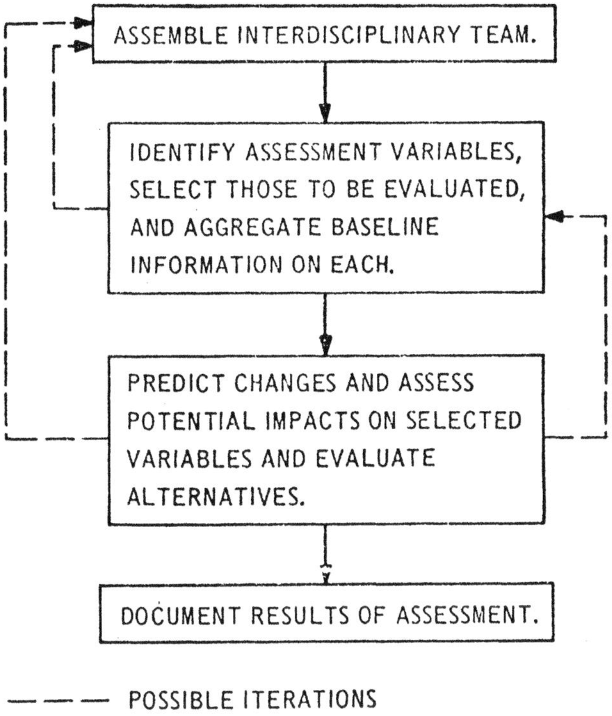

Figure 5: Elements of Water Resources Assessment Methodology (Reproduced by kind permission of U.S. Army Waterways Experiment Station)

Table 9: Delineation of Water Resources Assessment Methodology (Reproduced by kind permission of U.S. Army Waterways Experimental Station)

Element	Activities
Interdisciplinary Team	A. Selection
	1. Select members of interdisciplinary team.
	2. Designate team leader.
	B. Review and Familiarization
	1. Review study area history.
	2. Visit study area.
Assessment Variables	A. Selection
	1. Assemble list of mandatory* or critical** variables for each of the four national accounts (EQ, NED, SWB, and RD).+
	2. Use criteria questions or weighting portion of weighted rankings technique, along with professional judgment, to select additional relevant variables.
	3. Identify any resulting interactive or cross-impact variables or categories.
	B. Environmental Inventory
	1. Assemble baseline data for selected variables.
	2. Identify variables with data deficiencies.
	3. Use weighted rankings technique and other criteria to allocate manpower and funding resources to

Table 9 (continued)

Element	Activities
	data collection effort.
	4. Conduct field studies or assemble information on data-deficient input variables.
Impact Prediction, Assessment, and Evaluation	A. Prediction and Delineation
	1. Predict changes in each variable for each alternative plan and the no-action alternative using available techniques and/or professional judgment.
	2. Delineate potential impacts of alternatives.
	3. Highlight significant impacts and "red flag" any critical issues.
	B. Weighting and Scaling
	1. Use weighted rankings technique to determine relative importance coefficients (RIC) for each variable.
	2. Scale predicted impacts through development of alternative choice coefficients (ACC) or use of function graphs or linear scaling.
	C. Evaluation and Interpretation of Results
	1. Multiply RICs by ACCs to obtain final coefficient matrix. Sum coefficient values for each alternative.
	2. Use values in final coefficient matrix as basis for description of impacts of alternatives and trade-offs between alternatives.
	3. Discuss any critical issues and

Table 9 (continued)

Element	Activities
	predicted impacts.
Documentation of Results	A. Rationale
	1. Describe rationale for selection of decision variables.
	2. Describe procedure for impact identification and prediction, and rationale for weighting, scaling, interpreting results.
	B. Referencing of Sources of Information
	C. Decision on Environmental Impact Statement

*Mandatory = variables required by legislation or regulations.

**Critical = variables that are not mandatory but usually affected by water resources projects.

+ EQ = Environmental Quality
NED = National Economic Development
SWB = Social Well-Being
RD = Regional Development

Table 10: Illustration of Importance Weighting by Using Paired-Comparison Techniques
(Reproduced by kind permission of U.S. Army Waterways Experiment Station)

Variable	Assignment of Importance Values																					Sum	RIC
V1	0	1	1	0	0	1																3.0	0.14
V2	1						1	0.5	0.5	1	1											5.0	0.24
V3		0					0					0.5	0	0.5	1							2.0	0.09
V4			0					0.5				0.5				0.5	0	1				2.5	0.12
V5				1					0.5				1			0.5			1	1		5.0	0.24
V6					1					0				0.5			1		0		1	3.5	0.17
V7 (Dummy)						0					0				0			0		0	0	0.0	0
																					Total	21.0*	1.00**

*Check: $\frac{N(N-1)}{2} = \frac{7(7-1)}{2} = \frac{42}{2} = 21$

**Must sum to unity

Description:

Six hypothetical impacted variables are shown along with a dummy variable. The dummy variable, which by definition has no relative importance, is included to preclude any affected variable from being assigned a value of zero (no relative importance). Each variable is compared with every other variable in order to determine which of each pair is considered to be the most important for the study area. The variable considered to be more important is assigned a value of one, and a value of zero is assigned to the other. If a decision cannot be made regarding relative importance, or if both variables are considered to be of equal importance, a value of 0.5 is assigned to each.

The columns in the table under the heading "Assignment of Importance Values" represent the results of the variable-by-variable comparisons. The assigned values for each variable are summed, and each sum is divided by the total of all assigned values, 21, to determine the relative importance coefficient (RIC). Two checks concerning the accuracy of the mathematical calculations can be made using the totals in the tabulation: the sum column should total to N(N-1)/2, where N is equal to the number of variables considered (including the dummy); and the RIC column should total to unity.

The RIC column in the table shows the ranking of each variable considered. The dummy variable is zero (no priority) by definition, and the ordering of the remaining six, from highest to lowest priority, is V2 and V5, V6, V1, V4, and V3.

native on each factor (assessment variable). Table 11 contains an example of ACC assignments, with the basic approach similar to the assignment of importance weights in that a decision is required between two alternatives at a time as to which is best relative to their impacts on the variable.

The final matrix from the application of the WRAM is illustrated in Table 12. Five variables are identified along with their RICs. The ACCs for each alternative are displayed, with the final coefficient matrix representing the product of the RICs and ACCs. The best choice is the alternative with the highest product summation, thus in the example, alternative A would be the one of choice. The WRAM was developed with an orientation to water resources projects; however, examination of the elements as shown in Figure 5 and Table 9 indicates that this approach could be used for any project type, with the methodology providing more of a framework or structure than a specifically-developed approach limited to a given project type.

Table 13 illustrates a weighting-ranking checklist utilized in an environmental impact study of sites for a wastewater treatment plant. Two groups of importance weights were used, with factors assigned an importance weight of 2 being more important than factors assigned an importance weight of 1. Each of the three sites in the study were ranked on a scale from 1 to 3, with 1 denoting the worst site and 3 the best site relative to 13 decision factors. The best overall site based on a composite evaluation was site C.

Numerous other weighting-scaling or weighting-ranking checklists could be described. Table 14 provides a summary of some studies which used numerical systems for development of indices for baseline conditions and the alternatives being evaluated. Several impact scaling techniques have already been described along with the ranked pair-wise comparison technique and the paired-comparison technique for importance weighting. Additional approaches for importance weight assignments are available, including ranking, rating, several variations of paired-comparison techniques and successive comparisons (51). The Delphi technique can be utilized for solicitation of professional judgment and group decision-making on importance weights. The multi-attribute utility measurement technique can also be used for assigning importance weights to environmental factors (52). Finally, the nominal group process technique can be used, with this interactive group approach consisting of the identification and discussion of environmental factors followed by independent voting on their priority (53). Final group decisions are based on mathematical rank ordering.

Table 11: Illustration of Impact Scaling by Using Paired Comparison Technique
(Reproduced by kind permission of U.S. Army Waterways Experiment Station)

Problem

The impacts of four alternatives on Variable X have been predicted to be as follows:

Alternative	Predicted Impact
Plan A	Beneficial
Plan B	Beneficial
Plan C	Most Beneficial
Plan D	Detrimental
Dummy	Most detrimental by definition

Determine Alternative Choice Coefficients (ACCs) by using paired comparison technique

Solution

Alternative	Choice Assignment										Choice Total	ACC
Plan A	0.5	0	1	1							2.5	0.25
Plan B	0.5				0	1	1				2.5	0.25
Plan C		1			1			1	1		4	0.40
Plan D			0			0		0		1	1	0.10
Dummy				0			0		0	0	0	0.00
										Total	10	1.00

Table 12: Illustration of Final Product Matrix in Water Resources Assessment Methodology (Reproduced by kind permission of U.S. Army Waterways Experiment Station)

Variable	RIC	ACC of Alternative				Final Coefficient Matrix RIC × ACC			
		A	B	C	D	A	B	C	D
V1	0.20	0.25	0.25	0.40	0.10	0.05	0.05	0.08	0.02
V2	0.40	0.33	0.00	0.17	0.50	0.13	0.00	0.07	0.20
V3	0.10	0.30	0.30	0.20	0.20	0.03	0.03	0.02	0.02
V4	0.20	0.30	0.30	0.30	0.10	0.06	0.06	0.06	0.02
V5	0.10	0.50	0.17	0.33	0.00	0.05	0.02	0.03	0.00
					Total	0.32	0.16	0.26	0.26

Table 13: Weighting-Ranking Checklist for Wastewater Treatment Plant Site Evaluation (Reproduced by kind permission of L. A. Wilson).

Factor to be considered	Significance of Factor	Importance to decision (2 = most)	Ranking (Scale 1 = worst 3 = best) Site A	Site B	Site C
1. Construction cost	One-time cost with federal share	1	2	2.5	3
2. Operating cost	Ongoing cost; includes energy costs (all local share)	2	1	2	3
3. Non potable reuse	"Safe", economic benefit; key is proximity to users	2	1	2	3
4. Potable reuse	More long-range than 3; best sites are near well field or water plant; industrial sewage should not be present	1	3	2.5	1
5. Odor potentials	Assumes good plant design and operation	2	2	1	3
6. Other land use conflicts	Potential to interfere with agricultural/ residential land	1	2	1	3
7. Site area available	Future expansion capability, flexibility	2	3	2	1
8. Relationship to growth area	Assumes growth to State line (in 50 year time frame)	1	3	2	1
9. Construction impacts	Reworking of lines	1	1	2	3
10. Health of workers	Air pollution	1	3	3	2
11. Implementation capability	Land acquisition problems	1	3	1	2
12. Operability	One plant is better than two	1	1	2	3
13. Performance reliability	Assumes equal treatment plants	1	2	2	2
14. TOTAL	Ranking times Importance, added up Highest number - best site		34	32	40

Table 14: Examples of Studies Using Weighting-Scaling and Weighting-Ranking Checklists

Project Type	Reference and Comments
Transportation	(46) -- This methodology is a weighting-scaling checklist of 56 measurable environmental parameters. The actual values of alternative plan impacts on a parameter are normalized and expressed as a decimal of the largest impact (on that one parameter). These represent the scaling procedure. These normalized values are multiplied by a subjectively determined weighting factor, which is the sum of a weight for initial effects plus 10 times a weight for long-term effects. Uncertainty is compensated for by conducting several computer passes and incorporating a randomly generated error variation in both actual measurements and weights. This procedure also provides a basis for testing the significance of differences in total impact scores between alternatives. The methodology was used to evaluate eight highway project alternatives.
	(40) -- This weighting-scaling methodology was applied to a rapid transit system project in Atlanta. A total of 68 factors are included and grouped into ecological, physical/chemical, aesthetic, and human interest categories. Functional curves are used and 1000 importance weights (Environmental Quality Units) are assigned to the 68 factors. The Delphi technique was employed to reach a consensus on the optimum distribution of the 1000 units among the factors.
Waterway Navigation	(39) -- A weighting- scaling checklist methodology was used to evaluate the environmental impacts of 9 alternatives (8 waterway locational routes and the no-action alternative) for extending waterway navigation from Tulsa to Wichita. A total of 102 assessment variables grouped into 6

Table 14 (continued)

Project Type	Reference and Comments
	categories (biology, physical/chemical, regional compatability, archaeology, aesthetics and climatology) were used. A total of 1000 importance points were distributed to the categories and variables through use of the ranked pairwise comparison technique. Impact scaling was accomplished through the use of functional curves. Composite scores for each alternative were developed by summing the products of the importance points and scale values. A stochastically selected error term was included so as to allow the importance weights to vary by ± 50%. The entire methodology was computerized to allow for ease of calculation and the conduction of a sensitivity analysis.
Wastewater Treatment Plants	(38) -- This methodology can be classified as a weighting-scaling checklist although it also contains elements of matrix and network approaches. Areas of possible impacts are defined by a hierarchical system of 4 categories (ecology, physical/chemical, aesthetic, social), 19 components, and 64 parameters. An interaction matrix is presented to indicate which activities associated with wastewater treatment projects generally affect which parameters. Impact measurement incorporates both scaling and weighting. A set of ranges is specified for each parameter to express impact magnitude on a scale from zero (worst environmental quality) to one (best environmental quality). A unique part of the methodology is the use of an "environmental assessment tree". Specifically, the ranges of the parameters are aggregated into 19 components with 19 network trees which have been constructed to translate the relationship between the good and the bad values on individual parameters into good and bad values for a set of parameters. The advantages of the assessment tree are that it is

Table 14 (continued)

Project Type	Reference and Comments
	simple to use, requires less precise data about individual parameters, and places the single parameter in the context of the real-world interrelationships among similar variables. The disadvantage of the assessment tree approach is the potential loss of information about individual parameters. Weighting is accomplished by the interdisciplinary team assigning importance index values based on 100 units to each category and component. An overall score for any alternative can be obtained by multiplying each component scale by its weight factor and summing across components.
Nuclear Power Plants	(47) -- A formal methodology was developed for the selection of the best sites from among alternative suitable sites for a nuclear power station. The method is based on reducing the various variables affecting the decision to a single function that provides a metric for the level of site acceptability. The function accommodates for well known sites selection criteria as well as other factors such as public reactions to certain choices. The method is applied to the selection of a site from three acceptable alternative sites for Wolf Creek nuclear power station, Kansas.
	(48) -- A methodology was developed for environmental decision-making that combines societal and technical judgments. Eight factors that characterized the major economic and environmental impacts of nuclear power plant sitings were identified. These factors were used to construct "mini-environmental impact statements" for six siting alternatives. The impact statements formed the core of a survey questionnaire administered to three groups of respondents. Data analysis produced estimates of the relative importance of each factor. A procedure is described for using these estim-

Table 14 (continued)

Project Type	Reference and Comments
	ates of importance as weighting factors to be applied to techno-economic scores. These latter scores would be generated by technical experts and would represent the actual or anticipated impact of a plant siting upon the eight factors.
Miscellaneous	(49) -- Environmental problems usually involve multiple conflicting objectives, large uncertainties concerning the possible environmental impact, and several individuals or groups whose preferences are very different, but yet very important in choosing an alternative. The basic ideas of multiattribute utility are presented and placed in the context of a decision-making framework. The manner in which it aids the analysis of environmental issues is discussed. Several problems in which multiattribute utility was used are surveyed. These include siting nuclear power facilities, developing fisheries, controlling a forest pest, examining energy policy, and transporting hazardous materials.
	(50) -- An environmental impact assessment in its broadest interpretation requires the documenting and accounting for multiple conflicting objectives such as cost, protection of various components of the environment, and economic benefits to society in selecting from alternative courses of action. Multiattribute utility theory is a methodology that addresses the difficult issues present in environmental impact assessments in a systematic and theoretically sound manner. It provides techniques for defining the problem alternatives and the preferences of different interest groups in quantitative terms. These techniques allow the alternatives to be compared in a consistent fashion. The models used in the methodology can be implemented on the computer and thereby provide a general tool

Table 14 (continued)

Project Type	Reference and Comments
	for analyzing environmental impact assessment problems. This paper describes two applications that illustrate the use of the methodology and the computer tool. The first is a transmission line corridor analysis done for the U.S. Bureau of Reclamation. The second is an analysis of potential sites for oil terminals on the island of Kodiak done for the Alaska Department of Community and Regional Affairs.

NETWORK METHODOLOGIES

Networks refer to those methodologies which attempt to integrate impact causes and consequences through identifying interrelationships between causal actions and the environmental factors affected, including those representing secondary and tertiary effects. A linear network display for a dredging activity is shown in Figure 6. The initiating action is shown on the left with various other causal actions and affected factors shown in the phases of the network. Network analyses are particularly useful for identifying anticipated impacts associated with potential projects. Networks can also aid in organizing the discussion of anticipated project impacts. Network displays are useful in communicating information about an environmental impact study to interested publics. The primary limitation of the network approach is the minimal information provided on the technical aspects of impact prediction, and the means for comparatively evaluating the impacts of alternatives.

ENERGY SYSTEM DIAGRAM METHODOLOGIES

Energy system diagrams have been utilized to a limited extent in environmental impact studies. An energy system diagram shows the flows of all energies in the system, keeping track of the main components of the system such as the plants, animals, chemical processes, reservoirs of resource storage, flows of information, and outside actions that cause change (54). As shown in Figure 7, the natural environment system of any study area processes sunlight and other forms of energy (55). Sunlight is shown as an energy resource for the system, while air, water, land and biota are shown as subsystems. Each subsystem makes use of energies of various forms to produce and store a form of energy characteristic of that subsystem. In turn, that form of energy is one of the many forms used by other subsystems. For example, the water subsystem makes use of all of the forms of energy which nature has processed (shown as the pathway labeled "natural environmental goods and services") to achieve water evaporation, which provides a mechanical energy potential (due to elevation of the water) and a chemical energy potential (due to purity of the water). These potential energies get used (on the pathway labeled "use of water resources") along with other natural system energies, in a variety of natural environment processes. The mechanical energy potential gets used for such processes as building and eroding land forms and rocks, building and eroding soil, carrying nutrients to and from biota, and scavenging particulate matter from air. Figure 8 illustrates how energy from the sun becomes embodied in the goods and services of the natural environment, and subsequently in the goods and services of the human system. An environmental impact study utilizing energy system diagrams should consider relationships and changes in both the natural environment and human systems

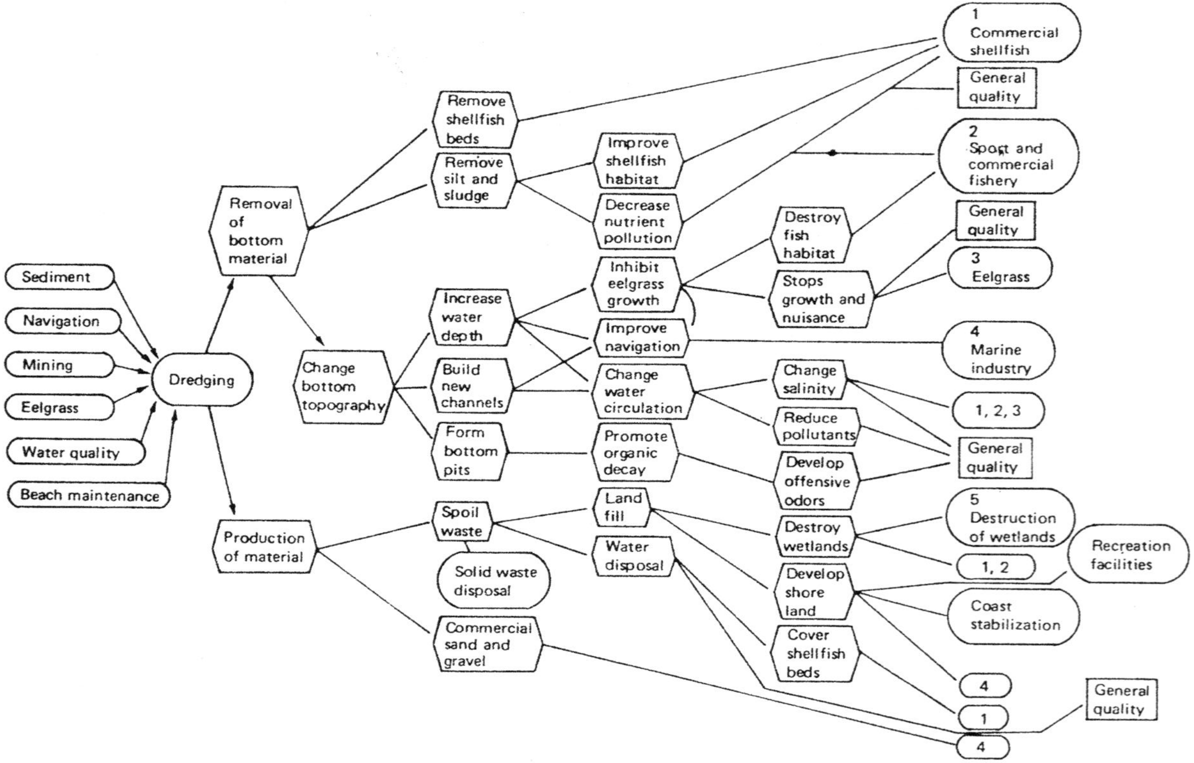

Figure 6: Network Diagram for Dredging Project
(Reproduced by kind permission of J. Sorensen)

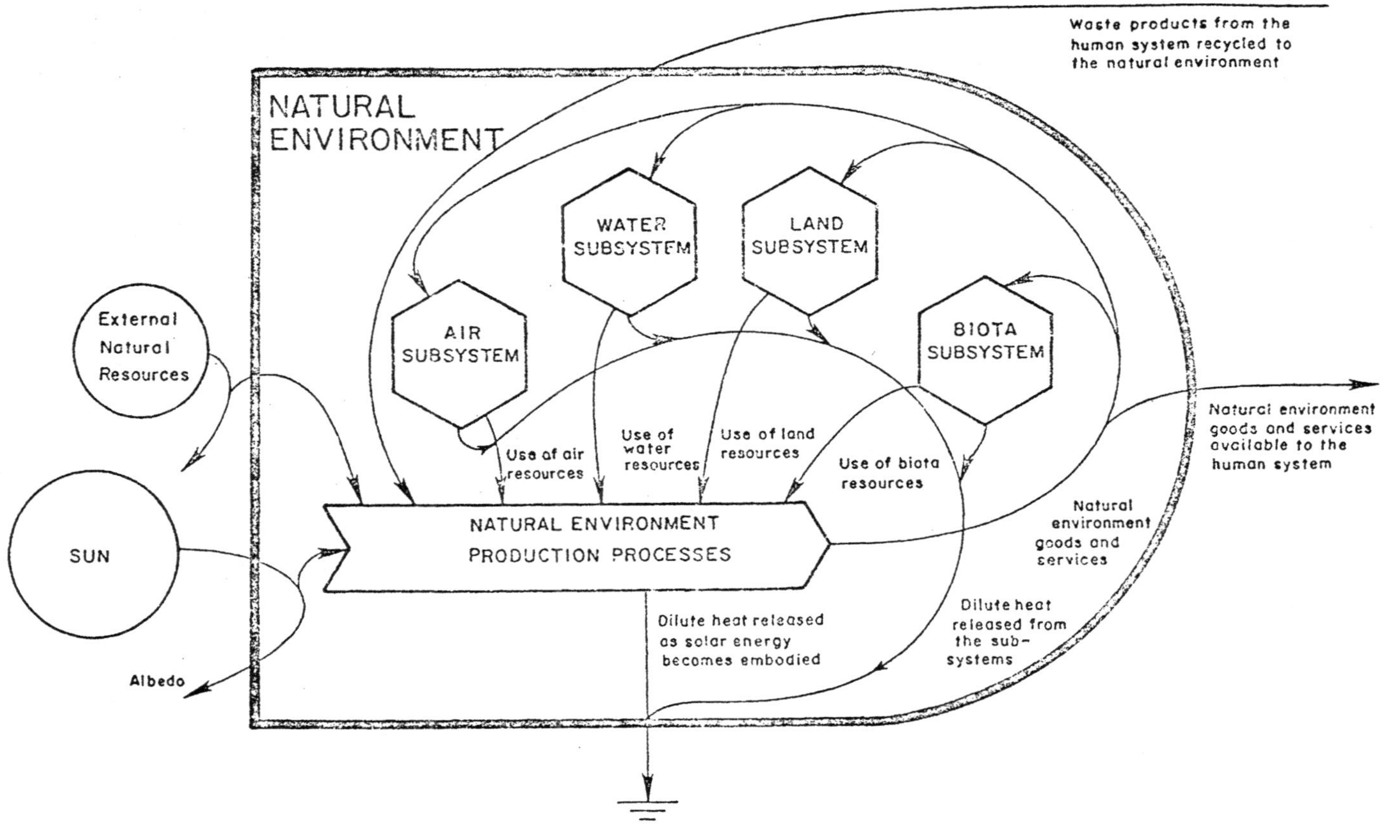

Figure 7: Concept of Natural Environment Energy System
(Reproduced by kind permission of Center for Environmental Research, Cornell University)

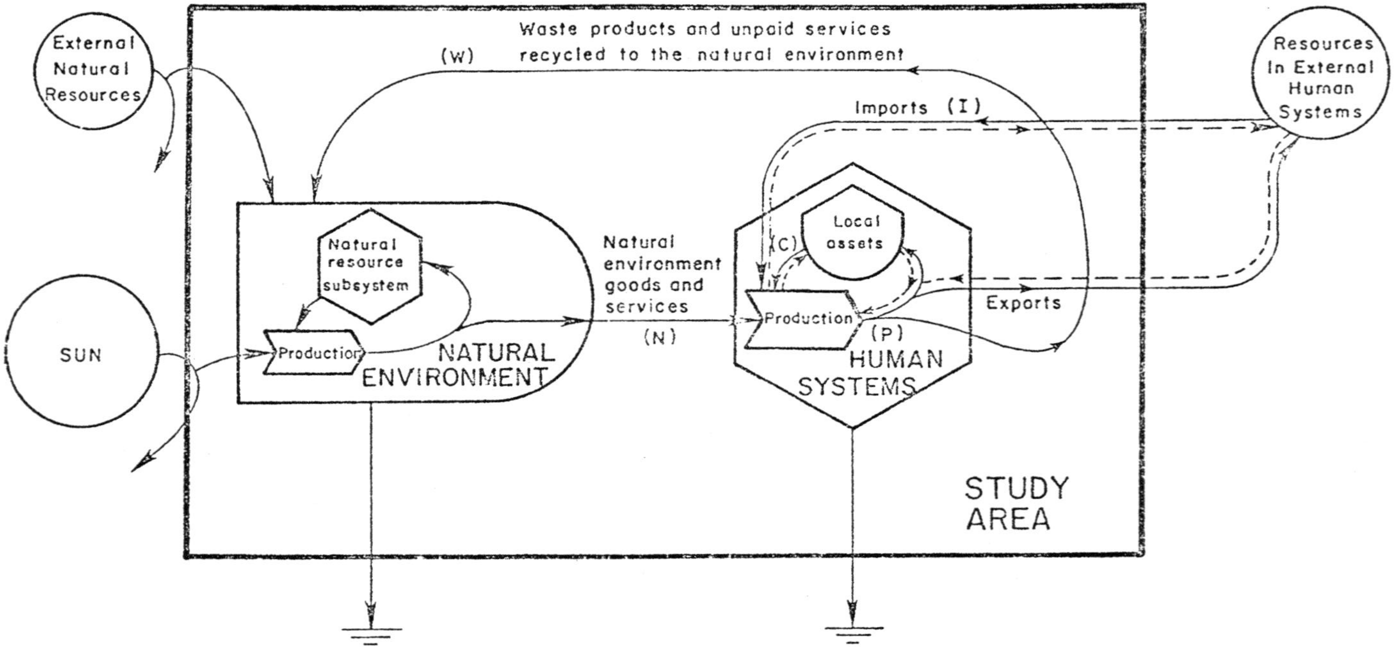

Figure 8: Concept of Study Area Energy System (Solid-line pathways indicate energy flows; broken-line pathways indicate money flows)
(Reproduced by kind permission of Center for Environmental Research, Cornell University)

within the study area.

The steps associated with developing an energy system diagram for an environmental impact study are listed in Table 15. The energy system diagram utilized to predict the impacts of estuarine development on a marine meadows community is shown in Figure 9. Table 16 provides brief summaries of three applications of energy system diagrams for environmental impact studies. Energy system diagrams are scientifically valid approaches for environmental impact studies; however, extensive data are needed for the successful usage of diagram approaches. Odum (54) identified the following two primary difficulties associated with use of this type of methodology:

(a) Many of the processes, and particularly the rates of change in the processes, are simply unknown.

(b) The use of an energy system diagram requires a high degree of technical sophistication as well as back-up resources such as computer facilities, and this type of resource availability may be limited for many environmental impact studies.

Despite these limitations, energy system diagrams offer potential advantages over traditional methodologies for environmental impact assessment. In energy system diagrams the evaluation of environmental impacts is based on a comprehensive assessment of all direct and indirect effects caused by the category of impact action being considered (55). For example, evaluation of the impact of introducing extra salt into the environment (due to road salt for de-icing) is based on a single comprehensive display of the total impact of all kinds of effects that will occur due to the extra salt. In contrast, traditional methods of impact assessment involve identifying and evaluating each of the known individual effects, such as increased corrosion of road surfaces and cars and other capital assets; decreased aquatic and terrestrial ecosystem productivity; and increased health costs due to contaminated drinking water supplies.

Energy system diagrams use a common measure (energy) to quantify all impacts, thus providing a basis for making quantitative comparisons among different categories of impacts as well as different alternatives. Finally, one other way in which the energy system diagram approach supplements traditional methods is by providing a basis for determining the monetary value of environmental impacts. Because the monetary value of economic production is proportional to the total embodied energy cost of the set of activities supporting that production, the energy values of impacts can be converted to monetary values. This enables natural system impacts to be compared with the traditional components of project

Table 15: Steps in Developing Energy System Diagram for Environmental Impact Study (Reproduced by kind permission of Marine Technology Society)

---

1. List the main properties of the system believed to be important in its operation and/or to man. These are the parts of the system called state variables. They are the tanks, or associated with the tanks, of the diagram (see Figure 10).

2. Make a list of the outside causal forces and energy sources including the proposed new ones associated with plans and actions of man. These are called forcing functions and are represented as circles.

3. Connect the symbols for forcing functions on the left to the components of the system with causal energy flows using appropriate intersection functions.

4. Each function has a different symbol:

   a. The intersections are additive when two properties may be used interchangeably.

   b. The intersections are multiplicative when both are different and required.

   c. The intersections are switching when there are thresholds of action that are either on or off.

   d. The intersections are integrative, and cumulative when there are storages that receive inflows and outflows developing balance between them.

   e. The intersections are autocatalytic, feeding storage back to act multiplicatively on the inflow, when the module represents some self-maintaining activity such as organisms, industries, and cities.

   f. One intersection type has the property of leveling itself due to a circular recycling of some material which is limited. This is called the Michaelis Menton property.

5. Having identified the pathways, try to put numbers on them to gain perspectives on importance and times involved for effects. Estimate the main effects believed to result in the system

Table 15 (continued)

from the change through visual estimation of the effects of the change in or diminishing flows that converge and diverge within the diagram.

6. Having identified the pathways and compartments most likely to be involved in the proposed change, develop a simplified diagram that includes the main change and the essence of the system in few enough modules to make simulation feasible.

7. Run a simulation to test the simpler version to determine if there are unexpected patterns emerging from the network not previously recognized in the diagrams or in experiences. Whereas the simpler model cannot simulate the real complexity, it can identify effects and complications. For example, the system may show if there is a tendency to oscillate or if effects are counter intuitive (go up when you expect them to go down, etc.).

8. Sum the total energy values including those covered by money and those of nature not covered by money flows. Use these energy flow values in overall value calculations.

9. Prepare written accounts to go with the diagrams to lead the reader to the points of main consideration in examining propositions for change.

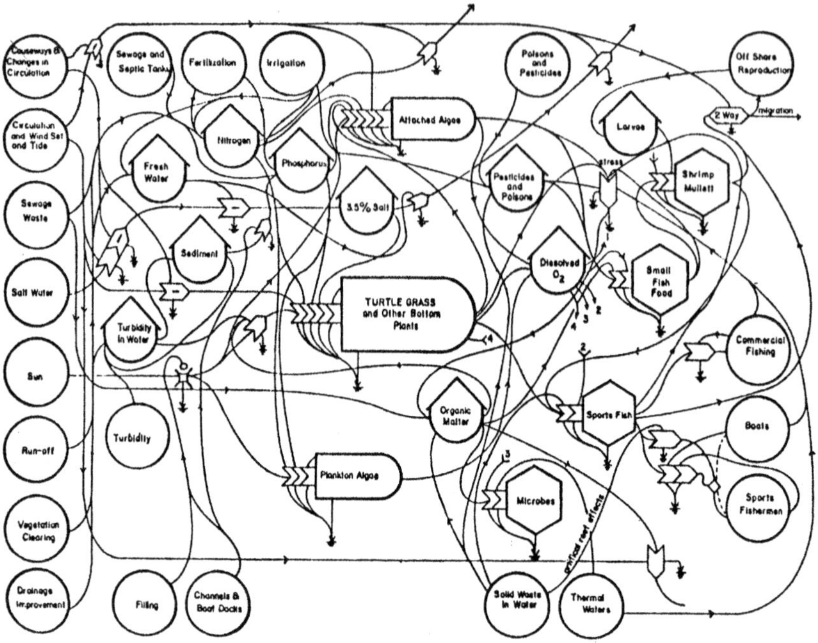

Figure 9: Energy System Diagram of Marine Meadows Community with Disturbances Imposed by Estuarine Development (Reproduced by kind permission of Marine Technology Society)

Table 16: Examples of Studies Using Energy System Diagrams

Project Type	Reference and Comments
Transportation	(56) -- This study consisted of a comparison of economic and energetic approaches for evaluating transportation systems. Methods for calculating the energy value of goods and services, the role of energy flows of natural systems in a regional or transportation system, and spatial energy theory for determining the competitiveness of different fuel source locations are presented. In particular, economic benefit/cost analysis as applied to transportation systems by the Corps of Engineers is compared to a comparable energy benefit/cost analysis. In order to illustrate the methodology of energetics several transportation systems are analyzed to calculate the energy cost of each. Both direct fuel energies for operation and indirect fossil fuel and natural energies for goods and services were considered. Some attempt was also made to measure the disruption of natural systems by an existing or planned transport system. Approximate analyses were made for barge transportation, railroad transportation, slurry pipelines, and electrical transmission lines. In particular, the direct and indirect energy costs of transporting coal were evaluated. Energy costs per ton-mile and energy yield ratios (energy transport ÷ energy costs) are presented. Some analysis of the direct and indirect energies associated with building barges, towboats, and locks and dams were also attempted. In order to show how energetics might be used at a regional scale of evaluation the problem of coal development and transportation in the Northern Great Plains is presented.
Transportation	(55) -- This study was focused on the theory and application of energy analysis approaches. The study report is in two volumes. Volume I documents and presents

Table 16 (continued)

Project Type	Reference and Comments
	the products of an effort to develop an energy analysis applications manual. The manual is intended for agency analysts to use in applying the energy analysis theories and procedures to environmental benefit/cost analyses of transportation actions. A step-by-step analysis procedure was developed and presented, describing energy analysis calculations and their application in transportation environmental benefit/cost analyses. The procedure's limitations and applications are discussed. Five case studies were developed and documented in Volume II, providing detailed information of the application of the procedure to various real transportation problems.
Missile Range	(57) -- The utility of system diagrams and of energy flow per unit time as the measure for environmental impact assessment is illustrated using results from the environmental impact statement for White Sands Missile Range, New Mexico. Procedures for developing and evaluating the diagrams are given and applied to White Sands. Both large-scale and small-scale environmental effects of the missile range were evaluated: on a macro scale, environmental stresses represented 1% of the natural energy flow through the system. At a more detailed level, a hydrological model used to assess effects of missile range water consumption on the aquifer showed that salt water intrusion into the aquifer was imminent, and identified two management strategies to prevent the intrusion.

benefits and costs (55).

OTHER METHODOLOGIES

Several new types of environmental impact methodologies are being developed, including more extensive usage of modeling and computer graphics. Table 17 summarizes some example references on new methodologies. The methodology by (66) includes the concept of risk analysis, and this is anticipated to become more important in future environmental impact studies.

SELECTION OF A METHODOLOGY

As indicated earlier, there is no "universal" methodology for meeting the environmental impact assessment needs for all project types in all environmental settings. Accordingly, selection of an existing methodology or portions thereof, or the development of a new methodology, is required in the conducting of an environmental impact study. A reasonable early selection should focus on methodologies developed for project types similar to the potential project being evaluated. In other words, if the environmental impact study is to be conducted for a highway project, methodologies which have been developed for usage on highway projects should be reviewed. The selection process can be aided by consideration of the following desirable characteristics of an impact assessment methodology or portions thereof.

An environmental impact methodology should be comprehensive in that it addresses appropriate environmental factors and impacts related to the project type being evaluated. Attention should be given to both beneficial as well as detrimental impacts. In addition, methodologies should encourage the highlighting of key environmental factors and impacts, with this concept being analogous to "red flagging" of significant issues. Methodologies should also be responsive to the legal and policy requirements of the governmental entity related to the impact study. For example, in the U.S., an EIS for an airport development project must address all land use, noise, and air pollution regulatory issues related to airport construction and operation.

Environmental impact methodologies should be dynamic in terms of included environmental factors and technologies used for impact identification, prediction and assessment. Since the environmental impact assessment field is only slightly over one decade old, rapid changes are occurring in study technology; accordingly, adaptation of methodologies for inclusion of new information is appropriate. Methodologies should also be adaptable to the unique environmental characteristics of the study area. The importance of environmental factors from one geographic location to another should be considered, and the flexibility provided to adapt the methodology to speci-

Table 17: Example References on Other Environmental Impact Methodologies

Reference and Comments
(58) -- Use of computer graphics for evaluation of impacts from conventional energy systems.
(59) -- Computer model used to trace flow of resources generated by various processes.
(60) -- Linear programming model for economic evaluations of wetlands.
(61) -- Resource allocation model for evaluation of wastewater management alternatives.
(62) -- Simulation model for determining the impacts of a series of reservoirs.
(63) -- Use of grid cell data banks in environmental impact assessment.
(64) -- Geographic-type environmental analyses using grid cell data banks.
(65) -- Human preference model developed for use with an environmental emissions model.
(66) -- Methodology for including risk analysis in environmental impact assessment.
(67) -- Simulation models for evaluating the impacts of alternative wildland use decisions.

fic environmental settings.

Another desirable characteristic is that the methodology should stress objective, quantitative analyses rather than purely subjective, qualitative approaches. It is recognized that it may be impossible to make quantitative calculations for many environmental impacts; however, the orientation of the methodology should be toward the quantitative approach. It is also recognized that even with quantitative approaches there is the necessity for subjective evaluation of objective information. The methodology should provide guidance on the value judgments and subjective evaluations associated with impact interpretation.

Finally, the selected environmental impact methodology should be implementable in terms of manpower, funding, and data and time requirements. Overly-sophisticated approaches may be scientifically valid, but not particularly usable within the constraints of a specific environmental impact study. Trade-offs will be required in the process of selecting between simpler methodologies which are implementable and more complex approaches which require extensive time and effort without yielding significantly improved information for impact assessment.

RESEARCH NEEDS

A number of research needs can be delineated within the general context of methodologies for environmental impact assessment. Examples of research needs according to impact identification, prediction, assessment, and integration and evaluation are as follows:

(a) Impact identification -- Information and methodologies are needed for defining project boundaries and identifying anticipated secondary impacts associated with project construction and operation. At the current time there is poor guidance on delineating the boundaries of a particular project, and also limited methodologies available for identifying anticipated secondary impacts.

(b) Impact prediction -- Better predictive tools are needed for various aspects of the environment, particularly as related to long-term forecasts of changes in environmental factors. Numerous modelling techniques exist for the physical-chemical, biological, cultural, and socio-economic environments; however, they generally focus on the short-term and provide little practical guidance for long-term forecasts.

(c) Impact assessment -- Information and systematic

methodologies are needed to improve the process of making necessary subjective judgments about objective information. Quantitative information on environmental factors and associated impacts must be professionally interpreted, and clearer guidance is needed on impact interpretation. This is particularly true for the biological environment due to the general absence of ecological standards.

(d) Impact integration and evaluation -- Methodologies are needed for addressing the composite or system impacts of projects. A common criticism of many existing methodologies is that they use a piece-meal approach by focusing on individual environmental factors. Methodologies are needed for addressing the cumulative impacts of a series of actions related to a project, and the relationship of the project to other existing or potential projects within the study area. Finally, methodologies are needed for making systematic evaluations of impact mitigation measures. In many instances, mitigation measures are proposed and become a part of project requirements without an evaluation of their cost-effectiveness.

SUMMARY

Numerous environmental impact assessment methodologies have been developed within the last decade. Methodologies can be useful in identifying anticipated impacts, determining appropriate environmental factors for inclusion in a description of the existing environment, providing information on prediction and assessment of specific impacts, and allowing for systematic evaluation of alternatives and the selection of a proposed action. Methodologies can be categorized into interaction matrices, checklists, and networks, with variations of networks including energy system diagrams. Interaction matrices are of greatest value in impact identification and display of comparative information on alternatives. Checklist approaches range from simple listings of environmental factors to complex methods involving assignment of relative importance weights to environmental factors and impact scaling for each of a series of alternatives on the environmental factors. Checklist approaches are useful for identifying environmental factors and providing information on impact prediction and assessment. Checklists are particularly valuable for displaying trade-offs between alternatives and their associated environmental impacts, thus they are useful in the selection of a proposed action. Network methodologies provide useful information for impact identification as well as valuable approaches for communicating information on interrelationships between environmental factors and anticipated project impacts. Energy system diagrams represent scientifically sophisticated approaches particularly useful for impact prediction and assessment. They are also of value in impact identification and can be a useful tool for impact display.

With the current emphasis on improving the scientific and technical quality of environmental impact studies, there is an increasing need to select and apply systematic methodologies for environmental impact assessment. It is anticipated that the usage of existing methodologies, or portions thereof, will increase in the future. Additionally, research and new methodologies are needed which will enable more systematic evaluation of project boundaries, long-term impact predictions, inclusion of value judgments in the process, analysis of system and cumulative impacts, and evaluation of mitigation measures.

REFERENCES

1. Canter, L.W. *Environmental Impact Assessment* (New York, New York: McGraw-Hill Book Company, 1977).

2. Dee, N. *et al.* "An Assessment of the Usage of Environmental Assessment Methodologies in Environmental Impact Statements". *DMG-DRS Journal* 10(1)(1975)4-9.

3. Canter, L.W. *Environmental Impact Statements on Municipal Wastewater Programs* (Washington, D.C.: Information Resources Press, Inc., 1979).

4. Council on Environmental Quality. "National Environmental Policy Act - Regulations". *Federal Register* 43(230)(1978)55978-56007.

5. Baram, R. and R.D. Webster. *Interactive Environmental Impact Computer System (EICS) User Manual,* CERL-TR-N-80 (Champaign, Illinois: U.S. Army Construction Engineering Research Laboratory, 1979).

6. Davos, C.A. "A Priority-Tradeoff-Scanning Approach to Evaluation in Environmental Management". *Journal of Environmental Management* 5(3)(1977)259-273.

7. Fischer, D.W. and G.S. Davies. "An Approach to Assessing Environmental Impacts". *Journal of Environmental Management* 1(3)(1973)207-227.

8. Kruzic, P.G. *Cross-Impact Simulation in Water Resources Planning,* IWR Contract Report 74-12 (Fort Belvoir, Virginia: U.S. Army Engineers Institute for Water Resources, 1974).

9. Leopold, L.B. *et al. A Procedure for Evaluating Environmental Impact,* U.S. Geological Survey Circular 645 (Washington, D.C.: U.S. Geological Survey, 1971).

10. Loran, B. "Quantitative Assessment of Environmental Impact". *Journal of Environmental Systems* 5(4)(1975)247-256.

11. Moore, J.L. *et al. A Methodology for Evaluating Manufacturing Environmental Impact Statements for Delaware's Coastal Zone* (Columbus, Ohio: Battelle-Columbus Laboratories, 1973).

12. Phillips, K.J. and J.A. DeFilippi. "A Matrix Approach for Determining Wastewater Management Impacts". *Journal Water Pollution Control Federation* 48(7)(1976)1759-1765.

13. Ross, J.H. *The Numeric Weighting of Environmental Inter-*

actions, Occasional Paper No. 10 (Ottawa, Ontario: Lands Directorate, Environment Canada, 1976).

14. Schlesinger, B. and D. Daetz. "A Conceptual Framework for Applying Environmental Assessment Matrix Techniques". *Journal of Environmental Sciences* 16(1)(1973)11-16.

15. Schlesinger, B. and D. Daetz. *Development of a Procedure for Forecasting Long-Range Environmental Impacts, Report to the Resource and Land Investigations (RALI) Program, U.S. Geological Survey,* Report No. USGS-LI-75-007 (Stanford, California: Department of Industrial Engineering, Stanford University, 1975).

16. Schwind, P.J. "Environmental Impacts of Land Use Change". *Journal of Environmental Systems* 6(2)(1977)125-145.

17. Sellers, J. and R.M. North. "A Viable Methodology to Implement the Principles and Standards". *Water Resources Bulletin* 15(1)(1979)167-181.

18. Tamblyn, T.A. and E.A. Cederborg. "Environmental Assessment Matrix as a Site-Selection Tool - A Case Study". *Nuclear Technology* 25(4)(1975)598-606.

19. Whitlatch, E.E. "Systematic Approaches to Environmental Impact Assessment: An Evaluation". *Water Resources Bulletin* 12(1)(1976)123-137.

20. Yorke, T.H. *Impact Assessment of Water Resource Development Activities: A Dual Matrix Approach,* FWS/OBS-78/82 (Kearneysville, West Virginia: U.S. Fish and Wildlife Service, 1978).

21. U.S. Federal Power Commission. *Implementation of the National Environmental Policy Act of 1969- Order 485* (Washington, D.C.: U.S. Federal Power Commission, 1973).

22. Carstea, D. *et al. Guidelines for the Environmental Impact Assessment of Small Structures and Related Activities in Coastal Bodies of Water,* MTR-6916 Rev. 1 (McLean, Virginia: Mitre Corporation, 1975).

23. Canter, L.W. and L.G.Hill. *Handbook of Variables for Environmental Impact Assessment* (Ann Arbor, Michigan: Ann Arbor Science Publishers, 1979).

24. Environmental Impact Center, Inc. *A Methodology for Assessing Environmental Impact of Water Resources Development* (Cambridge, Massachusetts: Environmental Impact Center, Inc., 1973).

25. Battelle-Columbus Laboratories. *A Methodology for Assessing*

Environmental, Economic and Social Effects of Dredge Soil Disposal on Marsh and Upland Areas, Draft Report (Columbus, Ohio: Battelle-Columbus Laboratories, 1974).

26. Warner, M.L. *et al. An Assessment Methodology for the Environmental Impact of Water Resource Projects*, EPA-600/5-74-016 (Columbus, Ohio: Battelle-Columbus Laboratories, 1974).

27. U.S. Department of the Army. *Handbook for Environmental Impact Analysis*, DA Pamphlet No. 200-1 (Washington, D.C.: U.S. Department of the Army, 1975).

28. Jain, R.K. *et al. Handbook for Environmental Impact Analysis*, Interim Environmental Planning Bulletin 11 (Washington, D.C.: U.S. Department of the Air Force, 1976).

29. U.S. Department of Transportation. *Environmental Assessment Notebook Series* (Washington, D.C.: U.S. Department of Transportation, 1975).

30. Schaenman, P.S. *Using an Impact Measurement System to Evaluate Land Development*, URI 15500 (Washington, D.C.: The Urban Institute, 1976).

31. Fitzsimmons, S.J. *et al. Social Assessment Manual - A Guide to the Preparation of the Social Well-Being Account* (Cambridge, Massachusetts: Abt Associates, Inc., 1975).

32. Adkins, W.G. and D. Burke Jr. *Social, Economic, and Environmental Factors in Highway Decision Making*, Research Report 148-4 (College Station, Texas: Texas Transportation Institute, Texas A and M University, 1974).

33. Voorhees, A.M. and Associates. *Interim Guide for Environmental Assessment: HUD Field Office Edition* (Washington, D.C.: A.M. Voorhees and Associates, 1975).

34. Duke, K.M. *et al. Environmental Quality Assessment in Multi-Objective Planning* (Columbus, Ohio: Battelle-Columbus Laboratories, 1977).

35. U.S. Environmental Protection Agency. *Draft Environmental Impact Statement, Wastewater Treatment Facilities, Twining Water and Sanitation District, Taos County, New Mexico*, EPA-906/9-81-002 (Dallas, Texas: U.S. Environmental Protection Agency, 1981).

36. Canter, L.W. *Water Resources Assessment - Methodology and Technology Sourcebook* (Ann Arbor, Michigan: Ann Arbor Science Publishers, Inc., 1979).

37. Dee, N. *et al. Environmental Evaluation System for Water Resources Planning, Final Report* (Columbus, Ohio: Battelle-Columbus Laboratories, 1972).

38. Dee, N. *et al. Planning Methodology for Water Quality Management: Environmental Evaluation System* (Columbus, Ohio: Battelle-Columbus Laboratories, 1973).

39. School of Civil Engineering and Environmental Science and Oklahoma Biological Survey. *Mid-Arkansas River Basin Study - Effects Assessment of Alternative Navigation Routes from Tulsa, Oklahoma to Vicinity of Wichita, Kansas* (Norman, Oklahoma: University of Oklahoma, 1974).

40. Smith, M.A. *Field Test of an Environmental Impact Assessment Methodology,* Report ERC-1574 (Atlanta, Georgia: Environmental Resources Center, Georgia Institute of Technology, 1974).

41. Solomon, R. *et al. Water Resources Assessment Methodology (WRAM) - Impact Assessment and Alternative Evaluation,* Contract Report Y-77-1 (Vicksburg, Mississippi: U.S. Army Waterways Experiment Station, 1977).

42. Ott, W.R. *Environmental Indices - Theory and Practice* (Ann Arbor, Michigan: Ann Arbor Science Publishers, Inc., 1978).

43. U.S. Army Corps of Engineers. *A Habitat Evaluation System for Water Resources Planning* (Vicksburg, Mississippi: U.S. Army Corps of Engineers, 1980).

44. U.S. Fish and Wildlife Service. *Ecological Services Manual - Habitat as a Basis for Environmental Assessment* (Washington, D.C.: U.S. Fish and Wildlife Service, 1980).

45. Dee, N. *et al. Literature Review to Identify Rationale for Developing Functional Relationships Between Environmental Parameters and Environmental Quality* (Columbus, Ohio: Battelle-Columbus Laboratories, 1978).

46. Odum, E.P. *et al. Optimum Pathway Matrix Analysis Approach to the Environmental Decision Making Process - Test Case: Relative Impact of Proposed Highway Alternates* (Athens, Georgia: Institute of Ecology, University of Georgia, 1971).

47. Ahmed, S. *et al.* "Formal Methodology for Acceptability Analysis of Alternate Sites for Nuclear Power Stations". *Nuclear Engineering Design* 51(3)(1979)361-388.

48. Burnham, J.B. *et al.* "Method for Integrating Societal and Technical Judgement in Environmental Decision Making". *Nuclear*

Technology 25(4)(1975)675-681.

49. Keeney, R.L. *Preference Models of Environmental Impact,* IIASA-RM-76-4 (Laxenburg, Austria: International Institute for Applied Systems Analysis, 1976).

50. Sicherman, A. "General Methodology and Computer Tool for Environmental Impact Assessment with Two Case Study Examples", pp. 638-642 in *Proceedings of the International Conference of the Cybernetics Society, Tokyo and Kyoto, Japan, Nov. 1978,* vol. 1 (New York, New York: IEEE, 1978).

51. Eckenrode, R.T. "Weighting Multiple Criteria". *Management Science* 12(3)(1965)180-192.

52. Edwards, W. *How to Use Multi-Attribute Utility Measurement for Social Decision Making,* SSRI Research Report 76-3 (Los Angeles, California: Social Science Research Institute, University of Southern California, 1976).

53. Voelker, A.H. *Power Plant Siting, An Application of the Nominal Group Process Technique,* ORNL/NUREG/TM-81 (Oak Ridge, Tennessee: Oak Ridge National Laboratory, 1977).

54. Odum, H.T. "Use of Energy Diagrams for Environmental Impact Statements", pp. 197-213 in *Proceedings of the Conference on Tools for Coastal Zone Management, Washington, D.C., Feb. 1972* (Washington, D.C.: Marine Technology Society, 1972).

55. Lavine, M.J. *et al. Energy Analysis Manual for Environmental Benefit/Cost Analysis of Transportation Actions* (Ithaca, New York: Center for Environmental Research, Cornell University, 1979).

56. Bayley, S. *et al. Energetics and Systems Modelling: A Framework for Energy Evaluation of Alternative Transportation Models in Comparison with Traditional Economic Evaluation* (Gainesville, Florida: Department of Environmental Engineering Sciences, University of Florida, 1976).

57. Gilliland, M.W. and P.G. Risser. "The Use of Systems Diagrams for Environmental Impact Assessment: Procedures and An Application". *Ecological Modelling* 3(3)(1977)183-199.

58. Babb, M.C. and H.R. Hickey. *Applications of Computer Graphics to Integrated Environmental Assessments of Energy Systems,* TVA/EP-78/10 (Chattanooga, Tennessee: Tennessee Valley Authority, 1978).

59. Bryant, J.W. "Modelling for Natural Resource Utilization Analysis". *Journal Operations Research Society* 29(7)(1978)667-676.

60. Hill, D. "A Modelling Approach to Evaluate Tidal Wetlands", pp. 105-117 in *Transactions 41st North American Wildlife and Natural Resources Conference* (Washington, D.C.: Wildlife Management Institute, 1976).

61. Hill, D. "A Resource Allocation Model for the Evaluation of Alternatives in Section 208 Planning Considering Environmental, Social and Economic Effects", pp. 401-406 in *Proceedings of the Conference on Environmental Modeling and Simulation,* EPA 600/9-76-016 (Washington, D.C.: Environmental Protection Agency, 1976).

62. Hodgins, D.B. *et al.* "A Simulation Model for Screening a System of Reservoirs for Environmental Impact". *Canadian Journal of Civil Engineering* 4(1)(1977)1-9.

63. Hydrologic Engineering Center. *Guide Manual for the Creation of Grid Cell Data Banks* (Davis, California: U.S. Army Corps of Engineers, 1978).

64. Hydrologic Engineering Center. *Resource Information and Analysis Using Grid Cell Data Banks, Computer Program 401-X6-L 7590, User's Manual* (Davis, California: U.S. Army Corps of Engineers, 1978).

65. Lincoln, D.R. and E.S. Rubin. "Cross-Media Environmental Impacts of Coal-Fired Power Plants: An Approach Using Multi-Attribute Utility Theory". *IEEE Transactions for Systematic Management of Cybernetics* 9(5)1979 285-289.

66. Rubenstein, S. and R.L. Horn. *Risk Analysis in Environmental Studies,* CONF-780316-8 (Richland, Washington: U.S. Department of Energy, 1978).

67. Schrender, G.F. *et al.* "A Computerized System for Wild Land Use Planning and Environmental Impact Assessment". *Computers and Operations Research* 3(2/3)(1976)217-228.

A METHOD FOR EVALUATING ENVIRONMENTAL IMPACTS FROM LAND DEVELOPMENT PROJECTS

Fotis Chatzimikes

Scientific Advisor, National Council for Physical Planning and the Environment, Ministry of Co-ordination, Athens.

1 INTRODUCTION

The US National Environmental Policy Act (NEPA) of 1969, brief as it was, began a new phase of response to the old ever-increasing problems of environmental deterioration from man's activities. One of the requirements of this act was the preparation of environmental impact statements (EISs) for proposed activities which were likely to significantly affect the existing environment. These EISs had to contain an account of the environmental impacts of the proposed activities. While the objectives of the Act appeared to be quite clear, the methods and procedures to be used in accomplishing them were less well understood. One of the problems related to the difficulty in deciding which projects should be subject to examination by an EIS.

Attempting to fill the void created between legislation and implementation, a series of various federal environmental statutes, executive orders and agency guidelines (1), including the Department of Housing and Urban Development (HUD) Environmental Review Procedures (2), was issued during the decade following the Act. While the various guidelines, and those issued by HUD in particular, were helpful in outlining the major steps and components of an EIS system, pertinent points relating to technical and philosophical approaches, methodologies, techniques and criteria for judgement, remained vague and subject to speculation and to the views of the practitioners. As both public and private sectors became busy in preparing EISs, this lack of clear understanding as to the subject's underlying structure was the main cause for the emergence of an abundance of problems, antagonisms and idealogical differences among the practitioners themselves, between the practitioners and NEPA, and between NEPA and other private and public interest groups.

As a result, there has been over the years a gradually unfolding judicial interpretation of NEPA that has caused considerable national controversy. Many federal agencies were involved in litigation over their environmental decisions. They were principally challenged by citizen groups and individuals, and were pressed to defend their decisions, strategies and methodologies case by case.

Throughout the past decade, one of the most frequently challenged elements of the EIS system has been the impact *evaluation* process. While substantial agreement has been reached on research methods for the identification of impacts, vigorous controversy still exists on how best to evaluate impact significance.

Starting with the fundamental argument of whether the assessment process should be quantitative or qualitative, the controversy continues as to the methodologies used for evaluation. It is generally agreed that each of the several methodologies which have been proposed and used, such as *ad hoc* evaluation, overlays, checklists, matrices and networks, has its merits and limitations. While the network method, for example, is excellent for identification and evaluation of indirect impacts, it has the disadvantage of being difficult to apply, requiring highly skilled staff and computer facilities. On the other hand, the checklist method is easy to apply and good for impact identification, but does not provide adequate evaluation means. Figure 1 gives a summary and evaluation of the methodologies most commonly used.

The US Housing and Community Development Act of 1974 enabled use of federal funds for "blocks" of development projects within a community rather than single and specific works. They were named Community Development Block Grant projects and required assessment of environmental impacts prior to releasing of funds. Since the overwhelming majority of these projects involved land development works there emerged an overabundance of impact studies for the development of land, using various methodologies with varying degrees of success.

In 1976, the North Texas Council of Government, in association with the University of Texas, and in collaboration with planning consultants in the North Texas region, attempted to formulate a comprehensive land development impact evaluation methodology which would comply with the spirit of NEPA. Since its development, the methodology has been tested, through frequent use, for Community Development Block Grant projects by Texas municipalities. The methodology is directed at selecting projects for examination

Type of Method	Addresses Impact								Ease of Application			Resource Requirement			
	Identification		Measurements		Interpretation		Evaluation					Staff Needed		Computer Needed	
	Yes	No	Yes	No	Yes	No	Yes	No	Difficult	Moderately Difficult	Not Difficult	Highly Skilled	Moderately Skilled	Desired	Not Needed
Ad Hoc	X			X		X		X			X		X		X
Overlays	XX			X		X		X		X		X			X
Checklists	XXX			X		X		X			X		X		X
Matrices	XXX		X		X		XX			X			X		X
Networks	XXX		XXX		XXX		XXX		X			X		X	

X = Fair   XX = Good   XXX = Excellent

FIGURE 1: Summary and Evaluation of Assessment Methodologies by EIS. While it is recognized that the methodology is not free of shortcomings, it offers several points of interest and may be viewed as a positive step in the continuing process of developing evaluation methodologies.

2 THE ENVIRONMENTAL REVIEW PROCESS FOR LAND DEVELOPMENT PROJECTS

The environmental review process is the step-by-step procedure through which an environmental assessment is conducted to identify and evaluate the impact of a man-made activity on the environment. More specifically the review process is a structure for collecting and interpreting facts with regard to the environmental consequences of a project. The manner in which the process is structured varies according to the particular approach used.

The North Texas Council of Government/University of Texas/ private consultants team therefore developed a recommended review process and evaluation methodology, the structure of which is in compliance with HUD's Environmental Review Procedures. It is shown in Figure 2. Of the five major components of Figure 2, it is the third component, which deals with the assessment and evaluation of impacts, which is addressed by this evaluation methodology.

3 IMPACT ASSESSMENT

As previously stated, the evaluation is central to the entire environmental review process. It provides a basis for determining whether or not a particular project has any significant impact on the quality of the human environment and whether that project, if implemented, will be environmentally acceptable.

3.1. Assessment Assumptions

Two interrelated assumptions have been used to guide the assessment and evaluation of impacts. One deals with the definition of the "environment" and the other with the national environmental policy expressed in NEPA.

3.3.1. First assumption. The term "environment" has a wide range of meanings for different people. Among the alternative definitions that can be applied to the term are the following:

a) restriction of the definition to the natural environment: air, water, noise, vegetation, soils, geology, etc..
b) broadening of the definition slightly to add energy, historic preservation, land use, open space, population

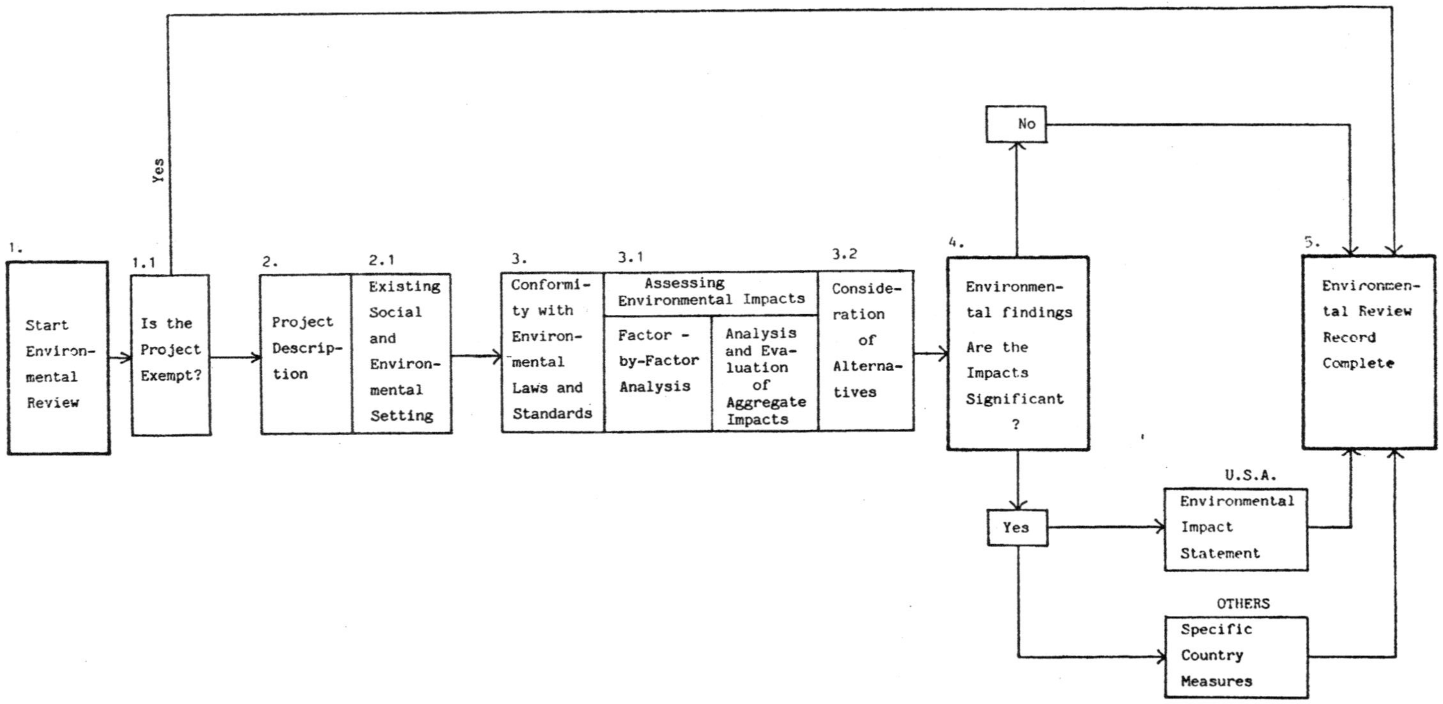

FIGURE 2: The Environmental Review Process for Land Development Projects

and wildlife preservation.

c) addition to the above factors of such elements as aesthetics, health, housing, mass transportation, recreation, infrastructure facilities etc..

d) extension of a) to c) above to include economic development, education, public safety and welfare, reflecting a "quality of life" scope.

It was this last definition of the environment which was adopted in the methodology to determine the impact significance of a project on the "quality of human environment". The adoption of this definition was based on the belief that NEPA's concept of the human environment reflects the "quality of life" scope, as further illustrated in the second assumption.

3.1.2. Second assumption. NEPA's aim is to introduce environmental factors into national decision-making, thereby protecting the environment from any action which may produce significant adverse effects. It was assumed that NEPA does not refer to the natural environment alone, but also includes man-made components. This may be derived from its policy objective in Section 101. This Section has been characterized as a "substantive mandate" and states that:

> "It is the continuing policy of the Federal Government .... to use all practicable means and measures ... in a manner calculated to foster and promote the general welfare, to create and maintain conditions under which man and nature can exist in productive harmony, and fulfill the social, economic and other requirements of present and future generations ..."(3).

Similarly, Section 101(b) contains sweeping language, referring to "high standards of living and a wide sharing of life's amenities", and "safe, healthful, productive, and aesthetically and culturally pleasing surroundings". In short, Section 101 of NEPA attaches to the concept of environment a wide meaning, including social and economic imperatives.

The preceding assumptions are important in determining the impact significance of a project for each individual assessment factor. They are also basic to establishing criteria for making trade-offs between and among the assessed impacts of a project on the human environment.

The use of the assumptions mentioned above was deemed necessary to establish common ground between aims and the existing political and philosophical framework, through which the desired objectives may be achieved. It is generally advisable that such assumptions relating to policy interpretation, definitions and existing philosophical imperatives be taken into account and used as guidance for the formulation of any assessment methodology.

3.2 Types and Significance of Impacts

Even though classification of impact types varies in the literature, there exists substantial agreement on the *nature* of impacts. A typical classification recognizes two basic impacts: direct and indirect. While direct impacts are those that a project directly generates on the environment, indirect impacts are those which may be generated by activities that are induced by a project. Indirect impacts are often known as induced impacts, or impacts of induced growth or secondary impacts.

A further classification of these two types of impacts can be made on the basis of their influences on other dimensions such as the time frame, the number and socio-economic characteristics of people and/or the size and characteristics of the area affected. In terms of time, an impact can be temporary (short-term) or permanent (long-term). In terms of geographic and population coverage, a project can affect its own site, its immediate vicinity, or the service area it is designed to serve, or even sometimes an entire city or region. Defined by these parameters, an impact may be significant (major) or less significant (minor).

The significance of an impact is also a function of the existing quality of the environment. Clearly, a small project may be capable of causing a significant impact if the base level of the existing environment is already low and vulnerable and has little or no tolerance left to absorb adverse effects from a project. On the other hand, a relatively large project may not produce any significant impact if the existing environment's tolerance to withstand adverse effects is high.

The significance of an impact, therefore, is a function of the following conditions:

a) Impact duration;
b) Population and/or geographic area affected;
c) The existing quality of the environment.

As will be shown later, the evaluation of each impact of a project will first be measured in terms of these parameters before establishing its level of significance.

3.3 Evaluation Methodology

The process of environmental assessment is both an art and a science. There are no universally applicable procedures for assessing impacts of any action or activity, including land development works. Any evaluation process is the product of numerous combinations of objective and subjective factors. Total objectivity is impossible, and it is informed subjectivity which often contributes most to the success of an evaluation effort. The methodology described below attempts to establish some criteria which, if studiously executed, may provide the pertinent information required for judgement on impact significance.

This methodology is simple, cost-effective and permits any number of environmental factors to enter the evaluation process. As a result, it can be made suitable for assessing secondary impacts of a project. After deleting and adding relevant factors, it can be adapted to allow simultaneous assessment of several projects in one format. It permits the consideration of the impact of project alternatives, including the no-action option, on the environment in a comparable way. Further, a variety of mechanisms are available to adjust the format to make the method responsive to almost any project, large or small.

3.3.1. Impact gauging scale. The scale developed in this method for measuring the impacts of land development projects consists of positive, neutral and negative values. As shown in Figure 3, the scale represents a possible impact range from highly beneficial to highly adverse. The "score" actually assigned to a project is dependent on the magnitude of the interactions between the various environmental factors and the given project, and the *significance of those interactions in terms of duration, the population and geographic area covered, and the existing quality of the environment*. The various segments of the scale correspond to the numerical values assigned to them within the scale.

3.3.2. Methodology application. Appendix I represents the structure within which land development projects may be evaluated in relation to their impacts on the environment. The table may take the form of a checklist if only one project is to be evaluated, or it may take the form of a matrix if a project and its alternatives or even several projects and their alternatives are to be evaluated simultaneously. The project or projects and the respective alternatives are shown as rows of the matrix while environmental concerns, community objectives, constraints and other factors are shown as columns. The first row of the matrix is reserved for evaluation of the existing condition of the environment. A project and its alternatives are evaluated against all the factors shown in columns 1 to 49 of the matrix. The rest of the columns are used for summary evaluation, including pertinent observations and qualitative remarks. The matrix in Appendix I describes the evaluation of a hypothetical project and illustrates the evaluation process.

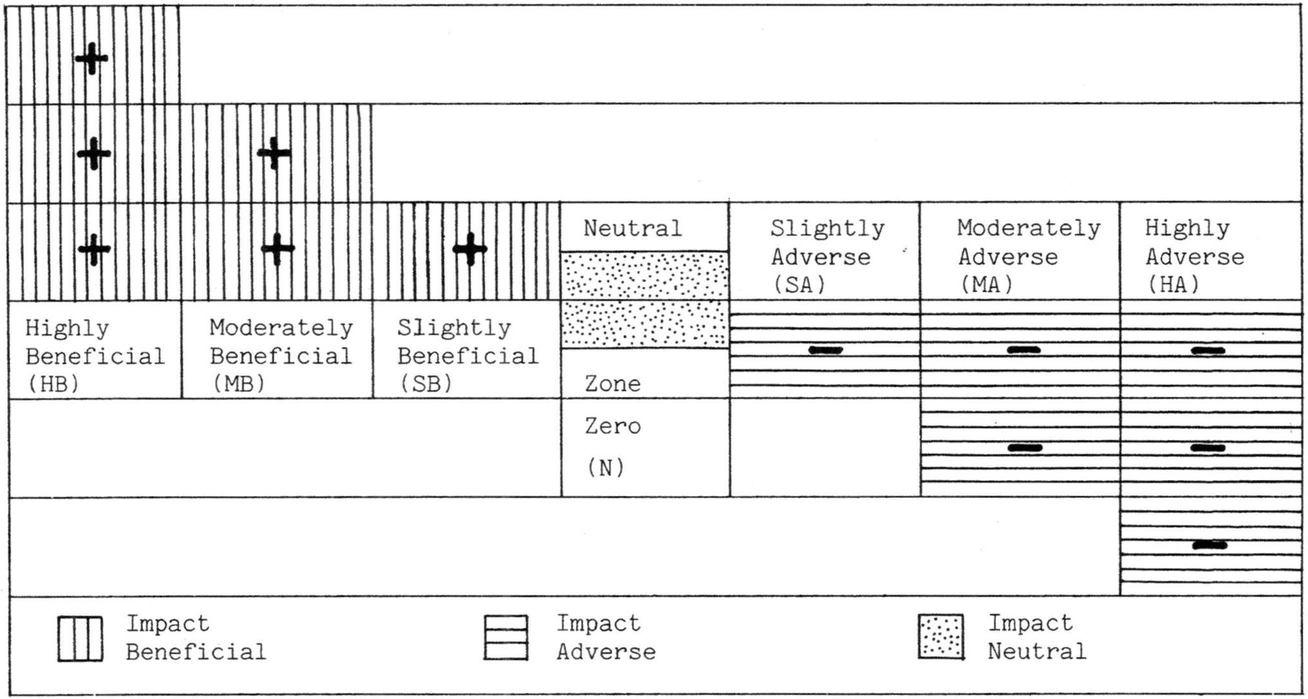

Highly Beneficial (HB) = +3
Moderately Beneficial (MB) = +2
Slightly Beneficial (SB) = +1

Slightly Adverse (SA) = -1
Moderately Adverse (MA) = -2
Highly Adverse (HA) = -3

Neither Beneficial Nor Adverse (Neutral) = 0

FIGURE 3: Impact Gauging Scale

3.3.3. Evaluation factors. One basic principle should be kept in mind when selecting project evaluation factors. The principle is *relevancy*. The factors selected should be socially, economically and environmentally relevant for a given project in a given area or community. The incorporation of additional factors not relevant to the specific environment upon which the project will be imposed, tends only to confuse the resulting assessment. Likewise, the omission of relevant factors ensures inadequate assessment.

The example in Appendix I considers representative factors which may be relevant in a typical land development project. It should be emphasized that these are only illustrative; there is no intention of suggesting that these are the most or only relevant factors in a land development project. This matrix is open and the number and/or type of factors to be considered for a given project will always vary in relation to legal, technical and philosophical approaches through which a community is prepared to confront the problem of environmental protection.

3.3.4. Summary of the evaluation process. The evaluation process involves the hypothetical imposition of a project on the existing environment and the estimation of the resultant changes (beneficial or adverse). Depending on the extent of changes and their significance, ratings are assigned to the factors by the use of the impact gauging scale. The same process is, of course, repeated for the alternatives of the project. Total scores are determined by the sums of the ratings assigned. The total scores, coupled with qualitative examination of other factors which may be of overriding importance, provide a basis for determining whether or not projects and alternatives will have a significant impact on the human environment.

3.3.5. Step-by-step rating procedures and criteria. The first task is to complete the matrix boxes in the evaluation table. Judgement is, of course, necessary in this task, but by following the available procedures and criteria, consulting related research and analysis, and considering the ultimate purpose of a given project, the staff of the pertinent agency conducting an evaluation can turn that judgement into a satisfactory tool for assessing the environmental impacts of a project.

The following steps are suggested for this purpose:

a) <u>Select one evaluation factor from the matrix table of Appendix I</u>.

b) <u>Assess the existing conditions of the environment in terms of that factor only</u>.

The rating may be done on the following basis:

Excellent	E
Good	G
Fair	F
Poor	P
Very Bad	B

In assigning one of the above values in the matrix box pertaining to the conditions of the existing environment in terms of that particular factor, the assessor may have to go back and consult the information gathered from research and analysis conducted to characterize the existing social and environmental setting (see step 2.1 in Figure 2).

c) Match the project and the particular evaluation factor being considered (the environment now is imagined as the existing environment plus the project), and identify if there would be any interaction between the existing environment and that factor. The answer may be noted as follows:

- Has no impact
- Has beneficial impact
- Has adverse impact

If the answer is that it "has no impact", then the evaluator will leave that box empty or write N (for Neutral). In that case, this will complete the rating for that project on that factor under consideration, and the evaluator may move to the next factor or move to rate an alternative solution of the project on the same factor. If, however, the answer involves the last two choices, the evaluator will proceed to examine the following.

d) Assess the significance of the effect of the interaction between the existing environment and that factor.

The effect will be either beneficial or adverse (see Figure 3).

Beneficial (Positive)	Adverse (Negative)
- Highly Beneficial (HB)	- Highly Adverse (HA)
- Moderately Beneficial (MB)	- Moderately Adverse (MA)
- Slightly Beneficial (SB)	- Slightly Adverse (SA)

The task from now on is to determine whether the particular matrix box will receive HB, MB, SB, or HA, MA, SA. This will be answered in the following steps.

e) The duration (t) of the effect may be:

- Permanent (HBt or HAt)
- Intermediate Duration (MBt or MAt)
- Temporary (SBt or SAt)

f) The effect on people (p) may be:

- Many People (HBp or HAp)
- Few People (MBp or MAp)

- No People (O)

g) <u>The effect on land area (a) may be</u>:
- A large area (HBa or HAa)
- A moderate-size area (MBa or MAa)
- A small area (SBa or SAa)

h) <u>Aggregation of results from steps e), f) and g)</u>:
Depending on the responses noted in these steps, a variety of combinations may result (either on the adverse side or beneficial side). From the twenty-seven possible combinations, four are shown below for the purpose of illustration.

Conditions	Aggregate Effect	Conditions	Aggregate Effect
HBt, HBp, HBa	HB	HAt, HAp, HAa	HA
HBt, MBp, MBa	HB	HAt, MAp, MAa	HA
MBt, SBp, SBa	MB	MAt, SAp, SAa	MA
SBt, MPp, MBa	MB	SAt, MAp, MAa	MA

The aggregate effect on any evaluation factor will result after the effect has been evaluated in terms of the three parameters (time, people, area). As shown in the example above, the governing criteria is that within any combination, the highest effect(s) will dictate the aggregate rating. Any one extreme effect, in other words, will be adequate to cause the entire score to be in that category. Prior to placing the aggregate effect score (HB, MA, HA, etc.) in the appropriate matrix box, the evaluator must check the box describing the existing condition of the environment in terms of the factor under consideration, and make a judgement as to the final score the particular matrix box should receive. If, for example, the aggregate effect on a particular evaluation factor turns out to be Moderately Adverse (MA) and the condition of the existing environment in terms of that factor happens to be Poor (P) or very Bad (B), the final grading of the matrix box may receive HA instead of MA. This is how the matrix boxes are completed for each environmental factor.

3.3.6. Aggregation of individual impacts. The aggregate ratings assigned to the factors against a project are totalled under the seven different categories (from Highly Beneficial to Highly Adverse, see the matrix, Appendix I, columns 51 to 57). The sum of the adverse effects scores and the scores of the beneficial effects will yield the net impact. This net impact becomes one of the major indicators of whether the environmental impact is significant. If the net impact is positive, it can be interpreted as a good indication that the project is satisfactory overall. In the example shown in the matrix of Appendix I, the hypothetical project received 30 neutral scores, 3 SBs, 8 MBs, 3 HBs, 2 SAs and 3 HAs. The total score is computed as follows:

30 (N)	=	30 x 0	=	0
3 (SB)	=	3 x 1	=	+ 3
8 (MB)	=	8 x 2	=	+16
3 (HB)	=	3 x 3	=	+ 9
2 (SA)	=	2 x -1	=	- 2
3 (HA)	=	3 x -3	=	- 9
			=	17 (shown in column 60)

The corresponding no-action alternative score is also shown in column 60 and the comparison between the two favours the implementation of the project as far as its environmental impact is concerned.

3.3.7. Evaluation of environmental impacts. As previously stated the major purpose of the environmental review process is to determine the *significance* of a proposed action on the human environment. The net impact derived from the evaluation described is one of the major factors for determining significance. Additional examination and qualitative evaluation is necessary, however, to avoid the pitfalls of a stereotyped, rigid, quantitative assessment. The following criteria may be used to aid the evaluation process and make the final determination on impact significance: the overall net impact derived from the matrix methodology, significant adverse impacts on natural systems components, and overriding factors or conflicts with environmental laws.

a) The overall net impact. This net impact derived from the matrix methodology is viewed as adequate in offering an idea of the overall impact significance of a proposed project. Further, it offers an important means of comparing alternatives, and of selecting the option least harmful to the environment. One of its major shortcomings lies in the fact that it uses the equal-weight principle in factor evaluation. Clearly all factors do not hold equal weight for all people, but rather their importance varies in accordance with national and local goals, philosophical and cultural imperatives and political systems. The additional criteria described below are intended to supplement the matrix evaluation by bringing into focus additional elements of a qualitative nature.

b) Significant adverse impacts on natural systems components. If the environmental assessment indicates that a project activity would have significant adverse impacts on any component of the natural systems (despite the project's overall net beneficial impact), then a further study may be required to identify alternatives and introduce mitigation measures to counteract such effects. In the case of the U.S. this additional study would constitute an EIA, while in other countries it might require close collaboration with the component authority(ies).

c) Overriding factors. As previously stated some of the environmental factors are more important than others. In fact some are so dominant that they can be classified as "veto factors". Any project that is environmentally in conflict with such factors would constitute a major action with significant effects on the human environment, and would require additional studies, alternative approaches and/or mitigation measures. Water and air quality are almost always major factors. Other overriding factors are historical and archaeological resources and endangered species. Clearly, there are other major factors, but they vary in accordance with regional differences in environmental quality. In some areas, for example, surface subsidence may be a major factor, while wetland protection may be a major factor in other areas. National and local criteria and laws are also overriding factors requiring compliance.

3.3.8. Environmental findings and conclusions. The final step in the evaluation process deals with the environmental findings and conclusions to determine whether the project would or would not have a significant impact on the quality of the human environment.

The possible need for additional studies and for mitigation measures is one component of these findings and conclusions. Generally speaking, the matrix will indicate which effects are highly adverse (HA), moderately adverse (MA), and slightly adverse (SA). Those which are highly adverse will always require mitigation measures; those which are slightly adverse will generally not require mitigation measures. The effects which are moderately adverse (MA) may or may not require any mitigation measures depending on the type of impact, the existing capacity of the environment to absorb such impacts, and the cost of undertaking programmes for their mitigation.

Impacts which cannot be avoided should also be identified and taken into consideration. These include effects which are normally associated with all physical development projects, such as machine and engine noise during construction phases, disruption of traffic flow during construction, air contaminants due to construction activities, disruption of natural earth and loss of habitat.

Certain irreversible and irretrievable commitments of resources will be required for the construction, operation and maintenance of almost any land development project. Resources such as building materials and fuels are essentially non-renewable. Lists of such materials should be prepared and considered in terms of their relative availability at a regional or national scale. A summary evaluation of whether the overall benefits to be gained by the depletion of these resources will adequately offset the costs associated with the development and maintenance of the project should also be made. These resources, plus land and manpower, will be needed for any man-made activity to achieve social, economic and environmental goals for a population. The resources are in essence traded-off against the higher levels of human need.

Following the brief discussion above, it should be clear that an environmental review without meticulous qualitative analysis is not realistic. The methodology described is not offered as a substitute to qualitative elaboration. Rather, it has a twofold purpose: first, to provide a general idea, through its quantitative impact figure, of the overall significance of the project; and second, to isolate significant adverse impacts for further examination through more detailed and qualitative means.

REFERENCES

1. U.S. Council on Environmental Quality. "Preparation of environmental impact statements : guidelines". *Federal Register* 37 (147) (1973) 20550-20562.

2. U.S. Department of Housing and Urban Development. "Environmental Review Procedures for the Community Development Block Grant Program, Corrections and Amendments". *Federal Register* 40 (137) (1975) 29991-29995.

3. U.S. Council on Environmental Quality. *Regulations for Implementing the Procedural Provisions of the National Environmental Policy Act* (Washington, D.C.: U.S. Government Printing Office, 1979).

APPENDIX I: Worked Example of Evaluation of a Land Development Project

LAND DEV. PROJECT(S) AND THEIR ALTERNATIVES		I. COMPLIANCE WITH LOCAL AND NATIONAL ENVIRONMENTAL LAWS AND STANDARDS													II. COMMUNITY AND/OR REGIONAL OBJECTIVES												
		CONFORMITY WITH GENERAL PLAN 1	CONFORMITY WITH UTILITY EXTENSION AND GROWTH POLICY 2	CONFORMITY WITH ZONING AND OTHER CODES 3	CONFORMITY WITH HISTORIC PRESERVATION ACT - 1966 4	EPA'S NOISE CONTROL ACT 5	CLEAN AIR ACT - 1970 6	FLOOD DISASTER PROTECTION ACT 7	FISH AND WILD LIFE COORDINATION ACT 8	9	10	11	12	13	SAFE AND DECENT HOUSING SUPPLY 14	IMPROVED HOUSING FOR LOW INCOME PEOPLE 15	MINIMIZING BLIGHT AND BLIGHTING IFLUENCES 16	INCREASE IN JOB SUPPLY 17	INCREASE IN HOME OWNERSHIP 18	INCREASE IN CULTURAL AND EDUCATIONAL FACILITIES 19	INCREASE IN ACCESS AND TRANSPORTATION 20	EQUITABLE DISTRIBUTION OF URBAN AMENITIES 21	MINIMIZATION OF CRIME 22	MAXIMIZATION OF PUBLIC HEALTH AND SAFETY 23	INCREASE IN PARKS, RECREATION AND AND URBAN AESTHETICS 24	25
EXISTING ENVIRONMENT		G		G																F	F	G	F	F	G	
PROJECT NAME	t	HB		MB																MB	HB	MB	MB	MB	HB	
	p	MB		MB																SB	MB	MB	MB	MB	HB	
	a	SB		SB																SB	SB	MB	SB	SB	MB	
		HB	N	MB	N	N	N	N	N						N	N	N	N	N	MB	HB	MB	MB	MB	HB	N
NO-ACTION ALTERNATIVE	t	SA		SA																		SA	SA			
	p	SA		SA																		SA	SA			
	a	SA		SA																		SA	SA			
		SA	N	SA	N	N	N	N	N						N	N	N	N	N	N	N	SA	SA	N	N	N

APPENDIX I (continued)

LAND DEV. PROJECTS AND THEIR ALTERNATIVES		III. IMPACT ON THE NATURAL ENVIRONMENT																		IV. IMPACT ON COMMUNITY FACILITIES						V
		SUBSURFACE SETTING				SURFACE SETTING			CLIMATE & ATMOSPHERE			BIOLOGICAL ENVIRONMENT														
		PALEONTOLOGY & ARCHAEOLOGY	SURFACE-SUBSURFACE GEOLOGY	AQUIFER RECHARGE AREA	UNDERGROUND WATER QUALITY	SOIL EROSION SENSITIVITY	FLOOD PLAIN AND DRAINAGE BASINS	SURFACE WATER QUALITY	WIND DIRECTION AND TEMPERATURE CHANGE	AIR QUALITY	NOISE LEVEL	VEGETATION & PLANT LIFE	AQUATIC LIFE	ENDANGERED PLANTS	WILDLIFE	INSECT LIFE	ENDANGERED SPECIES	BIRD LIFE		WATER AND SEWER	PARKS AND SCHOOLS	POLICE AND FIRE PROTECTION	SOLID WASTE DISPOSAL	HEALTH AND OTHER SOCIAL SCIENCES	TRANSPORTATION	IMPACT OF ENERGY
		26	27	28	29	30	31	32	33	34	35	36	37	38	39	40	41	42	43	44	45	46	47	48	49	50
EXISTING ENVIRONMENT			F			F	P	F				G		P	G	G		G			F			G		
PROJECT NAME	t		SA			SB	MA	HA				SA		HA	MB	SB		MB			MB			SB		
	p		SA			SB	SA	SA				SA		SA	SB	SB		MB			MB			SB		
	a		SA			SB	HA	MA				SA		SA	SB	SB		SB			SB			SB		
		N	SA	N	N	SB	HA	HA	N	N	N	SA	N	HA	MB	SB	N	MB		N	MB	N	N	SB	N	N
NO-ACTION ALTERNATIVE	t													SB							SA					
	p													SB							SA					
	a													SB							SA					
		N	N	N	N	N	N	N	N	N	N	N	N	SB	N	N	N	N		N	SA	N	N	N	N	N

APPENDIX I (continued)

LAND DEV. PROJECTS AND THEIR ALTERNATIVES	VI. ANALYSIS AND EVALUATION OF PREDICTED IMPACTS										VII. DOES THE PROJECT HAVE SIGNIFICANT IMPACT?		VII. REMARKS
	FREQUENCY							TOTAL		NET IMPACT			OBSERVATION REMARKS AND RECOMMENDATIONS ON OVERRIDING FACTORS AND THEIR SIGNIFICANCE
	HA	MA	SA	N	SB	MB	HB	A	B		YES	NO	
	51	52	53	54	55	56	57	58	59	60	61	62	63
PROJECT NAME	3	0	2	25	3	8	3	-11	28	17		X	Alternative solutions and/or mitigation measures required to remedy the highly adverse impacts on soil erosion, endangered plants and surface water quality.
NO-ACTION ALTERNATIVE	0	0	5	38	1	0	0	-5	1	-4		X	

SOCIAL IMPACT ASSESSMENT: METHODOLOGICAL OVERVIEW

C. P. Wolf

Department of Social Sciences
Polytechnic Institute of New York
Brooklyn, New York

While environmental impact assessment has focused mainly on the physical and biological properties of natural environments, social impact assessment (SIA) is concentrated on the distinctively human side of human environments. SIA is about "people impacts" --what we are doing to folks (or failing to do for them) where they live, in families and communities, as a consequence of formulating policies, instituting programs and building projects. Its aim is to predict and evaluate those impacts before they have happened. Unlike the more familiar "evaluation research," which gauges the effectiveness of policies, programs and projects already in operation, the task for SIA is anticipatory research. It seeks to place the expectation and attainment of desired outcomes on a more rational and reliable basis.

While the impetus for SIA came largely through passage of the National Environmental Policy Act of 1969 (NEPA), it is gaining prominence in such diverse fields as technology assessment and cultural resource management. Among the topical areas of SIA interest are:

Community development	Weather modification
Industrial development	Highways and mass transit
Coastal zone management	Parks, forests, wildlands
Water resources	Buildings and housing
Energy development	Health and human services
Education and manpower	Toxic substances

The list could be extended to all forms of technology development, environmental modification, and social intervention. Indeed, there is a nontrivial sense in which all impacts are "social." François

Hetman's phrasing of "technology assessment" as "the social control of technology" is the correct one. Since "hyphenated" impact assessments (socio-technical, socio-economic, socio-environmental) have tended to be more technical, economic or environmental than "socio," however, it is necessary to identify a "distinctively" social (or, in Bill Freudenburg's (1) phrase, "social social") content for SIA. One clear example of social impact is the broad mandate for public involvement.

In the past, most SIA studies were focused on physical planning projects in rural areas, e.g. water resources. Today greater emphasis is being placed on social planning in "built" (e.g. urban) environments. (Naturally these too have physical, economic and environmental components, as in the construction and financing of housing units and the conversion of agricultural land to residential use.) "Software" applications, e.g. to health care delivery systems, demand and deserve increased assessment efforts. Although SIA began at the federal level, its gradual extension to regional, state and local levels is both feasible and desirable. Counterpart to those on other levels of the federal system, community-based assessments are receiving greater attention. In fact, SIA can and must be conducted at every appropriate scale, from local to global. International development of SIA is an especially pressing concern. As well as community involvement, corporate involvement is a promising trend and a practical necessity.

The analytic problem of SIA can be modestly stated as learning to make public (and private) decisions that will look good in 50 years, after the evaluative criteria by which they are judged have changed. This "need to know" has arisen because of the unwanted and untoward impacts (side effects and spillovers) of planning approaches that looked good at the time but didn't turn out so good--combined storm and sanitary sewers, the bulldozer approach to urban renewal, and so on and on. "Nobody told us it would be like this!" is the familiar refrain. Knowing now, as best we can, what we would know then is another way of stating the problem. How rational we can be remains to be seen; although we can never know the future with certainty, we can know better than we do. Who the "we" is also needs clarifying, since it turns out that there are multiple publics holding multiple objectives and employing multiple criteria of evaluation. A further complication is that all these can be expected to change over time.

The "bottom line" question for SIA is: "Who benefits and who loses?" (were a proposed action to be implemented). Since often these are not the same people, the dissociation of costs (or risks) and benefits creates a problem of equity. Just what the incidence and distribution of social costs and benefits may be in any particular impact situation is a matter for assessment. Getting them together, so that those who reap the benefits also bear the costs, is a matter of social justice and social policy. SIA cannot prescribe what social values "ought" to be in this instance, any more than expert judgment generally can be said to reflect public preference. It can only attempt to determine what the equity impacts would be were one or another planning alternative to be adopted. Here again, rationality must include values, although there can be no purely rational means for deciding what rank and weight should be assigned to various publics and their preferences.

Allowing for value changes, perhaps the best guide to the future is what we want it to be. If getting "from is to ought" is a commission of the "naturalistic fallacy," the reverse proposition ("from ought to is") is a legitimate exercise in normative planning. Why are we doing or proposing these things?--whether resource development, environmental protection, historic preservation, neighborhood revitalization, or whatever is at issue. The only valid answer can be: because they are deemed "good" under some criteria of social value applied to some recognized social need. This social goal orientation furnishes the justification for public decision and action. In the case of SIA, the "why" question resolves to some notion of "social well-being" or "quality of life"--what it is, how do we know when we have it, and how can we get more of it. Such conceptual understanding and operational measurement implies attention to social purposes as well as social consequences. The underlying reason for doing SIA, then, is to address and express these social concerns. Although SIA partly grew out of the sense of frustration over plans gone wrong or awry, its constructive use in achieving desirable futures is its chief social value.

In this sense, the goal of SIA is not really different from that of social science traditionally: to understand the causes, conditions and consequences of social life. As such, it can be argued that SIA is "what we have always done" as social scientists. There are social impacts of ordinary social life and social change, e.g. of large-scale shifts in demographic structure, such as rural depopulation. While the content of social knowledge may not differ--and SIA draws freely on all social science disciplines--the occasions and conditions for its application are distinctive. SIA is usually invoked in cases of purposive or planned social intervention, to provide "use knowledge" for impact mitigation and management.

What you do in SIA is assess the social impacts. If this seems to belabor the obvious, it must be said that how assessments are done varies widely. SIA is a multimethod approach, and its analytic tasks (see Table 1) require assessors to draw selectively from the full range of social research methodologies and techniques. Moreover, every impact situation has unique features, and general methodologies must be tailored to their dimensions.

Table 1: Social Impact Assessment: The Main Pattern

Assessment Steps	Analytic Operations
1. SCOPING How big a problem is it? How much is enough?	1. Set level(s) of assessment (policy/program/project) 2. Determine impact area boundaries 3. Establish time horizons 4. Develop study design
2. PROBLEM IDENTIFICATION What is the problem? What is causing it?	1. Formulate policy goals, planning objectives 2. Identify publics and concerns 3. Perform needs assessment 4. Determine evaluative criteria
3. FORMULATION OF ALTERNATIVES What are the alternatives?	1. Define set of "reasonable" alternatives (corresponding to identified concerns) 2. Determine change agents, instruments 3. Characterize and describe technical system; analyze for social (institutional/behavioral) components and correlatives 4. Analyze economic and environmental impacts for secondary social impacts
4. PROFILING Who is being affected and how?	1. Dimension impact categories 2. Select impact categories 3. Assign impact indicators 4. Perform indicator measurements 5. Compile social profile
5. PROJECTION What is it causing?	1. Explicate "state of society" assumptions 2. Perform trend impact analysis 3. Construct dynamic system models 4. Estimate impact indicator values for alternative plans ("with and without" implementation)
6. ASSESSMENT What difference does it make?	1. Perform sensitivity analysis for alternative outcomes of alternative plans 2. Perform cross-impact analysis 3. Describe and display "significant" impacts

7. EVALUATION How do you like it?	1. Reidentify publics and concerns 2. Reformulate evaluative criteria 3. Rank and weight preferences for alternatives 4. Perform tradeoff analysis 5. Identify preferred alternative
8. MITIGATION What can you do about it if you don't like it?	1. Review unavoidable adverse impacts 2. Identify possible mitigation measures 3. Perform sensitivity analysis of possible measures
9. MONITORING How good are your guesses?	1. Measure actual vs. predicted impacts 2. Feed back measurements to policy-makers and publics
10. MANAGEMENT Who's in charge here?	1. Devise management plan 2. Adjust planning objectives, operating procedures, design specifications
X. (BOTTOM LINE) Who benefits and who loses?	(All of the above)

Now, associate and integrate general methodologies, specific techniques, and relevant data.

While no "one best way" has (or can) be devised to fit all circumstances and cases, there is growing professional consensus and methodological convergence on what may be described as the "main pattern" of assessment steps (Table 1). Essentially this is a "rational problem solving" schema closely resembling many others in technology assessment, decision analysis and related fields.

Instrumenting and implementing this series also requires the meshing of analytic and institutional systems. For example, public involvement enters the series at every step in various ways, decisively at some points (e.g. impact evaluation). Refining this pattern to a high level of detail and testing it against diverse case experience will be the work of many practitioners over many years. In what follows here I will briefly sketch the rationale for this series in general methodological terms, indicating some methodologies and techniques that can be usefully employed and illustrating their application by reference to some representative studies.

1 SCOPING

The leading questions here are "How big a problem is it?" and "How much is enough?" In general, the size of the study should reflect the magnitude of the problem, and that cannot be known in advance. The correct scope of assessment cannot be finally decided until after its completion--too late, for most immediate purposes. Although how big a problem it is (whatever "it" is) cannot be predetermined, nevertheless some of its major proportions and dimensions must be estimated in order to get on with the assessment work. This initial closure must remain provisional and flexible, however, to allow for the learning that is bound to occur if the study is competently done.

In practice, this flexibility is difficult to maintain; deadlines are set and budgets are fixed before the answers are in. These constraints on study resources provide very practical answers to how much "enough" is. In any case, no amount of assessment is ever really enough--there is practically no end to where studies might lead. The real proposition is not how to avoid complexity but how to manage it. At the same time, no single study can do everything; there must be limits, and effort must be scaled to what is possible within them. Whatever else is done in an impact study, the greatest attention should be paid to impacts that appear to be the most "significant." In turn, that implies some criteria of significance and some methods for their agreement and application. More will be said of this later.

Since SIAs are usually not self-initiated on the part of the assessors, moreover, consideration must also be given to the interests of the sponsor or client. They will have something in mind, otherwise the occasion for the study will not have arisen. But often they will not have much in mind in the way of specific guidance; they will need to learn along with the assessors, and communication between the two--and with others involved, such as potentially affected publics--will be crucial at many points. Again, the sponsor or client may be mistaken in their impression of what the "real" problem is, viewed from the assessor's side. It is then a matter of professional responsibility to secure the widest sharing of understanding as to its "true" nature.

1.1 Set Level(s) of Assessment (Policy/Program/Project)

The level of assessment is initially given in the nature of the problem under study. In the case of a development project--a highway, for instance--this would normally dictate that program- and policy-level impacts receive only secondary emphasis, if that. Yet there is a relativity among these levels that cannot be escaped; projects are the program actions designed to implement policies (2). Project construction will have program and policy impacts that should not be slighted even though falling outside the study guidelines. How to get a hearing for these, if that is important, must be negotiated between the assessor and sponsor. The situation becomes more complicated where other parties at interest, e.g. community groups, insist on raising policy issues in the context of a project development. Usually these are outside the authority of the project manager and must be referred to higher decision levels, where local interest must have a different, and often more difficult, access. Asking the right question on the right level is partly a matter of institutional analysis.

1.2 Determine Impact Area Boundaries

There are two main kinds of impact area boundaries for projects, on-site and off-site. Boundaries should be chosen to enclose the area likely to receive the greatest project impact, although secondary impacts and areas should not be ignored. While usually this will be adjacent to the project site, this is not always the case. For example, according to Stan Hirst, the site-specific impacts of the Liard River Hydroelectric Project in northeast British Columbia will probably be far less significant than those 1200 miles downstream, in the Mackenzie Delta. Similarly, project benefits in the form of electric power will be transmitted 1000 miles to the southwest, in the Vancouver region. In this case the impact area extends 2400 miles along river and transmission line corridors, and different types of impact will occur at different points along this route.

Another factor in delineating impact areas is the availability of data at different geopolitical scales, e.g. county level. The sensitivity of aggregate data to highly localized impacts and the match between areal units and impact gradients are methodological concerns here. Impact intensities tend to be attenuated over larger areas. Thus, while admitting severe local and regional impacts, the U.S. Air Force could argue that the MX missile mobile basing mode was justified in terms of the national interest. Special impacts on racial and cultural minorities may also be submerged in aggregate data.

1.3 Establish Time Horizons

Temporal as well as spatial considerations enter into impact area delineation. The question here is not where is an impact, but when. Although prompt impacts may command greatest attention, in cases such as Three Mile Island the delayed (post-accident) impacts were far more potent (3). If the expected lifetime of a project is used as a convenient time line, special attention should be paid to the "back end," as in nuclear reactor decommissioning and radioactive waste storage. In most boomtown studies, the "bust" side of the boom-bust cycle has been neglected.

1.4 Develop Study Design

The problem at this point is to structure the impact study in such a way as to capture the major impacts with the greatest economy of resources. There is now a wide assortment of approaches, frameworks, paradigms and models for conducting impact studies, although their systematization and codification have lagged behind field practice. The "main pattern" of assessment steps presented here embodies a systems approach, cutting across and combining institutional systems (such as government agencies), sociotechnical systems (such as building technologies), and social systems (such as communities). Much of the reason for urging a systems view is that planning efforts in the past have been piecemeal in their conception and operation, to the neglect of such phenomena as social networks (4). Tracing out the consequences of a planned intervention means viewing them in mutual relation.

If it is true that in some sense "everything is connected to everything," it does not follow that all impacts on all parts of the system are the same. Sameness or difference is partly determined by the degree of "systemness"--how structurally and functionally differentiated and integrated the systems in question are and may become. For example, "community cohesion" as an impact category says something about the connectedness of community systems as well as the quality of community life affected by institutional completeness and performance. A study design should outline not only what order of impacts we are looking for, where and when, but also how we intend to go about estimating their magnitude on relevant social units. The systems approach implies both a conceptual model and a measurement model. The conceptual model is embedded in the very notion of "system"; the measurement model gauges the state of the system in its relevant dimensions and estimates its transformation under impact conditions.

2 PROBLEM IDENTIFICATION

To do impact assessment, there must be something--a proposal for planned intervention--to assess. Our motto is: "No effects without causes." The prior question must also be entertained, however: Why are we doing or proposing to do this thing? What is the problem? What is causing it? Already in the scoping step there are some preconceived answers, given in the proposal under review and the goals or needs to which it is addressed. Nevertheless, these preconceptions may be misconceptions and in any case they must be analyzed and sharpened. What "the problem" is may be a foregone conclusion from someone's standpoint; the task at hand is to open it to question. Upon examination, "the problem" turns out to be a good deal more complicated, and its identification is variable upon the parties doing the defining. For example, before the 1973 OPEC embargo, there was only one solution to the problem of energy (and other resource) interruption or shortfall: increase supply. Since then another possibility has emerged: decrease demand. Correspondingly, the "energy problem" may be reidentified, from underproduction to overconsumption; alternatives for its solution are likewise altered, e.g. to conservation measures.

Another example is the case of local flooding I encountered in 1970 around Athens, Ohio (5). One way of putting the problem--the one used by Ohio University and the U.S. Army Corps of Engineers--was to keep the water away from the people, which they did by the structural (physical) alternative of channelization. Another way--mine--was to identify the problem as keeping the people away from the water, in which case the appropriate (non-structural) measure would have been a system of flood plain management. Again, whereas the Ontario Ministry of the Environment identified the disposal of hazardous and toxic substances as a problem of finding suitable (least objectionable) sites around the Province, my problem was why we had to have them at all--opening up alternatives of materials recovery and recycling through industrial process redesign or, at least, on-site storage. Interpretive structural modeling (6) is one formalized, computer-assisted method for problem identification.

2.1 Formulate Policy Goals, Planning Objectives

What are we "solving for" in SIA? A situation, such as a flood hazard, may be viewed as problematic; its removal then becomes the solution. Conversely, the proposed solution may itself appear to pose a problem--as in the Athens case--from the standpoint of habitat destruction and human occupancy. Proposals originate in stated policy goals and planning objectives; these are official definitions of the problem situations and of solutions addressed to them, embodied in legislation such as the National Environmental Policy Act of 1969 (NEPA) and the mission statements of agencies such as the Corps of Engineers. In addition, there are the interests of private proponents, whether of superport development in Palau or of housing estates in Podunk. Finally, there are the assessed needs of local publics. Documenting these statements of purpose and their elaboration in programs establishes the ends in relation to which proposals stand as instrumental means.

2.2 Identify Publics and Concerns

A wide array of parties and interests are present in any typical impact situation. They provide alternative viewings of identified problems and proposed solutions, and of evaluative criteria by which to judge them. How the various publics and positions are represented and considered depends on the institutional arrangements for reaching and reviewing decisions (7). From this participation we can discern some relevant impact categories and potentially affected publics an impact study must consider. In the Beaufort Sea Hydrocarbons Development case, for instance, public hearing transcripts record the various concerns voiced by participants. These were compiled into an issues list from which guidelines or "terms of reference" for the impact study were drawn. Identifying publics and their concerns can thus serve as a basis for scoping impact studies (8). Not all publics and concerns are immediately and systematically recognizable, however, even by the parties themselves.

2.3 Perform Needs Assessment

In needs assessment, the interest of the clientele of a public agency or of the community at large provides the starting point. Statutory requirements may dictate that community interests play a part in identifying problems and the formulation of alternatives for their solution. For example, the Community Development Block Grant program of the U.S. Department of Housing and Urban Development prescribes that funds be allocated to projects designed to serve residents in greatest need. Local citizen groups and planning agencies may have already set these priorities, but in most instances some original data collection will be required through the use of attitude surveys or some other method (9).

2.4 Determine Evaluative Criteria

From the preceding we can derive evaluative criteria for judging the consequences of carrying out (or failing to carry out) a course of action, and of calculating the social impacts--benefits and costs--of conditions staying the way they are or changing. These criteria are employed later, in the evaluation step. They represent the different value judgments of the various parties at interest as to the detrimental and beneficial consequences of a planned intervention. These criteria can be expected to change over time--e.g. the emphasis on equity concerns (10). The prediction of value change is part of the projection step.

3 FORMULATION OF ALTERNATIVES

By definition, rational solutions are relative to problem definitions. The various ways of identifying problems generate a correspondingly wide range of planning alternatives. These may include alternative routes and sites, alternative technologies and project designs, and alternative rates of development (such as construction workforce buildup) and uses for a fixed resource such as land.

3.1 Define Set of "Reasonable" Alternatives (Corresponding to Identified Concerns)

One the theory, "If it isn't broken, don't fix it," the simplest choice of alternatives would be to exercise the "do nothing" or "no build" option. Indeed, there are situations such as beach erosion in which inaction appears the most prudent course of action. Proposals are formulated and impact studies are initiated however because "doing nothing" seems unreasonable (11). Even so, it is worthwhile to weigh the expected benefits of a proposal against its projected costs. The "reasonableness" criterion is introduced to limit the range of conceivable alternatives to realistic proportions. One "rule of reasonableness" advanced in the U.S. Department of Housing and Urban Development (12) was that "Alternatives need not extend beyond those which serve the same general purpose of the project being considered."

> For example, if the purpose of the project is to provide housing to a community, then alternatives which provide something other than housing need not be treated. . . . In other words, for projects whose purpose is associated with a specific land use (e.g. housing/residential land use), other land uses (e.g. industrial, open space) need not be considered as "alternatives". . . .

The question naturally arises whether the intended purpose of the project is the one best calculated to meet community "needs," who is to make that determination, and how it can serve to orient the actions of project proponents.

It is reasonable however to associate a preferred alternative with each major interest identified in the previous step. Thus one Corps "display of effects" compared four alternative plans, each of which emphasized a distinct planning objective: national economic development, environmental quality, non-structural, and non-federal. To these could be added some others, such as "desirable" regional development--whoever is defining "desirable"--and social well-being or "quality of life." These last have not attained the rank of an officially designated planning objective for water resources projects, however, suggesting that their reasonableness has not been fully demonstrated to the satisfaction of some policy-makers.

3.2 Determine Change Agents, Instruments

If an alternative is to be implemented, someone must do it, using some available or potential means. Besides resources and equipment, energy development requires some agencies or companies and capital investment, among other things. Which ones, and what their managerial, technical and financial capability, are matters for institutional analysis. If the capability does not presently exist, further impacts may be incurred in the process of creating it. Just as the "receiving" communities are characterized in the profiling step to follow, the proponents should be analyzed in the present one--drawing corporate profiles for sponsors of private development proposals, for example (13).

3.3 Characterize and Describe Technical Systems; Analyze for Social (Institutional/Behavioral) Components and Correlatives

Although exact specifications of a technology-based proposal may not be known until late in the design phase, after an approvals process has been completed, its salient features can usually be discerned in sufficient detail for general assessment (14). These can then be composed into a "project profile." A convenient format for characterizing energy systems is the fuel cycle model, from exploration and extraction to end use and residuals management. Institutional arrangements and human activities are implied at each stage--for example, the construction workforce required and their probable social characteristics, such as number of in-migrants with dependents (15). From these attributes potential infrastructure impacts can be inferred and compared to existing and changing local conditions of supply and demand.

3.4 Analyze Economic and Environmental Impacts for Secondary Social Impacts

Other categories of impact besides social--notably economic and environmental--generally accompany physical planning projects. These have potential indirect or "secondary" social impacts which should be assessed, and which in any case require a social interpretation--employment, for instance. Detailed planning information of this kind may not become available until later in the impact study, but it should be sought out at the earliest time possible, at least in broad outline. The interaction among impact categories is also a consideration in regard to cross-impact and tradeoff analysis.

4 PROFILING

Having now gained some impression of what it is that is potentially having an impact, we turn to the question of what it may have an impact on. This is compiled and presented in the form of a social profile at some appropriate scale(s), most often the community but also, on occasion, the family or region. The question we are asking here is: how much of what kind of impacts on whom?

The answer must begin with a "baseline" measure of the system before proposal-related impacts have occurred. The procedure for profiling is fairly straightforward: first we establish some dimensions and categories of social impact, then we perform some appropriate measures of their current values among various segments of the population.

4.1 Dimension Impact Categories

Before deciding what content categories of social impact to examine, it is useful to inquire what about the kinds of change that may occur in them. These are impact dimensions, such as incidence and distribution, timing and magnitude of potential impacts. NEPA is the source of some of these--directness, intensity, duration, reversibility and cumulation, for example.

4.2 Select Impact Categories

One way to proceed is to list all conceivable impacts that may impinge on people and then to cluster them according to common properties: health and safety, legal, and so on (16). The resulting typology will provide broad coverage across most institutional areas: family, economy, polity, education, religion, etc. Identification of functional groups in these areas is a means of tracing organizational and cultural impacts as well (17). In certain instances the social impact categories of primary interest are mandated by law or regulation. For example, the U.S. Water Resources Council's "social well-being account" (18) includes: life, health and safety; emergency preparedness; demographic characteristics; community organization; income, educational, cultural and recreational opportunities; noise; and esthetic value. Impacts of greatest interest to local decision-makers tend to be in the categories of community services and facilities (sanitation, schools, recreation, fire and police protection, etc.), and fiscal and land use impacts.

4.3 Assign Impact Indicators

Once impact categories have been selected, the next task is to assign measures to their salient features. Social indicators research (19) affords many alternatives for measuring an impact category, and indicator selection may be based on the availability, quality and cost of impact data (20). Where direct measures are unavailable, "proxy" measures can be used, e.g. "community cohesion" as a function of the percent of home owners versus renters. Here an assumption is being made that renters are less cohesive community people, and that a higher percentage of renters is an indication of lower community stability. That assumption needs to be tested against independent research. As much as data gathering and analysis, data interpretation is a major responsibility of the assessor. Institutional standards such as those for measuring the quality of health care may be of assistance here. In many cases these will not be available, however, and interpretation must rest on professional judgment.

4.4 Perform Indicator Measurements

A method of "triangulation" is recommended for measuring impact categories, drawing on attitudinal, observational and statistical data. The first of these is gathered by interviewing or survey techniques and yields "subjective" measures such as perceptions of community satisfaction and possible threats or supports to that were a proposal to be implemented. Observational (or behavioral) data provide a reliability check on the consistency of attitudes and actions, e.g. whether reported "fear of crime" is followed by the curtailment of social activities. Statistical data, such as census materials, furnish an aggregate baseline for normalizing or contrasting individual and group measures.

Data sources are correspondingly varied: archival materials, official statistics, unobtrusive measures of social interaction, and the like. In any case, suitable data are those that satisfy the requirements established in profile analysis; in all cases, their value is relative to an understanding of the impact situation at hand. It is also desirable to obtain time series data, since the existing state of the system cannot be interpreted without insight into its recent history.

4.5 Compile Social Profile

A matrix approach is convenient for summarizing data gathered in the profiling step. In one format (21) the matrix rows are "quality of life" indicators such as job satisfaction and pollution exposure and the columns are "population segments" in various categories--socioeconomic status, racial and ethnic composition, sex, age, place of residence and so on. Cell entries are the anticipated impacts of a proposal for each indicator and population segment identified. While future states of the system (after impact) are projected in the next step, current values (before impact) can be depicted in this same fashion. The baseline condition is established to separate "normal change" within the population and community from that specific to the proposal; this is the "with and without" (proposal) distinction. The matrix is also useful as a scanning device for detecting the possibly most significant impacts of a proposal, and for diverting study efforts to their closer examination.

In the next step, what is projected are the current values of social profile features under various future conditions.

5 PROJECTION

The question at the projection step is what kind and how much change in profile features would occur were one or another planning alternative to be implemented. This induced change must be estimated against a background of "normal change," of course--the "without" condition mentioned above. One way of controlling for this is to perform the mental experiment of asking what the social system would be like in the absence of the proposal's implementation. Social systems are dynamic, naturally, and the "without" condition is not one of "no change" or even linear extrapolation.

5.1 Explicate "State of Society" Assumptions

"State of society" assumptions are predictions about general trends believed to influence social conditions over the period considered by the study. They must be taken into account when estimating changes in the social profile that would occur regardless of the proposal (22). Such trends as urbanization are usually presented in a highly aggregated form. Because many social impacts are site-specific, these general trends must be disaggregated in order to make contact with local conditions (23). Adding to the uncertainty of these forecasts are trend discontinuities, such as the recent "turnaround" in migration from rural to urban areas. To allow for these deviations in trend lines, a range of alternative futures or "scenarios" can be postulated, as in the various population projection series computed by the Bureau of the Census. In such cases, the predictive question is what profile changes would result were one or another of these levels to be attained.

5.2 Perform Trend Impact Analysis

"Normal change" can interact with proposal-induced change in ways that either intensify or moderate social impacts. Hence the task for trend impact analysis is not only to estimate the influence of general on local conditions "without" the proposal being implemented but also "with" the proposal. For example, a reservoir development may be proposed in order to meet the future water supply requirements of a growing population. Reservoir development itself may then induce further population growth attracted by recreation opportunities, environmental amenities, or water-using industries. Situations of this kind are sometimes referred to as "self-fulfilling prophecies"--efforts to meet future needs themselves become the stimulus to increased demand (24).

Another form of trend impact analysis is normative forecasting, in which the planning problem is not merely to provide for anticipated requirements but to realize some vision of a desirable future. For instance, the California Tomorrow Plan (25) sees two future states of the state, California 1 and California 2. The former is the most probable future, given the continuation of existing trends; the latter is a desirable future in which adverse trends such as land conversion from agricultural to residential use are abated or eliminated. The problem then is to deflect the trajectory of future development from California 1 to California 2.

5.3 Construct Dynamic System Models

To determine future states of the system, not only do we project social trends; we must also gain an understanding of the system dynamics that generate them and analyze their principal components. For example, in the surrogate site study of the Upper St. Lawrence Nuclear Energy Center (26), projected population growth was assumed to be a partial function of labor force within the commuting area and distance to the construction site. The hypothesis was that willingness to commute is inverse to the distance of the journey to work. When validated by empirical research, part-models such as this can offer at least some plausible basis for impact prediction. The use of causal models (27), of which the most familiar are the "system dynamics" models (Forrester and his associates) (28), can be used for predicting discontinuities--e.g. in the absence of negative feedback loops--as well as continuous change.

5.4 Estimate Impact Indicator Values for Alternative Plans ("With and Without" Implementation)

The projection step is completed when the baseline social profile is estimated for the range of trend assumptions and systematically varied with the predicted impacts for each planning alternative. The comparison of these alternative outcomes of alternative plans takes us to the next step, assessment.

6 ASSESSMENT

The task for assessment is to compare the potential impacts of the full set of reasonable alternatives under the range of assumptions about future conditions. This is done by predicting differential changes in the current values of impact indicators under alternative trend and plan assumptions, including second-order differentials (changes in the rate of change) caused by cross- and cumulative impacts.

6.1 Perform Sensitivity Analysis for Alternative Outcomes of Alternative Plans

Covariation is the basic notion in sensitivity analysis, a

technique well developed in economics. How much of X "goes with" a given amount of Y, and what change in X would occur with a certain change in Y? Since in SIA we are dealing with causal analysis, covariation is only the beginning of wisdom. Some indication of causal relations should have been acquired in the modeling phase of projection, however. As in projection, even if we cannot know the value of Y with certainty, we can assume a range of values Y might take and then look for systematic variation in X at different points within that range. Sensitivity analysis is also used in the next step for the evaluation of alternatives (29).

6.2 Perform Cross-Impact Analysis

"Cross-impact analysis strives to identify interactions among events or developments by specifying how one event will influence the likelihood, timing, and mode of impact of another event in a different but associated field" (30)--the reciprocal impacts of infrastructure development and land use, for example (31). We are typically dealing with mutual causal relations in long chains, where one impact can amplify or dampen the effects of others. Second- and higher-order impacts come into consideration here, as do the compound effects, or "synergisms," of cumulative impacts (32, 33).

6.3 Describe and Display "Significant" Impacts

We are now in a position to submit the assessed impacts of alternative plans for evaluation by the various parties at interest. While there are political reasons why assessment results are often poorly utilized, their prospects are not improved by the use of presentational techniques unintelligible to either decision-makers or publics. For purposes of public communication, the product is only as good as its packaging. Qualitative analysis of quantitative data is essential; tables must be converted into more legible charts, for example. A wide variety of display techniques is available in social graphics (34), e.g. computer mapping (35). More serious is the question of deciding which impacts are to be regarded as "significant." Usually this is left to the judgment of evaluators, but even the selection of results for presentation can bias that judgment. Evaluative criteria established in the problem identification step provide a basis for preliminary determination of significance.

7 EVALUATION

The distinction between "assessment" and "evaluation" is roughly that between fact and value, except that the facts in question are "future facts" that have no empirical existence in the present. Whereas in assessment the task is to estimate the possibility of alternative futures, in evaluation their desirability is the issue. Since assessors have no unique privilege in this regard, their role in evaluation is mainly that of facilitators.

7.1 Reidentify Publics and Concerns

This is an iteration of step 2.2. Over the course of the study new interests and interest groups may have emerged and some old ones may have been missed. Although the series of assessment steps has been presented in linear fashion, in practice there is --and ought to be--a good deal of interplay among them. While they form a logical series, psychologically it is difficult to keep them distinct. (The formal report of study results should more or less follow the outline, however.)

7.2 Reformulate Evaluative Criteria

Another iteration, this time informed by assessment data gathered throughout the study. For example, if opinion data have been collected from interviews or questionnaires, the value positions of various groups can be plotted and compared. It may turn out that smaller differences exist between them than was originally supposed. It should be emphasized however that preferences expressed at one point in time are not necessarily valid indicators of "final" positions. Values are learned with experience, and it matters at what point in the learning process the sample is taken. The educational effect of the study may itself have altered value positions. This refers to relative placement on some value dimension, of course; evaluative criteria are the dimensions themselves.

7.3 Rank and Weight Preferences for Alternatives

Ideally preferences would be expressed on a single dimension in equivalent units. The analytic situation we face in evaluating alternatives, however, is that of "comparing incommensurables"--apples and rabbits. Hence it is necessary to resort to some ranking and weighting schemes to evaluate different alternatives for different publics. Among the candidate metrics for weighting are the numbers of people potentially affected and the predicted magnitude of impact. In multiattribute utility theory (MAUT), intensity of concern is combined with group size to produce a composite weight (36). Various such schemes have been reviewed by Finsterbusch (37). In general, there is no "felicitous calculus" for performing this operation; the process must be value-*added*. In turn, that raises further questions as to *whose* values should be considered and what weight should be given them. At this point too the relative decision power of the parties at interest comes into play.

7.4 Perform Tradeoff Analysis

To reach this stage there must be at least tentative agreement on the evaluative criteria to be employed and the relative merit of planning alternatives ranked according to them (38). Techniques of tradeoff analysis such as policy capture (39) can be used to clarify the subjective preferences of the parties at interest. The question then is how these respective positions can be unified to produce a single "preferred" alternative. In arriving at this judgment, many concerns not explicitly addressed in the study will enter into consideration, such as the institutional decision rules which govern final selection. The context in which this decision is made is often one of multiobjective planning, in which all desirable features of a proposal cannot be maximized simultaneously. It is at this point also where a balancing among economic, social and environmental impacts must take place (40) to yield a comprehensive and integrated assessment.

7.5 Identify Preferred Alternative

The result of tradeoff analysis is the designation of one plan as the "preferred" alternative. Preference ideally would be given to the plan containing the most favorable net balance of beneficial to adverse impacts; no alternative, however preferable, is likely to be free of some disbenefits. How preferable the preferred alternative is therefore depends also on the availability of mitigation measures for avoiding or alleviating these adverse impacts. Who is doing the preferring--proponents or opponents, or some coalitions among them--is again a matter for institutional analysis.

8 MITIGATION

Although preponderance should clearly be on the positive side, negative impacts of the approved proposal must be accounted for as well. In the past these were simply imposed as the price of what passed for "progress." Now it is understood that just compensation is due those who bear these social costs without sharing equitably in the benefits. What is compensable, and to what extent, is partly given in legislation; partly it is arrived at through direct negotiation among the parties. "Fair market value" as the standard of compensating displaced persons has been widely adopted, although the cost of replacement housing may substantially exceed that figure and actual moving costs may not be covered; locational preference is also disregarded. "Grieving for a lost home" (41) and the real economic losses which disruption of social networks often entails do not appear on this balance sheet. Relocatees are regarded as economic individuals having equal access to the market, not as family members vulnerable to economic discrimination. For all these reservations, the principle of mitigation, to communities (e.g. in the form of "impact funds" for energy development projects) as well as persons, is now well established (42).

8.1 Review Unavoidable Adverse Impacts

The negative impacts of the preferred alternative should have been identified in the evaluation step. Now they must be considered from the standpoint of mitigation. (Modifications to the plan will have taken care of the avoidable ones.)

8.2 Identify Possible Mitigation Measures

Appropriate mitigation measures are naturally relative to the nature and extent of the negative impacts. In the case of habitat destruction, for example, the U.S. Fish and Wildlife Service requires that mitigation lands be acquired. (These need not be of the same ecosystem type, however, so the tradeoff here is between species--lake vs. riverine, bottomlands vs. uplands, etc.) Potentially impacted communities may also seize upon mitigation provisions as an opportunity for realizing goals unrelated to the specific nature of the proposal. There is no particular reason why mitigation should be in kind, however; impact funds have frequently be allocated to highway construction and other capital improvements, for instance (43).

8.3 Perform Sensitivity Analysis of Possible Measures

As in 6.1, the uncertainty inherent in mitigation suggests the advisability of testing across a range of assumed values. The difference is that here we are trying to anticipate the potential effectiveness of mitigation measures. Of course, the real test can only come after the fact. Besides effectiveness, cost will also be a factor in deciding to commit organizational resources to impact mitigation. One major source of uncertainty in mitigation is the institutional capability of the performing agency; affixing that responsibility is a matter of institutional analysis and, where current provisions are inadequate, of institution-building.

9 MONITORING

The purpose of monitoring is to compare expected against actual impacts; it provides the "after" measure in before-after designs. "Post audit" studies (44) contribute to this, but often too long after the fact for use in impact mitigation and management. The same is true of evaluation research, in which there has been considerable public investment but slight application to forward planning (45). Monitoring is routinely done for environmental quality, but only occasionally for quality of life. A notable exception is the Tennessee Valley Authority's socioeconomic monitoring program for Hartsville, Phipps Bend and Yellow Creek Nuclear Plants. B.C. Hydro and Power Authority has had an experimental program for Revelstoke Dam (46). The Huntly Social and Economic Monitoring Project in New Zealand (47) is so far mainly a research effort.

9.1 Measure Actual Vs. Predicted Impacts

To ensure comparability of measures in comparing anticipated with actual impacts, the same indicators selected in profiling and then projected should be used as the basis for monitoring. Ideally, the same population would also be sampled, although where population displacement has occurred, e.g. in reservoir development, the movers are sometimes impossible to locate. Whether the comparison can be done naturally depends in the first place on the impact predictions hypothesized in the projection and assessment steps. When discrepancies are detected between expected and actual values in indicator performance, their "significance" must be interpreted, again using the same criteria as employed for evaluation.

9.2 Feed Back Measurements to Policy Makers and Publics

Having determined the correspondence of actual and predicted impacts, these results should be reported both to responsible officials and to affected publics. Policy makers can use these to guide any policy adjustments that appear warranted and to improve future program and project designs. Publics may want to reevaluate their preferences on the basis of these results; this may prove influential in considering future expansion of a completed project, for instance. An excellent example of a community-based monitoring system was the Impact Information Center (now the Community Information Center) in Fairbanks, Alaska, established to monitor local impacts of construction and operation of the Trans Alaska Oil Pipeline (48). Experience gained from that effort can now be applied to proposals for a parallel gas pipeline.

10 MANAGEMENT

SIA is not done for its own sake but for its utility in conceiving and conducting socially sound development planning. Assessment results are what managers have to manage. Impact management can be viewed from the perspective of impact- vs. project-based assessment. The latter is the more usual; projects are proposed and then their impacts are assessed and mitigated. An equal and opposite viewing is to take social impacts as the starting point--a social goal orientation--and to ensure project development takes place in conformance with those social criteria. If a high rate of construction workforce buildup is expected to have destabilizing community impacts, for instance, it is possible to reduce that rate below what is assumed to be optimum from the standpoint of project management. Emphasis on social impact management has been especially pronounced in the work of the Foundation for Urban and Neighborhood Development (49).

10.1 Devise Management Plan

An impact management plan should establish oversight responsibility for the "back end" of the assessment cycle--mitigation and monitoring. Since a shift in the locus of control typically occurs between the construction and operation phases of project development, determining responsibility and capability are often not simple matters. In this process too further institutional impacts occur, e.g. in the planning capability of local authorities charged with managing an impact situation. This capability can be enhanced by outside assistance, such as the provision of impact funds and the authorization to impose "impact fees" on private developers. An institutional needs assessment may be useful in determining resource requirements and sources (50).

10.2 Adjust Planning Objectives, Operating Procedures, Design Specifications

The utilization of SIA in policy formation, program development and project implementation are its ultimate justification. Based on hindsight, policy adjustment, program redirection and project modification may now appear possible and desirable courses of action.

A FINAL WORD

Not all the steps and tasks outlined above will prove necessary in every assessment. What I have tried to present is a methodologically complete series to guide in their selection. This schema is still undergoing revision and elaboration, although the underlying logic seems about right. By concentrating more on the "why" than on the "how" questions, I hope to have laid a firmer basis for the enlightened use of assessment techniques. Stated in the general terms used here, the discussion is still at one remove from the specification of an operational methodology. It remains for assessors now to "associate and integrate general methodologies, specific techniques, and relevant data" adapted to the special circumstances of particular impact situations. This chapter may serve as a point of departure towards that end.

REFERENCES

1. Freudenburg, W.R. *A social impact analysis of a Rocky Mountain energy boomtown*, Paper presented at the 73rd Annual Meeting of the American Sociological Association, San Francisco, California, 1978.

2. Wolf, C.P. "Getting social impact assessment into the policy arena". *Environmental Impact Assessment Review* 1 (1980) 27-36.

3. Flynn, C.B. *Impact discontinuity: The case of Three Mile Island*, Paper presented at the 76th Annual Meeting of the American Sociological Association, Toronto, Ontario, 1981.

4. Leinhardt, S., ed. *Social Networks - A Developing Paradigm* (New York, New York: Academic Press, 1977).

5. Wolf, C.P. "They are filling in the river! Ecology action in Athens, Ohio", pp 407-433 in A.B. Shostak, ed. *Our Sociological Eye: Personal Essays on Society and Culture* (Port Washington, New York: Alfred, 1977).

6. Kawamura *et al.* "Probing complexity in social systems through interpretive structural modeling", pp. 335-342 in K. Finsterbusch and C.P. Wolf, eds. *Methodology of Social Impact Assessment*, 2nd edition (Stroudsburg, Pennsylvania: Dowden, Hutchinson and Ross, 1981).

7. Willeke, G.E. "Identifying publics in social impact assessment", pp. 305-311 in K. Finsterbusch and C.P. Wolf, eds. *Methodology of Social Impact Assessment,* 2nd edition (Stroudsburg, Pennsylvania: Dowden, Hutchinson and Ross, 1981).

8. Priester, K. "The issue - centered approach to SIA: From assessment to management". *Social Impact Assessment* 69/70 (1981).

9. Butler, L.M. and R.E. Howell. *Coping with growth: Community needs assessment techniques*, WREP 44 (Corvallis, Oregon: Western Rural Development Center, Oregon State University, 1980).

10. Lampman, R.J. "What does it do for the poor? - A new test for national policy". *The Public Interest* 34(1974)66-82.

11. Lane, J.S. *et al. Impact Assessment Guidelines: The Role of the No-Build Alternative in the Evolution of Transportation Projects*, NCHRP 8-11 (Washington, D.C.: Transportation Research Board, National Academy of Sciences, 1977).

12. Office of Environmental Quality. *Environmental reviews at the community level: A program guide* (Washington, D.C.: Office for Community Planning and Development, U.S. Department of Housing and Urban Development, 1975).

13. Michalenko, G. "The social assessment of corporations", pp. 168-179 in F.J. Tester and W. Mykes, eds. *Social Impact Assessment: Theory, Method and Practice* (Calgary, Alberta: Detselig, 1981).

14. Technology Assessments Division. *Technology Characterizations: Environmental Information Handbook*, DOE/EV-0072 (Washington, D.C.: U.S. Department of Energy, 1980).

15. Mountain West Research, Inc. *Construction Worker Profile*, 10 volumes (Washington, D.C.: Old West Regional Commission, 1975).

16. Liebermeister, U. "Universal checklists in the concept of 'impact trees'", pp. 133-158 in D.F. Burkhardt and W.H. Ittelson, eds. *Environmental Assessment of Socioeconomic Systems* (New York, New York: Plenum, 1978).

17. Finsterbusch, K. "The potential role of social impact assessments in instituting public policies", pp. 2-12 in K. Finsterbusch and C.P. Wolf, eds. *Methodology of Social Impact Assessment*, 2nd edition (Stroudsburg, Pennsylvania: Dowden, Hutchinson and Ross, 1981).

18. Andrews, W. *Evaluating Social Effects in Water Resources Planning: First Steps* (Washington, D.C.: U.S. Water Resources Council, 1981).

19. Gilmartin, K.J. *et al*. *Social Indicators: An Annotated Bibliography of Current Literature* (New York, New York: Garland, 1979).

20. Sullivan, J. "Multiple indicators: Some criteria of selection", pp. 243-269 in H.M. Blalock Jr., ed. *Measurement in the Social Sciences: Theories and Strategies* (Chicago, Illinois: Aldine, 1974).

21. Finsterbusch, K. and P.A. Weitzel-O'Neill. *A Methodology for the Analysis of Social Impacts* (Vienna, Virginia: Braddock, Dunn and McDonald, Inc., 1974).

22. Spilerman, S. "Forecasting social events", pp. 381-403 in K.C. Land and S. Spilerman, eds. *Social Indicator Models* (New York, New York: Russell Sage Foundation, 1975).

23. Leistritz, F.L. *et al.* "A model for projecting localized economic, demographic, and fiscal impacts of large-scale projects" *Western Journal of Agricultural Economics* 4(1979)1-16.

24. Hollis, J. and J. McEvoy. "Demographic effects of water development". *Impact of Water Resources Development* 1(1973)24-40.

25. Heller, A., ed. *The California Tomorrow Plan* (Los Altos, California: William Kaufmann, 1971).

26. Merry, P.A. *et al. Nuclear Energy Center: Upper St. Lawrence Region*, ANL/ES-57 (Argonne, Illinois: Division of Environmental Impact Studies, Argonne National Laboratory, 1977).

27. Blalock, H.M., Jr., ed. *Causal Models in the Social Sciences* (Chicago, Illinois: Aldine, 1971).

28. Sage, A.P. "Methodologies for socioeconomic resource system modeling". *Journal of Interdisciplinary Modeling and Simulation* 1(1978)1-95.

29. Henderson, J. "Evaluation and Sensitivity Analysis Program (EASP)". *Environmental Impact Assessment Review* 1(1980)422-426.

30. Mitchell, A. *et al. Handbook of Forecasting Techniques,* IWR Contract Report 75-7 (Fort Belvoir, Virginia: U.S. Army Engineer Institute for Water Resources, 1975).

31. Dajani, J.S. and L. Ortolano, eds. *Methods of Forecasting the Reciprocal Impacts of Infrastructure Development and Land Use*, Report IPM-11 (Stanford, California: Department of Engineering, Stanford University, 1979).

32. Grigsby, J.E. *Cumulative impact assessment: A case study of the aggregated effects of diversified federal projects on Santa Barbara County, California*, Ppaer presented at the 76th Annual Meeting of the American Sociological Association, Toronto, Ontario, 1981.

33. Husky, L. *Analysis of Cumulative Impacts, Western Gulf of Alaska Impact Analysis* (Anchorage, Alaska: Institute of Social and Economic Research, University of Alaska, 1979).

34. Feinberg, B.M. and C.A. Franklin, eds. *Social Graphics Bibliography* (Washington, D.C.: Bureau of Social Science Research, 1976).

35. Steinitz Rogers Associates, Inc. *The Santa Ana River Basin: An Example of the Use of Computer Graphics in Regional Plan Evaluation* (Fort Belvoir, Virginia: U.S. Army Engineer Institute for Water Resources, 1975).

36. Keeney, R.L. and H. Raiffa. *Decision Analysis with Multiple Objectives: Preference and Value Tradeoffs* (New York, New York: John Wiley, 1976).

37. Finsterbusch, K. *Methods for Evaluating Non-Market Impacts in Policy Decisions with Special Reference to Water Resources Development Projects*, IWR Contract Report 77-8 (Fort Belvoir, Virginia: U.S. Army Engineer Institute for Water Resources, 1977).

38. Central Arizona Water Control Study, *Summary factbook* (Phoenix, Arizona: Central Arizona Water Control Study, 1981).

39. Johnson, J.M. "Policy capturing in energy", pp. 329-334 in K. Finsterbusch and C.P. Wolf, eds. *Methodology of Social Impact Assessment*, 2nd edition (Stroudsburg, Pennsylvania: Dowden, Hutchinson and Ross, 1981).

40. Blood, D.M. *Techniques for assessing tradeoffs among social, economic and environmental effects of public investments in highways*, Highway Research Board Special Report 128 (Washington, D.C.: National Academy of Sciences, 1973).

41. Fried, M. "Grieving for a Lost Home: Psychological Cost of Relocation", pp. 151-171 in L.J. Duhl, ed. *The Urban Condition* (New York, New York: Basic Books, 1963).

42. Peelle, E. "Migrating Community Impacts of Energy Development: Some Examples for Coal and Nuclear Generating Plants in the U.S." *Nuclear Technology,* 44(1979)132-140.

43. Moore, K.D. *et al. Mitigating Adverse Socioeconomic Impacts of Energy Development* (Washington, D.C.: U.S. Department of Energy, 1978).

44. Bisset, R. "Problems and issues in the implementation of EIA audits". *Environmental Impact Assessment Review* 1(1980)379-396.

45. Meidinger, E. and A. Schnaiberg. "Social Impact Assessment as Evaluation Research: Claimants and Claims". *Evaluation Review* 4(1980)507-535.

46. Kopas, P. "The structure and functions of the Revelstoke impact monitoring program". *Social Impact Assessment* 51/52 (1980) 3-10.

47. Fookes, T.W. *Monitoring Social and Economic Impact: Huntly Case Study. A Report for Administrators, Politicians and Others Not Closely Associated with the Huntly Monitoring Project, on the Findings of the Project through to 1977* (Hamilton, New Zealand: Huntly Social and Economic Impact Monitoring Project, University of Waikato, 1979).

48. Dixon, M. "Impact Information Center, a community-directed research model". *Environmental Sociology* 7(1975)25-31.

49. Kent, J.A. *et al. An Approach to Social Resource Management. Report to the Surface Environment and Mining Program* (Washington, D.C.: U.S. Department of Agriculture Forest Service, 1979).

50. Energy Research and Development Administration. *Managing the Socioeconomic Impacts of Energy Developments: A Guide for the Small Community* (Washington, D.C.: U.S. Energy Research and Development Administration, 1977).

THE THEORY AND APPLICATION OF MODELLING IN ENVIRONMENTAL IMPACT ASSESSMENT

R.E. Munn

Professor, Institute for Environmental Studies, Toronto

1 INTRODUCTION

Models are representations of systems whether these are real or imagined. They have three characteristics:

a) Models are simplified representations of the system being studied.

b) Models contain explicit assumptions about the behaviour of systems being studied.

c) Models are imperfect, particularly when applied out of context.

Models are of four main types:

a) conceptual;
b) empirical/statistical;
c) physical (for example, wind-tunnel scale models);
d) numerical/mathematical.

In the paragraphs that follow, attention will be directed to types a) and d). Empirical regressions (category b) are not very useful in EIAs; they are best applied to existing well-behaved systems for which an extensive data base is available, rather than to proposals for new projects. Applications in category c) are sometimes made; for example, chimney height is sometimes determined by tests with a scale model in a wind tunnel. However, physical modelling will not be discussed in this paper.

Models serve three purposes:

a) Models describe:- The descriptions are often in convenient shorthand form that can, for example, be stored in a computer.

b) Models explain:- No model is able to explain all the intricate details of a complex system but the user can compare models, deciding which is better under specified conditions. By simplification, a model can in fact be useful in identifying processes of particular importance and in explaining their behaviour. Models need to be verified/validated with real data. In fact, modelling and monitoring are iterative processes: with increased understanding of a system, the monitoring requirements become more precisely specified.

c) Models predict:- The predictions are of three types:

(i) interpolation; given a series of values at discrete time intervals, estimation of the value at some intervening time. Or given observations at a network of stations, estimation of the value at a point.

(ii) extrapolation; predictions of the behaviour of a system at some future time.

(iii) "what if?" predictions; predictions under some hypothesized set of conditions that have been observed. This is called a simulation. Its main property is that it cannot be validated with data available at the time the prediction is made. (Note that if several alternatives are being compared in an EIA, only the alternative that is finally chosen can ever be validated).

Simulation models are used successfully within single disciplines; for example, the air pollution impacts of several different energy strategies can be compared with a simulation model. However, attempts to develop large interactive biogeophysical-socioeconomic models as EIA tools have generally failed. (One of the problems is that the output format is too complex, creating credibility gaps between the modeller and the environmental manager). However, big models have instructional value and sometimes affect the ways in which people think. For example, the Limits-to-Growth models led to calls for a society which conserved natural resources. Environmental simulation models have been discussed in HRI (1), Holling (2) and Munn (3).

Many books on EIA contain a section on modelling, but it is usually stated that the data requirements are too stringent for most applications. In fact the authors have missed the point. Simulation models are most useful in the early stages of an investigation, when there are few data and considerable uncertainty about the nature of the relationships amongst the variables. To emphasize this point of view, the chapter on environmental modelling in the second edition of SCOPE 5 (3) was re-titled "A Conceptual Framework for an EIA". There is need at the outset of an EIA study to conceptualize the system, providing for the subsequent investigation, and guidance for the baseline data collection phase. Otherwise, those involved in EIA may produce models for particular subsystems which cannot be linked together at a later date.

In this paper, an example will first be given of a simulation model, viz. the Gaussian diffusion model that is used frequently for a component part of an EIA. Then the principles to be followed in developing an EIA conceptual framework will be discussed, and an example will be given (CO_2 climate warming).

2 THE GAUSSIAN PLUME MODEL

The Gaussian air pollution plume model is a good illustration of a model widely used in EIA. Embedded beneath a handbook formula or component program for diffusion, there is a complex and almost intractable problem in fluid mechanics. It is surprising that the method works at all! Here it should be mentioned that the calculated pollutant concentrations are very small, of the order of parts per million, or parts per hundred million.

The Gaussian plume model was first hypothesized by Sir Graham Sutton (6) in the UK. It assumes a series of particles of fluid elements moving downstream in a turbulent fluid. The position of an individual particle is unpredictable. However, a cross-section taken at right angles to the wind direction would reveal a frequency distribution of particles that approximates the random Gaussian bell-shaped curve used in statistics. So at any downwind distance x, the crosswind distribution is defined by a standard deviation s_y, where s_y increases as x increases. Similarly in the vertical direction z, there is a standard deviation s_z, where s_z need not be the same as s_y. An example of an expanding plume is shown in Figure 1 (7).

The following derivation of the Sutton plume model is from Munn (8). The Gaussian probability distribution is given by

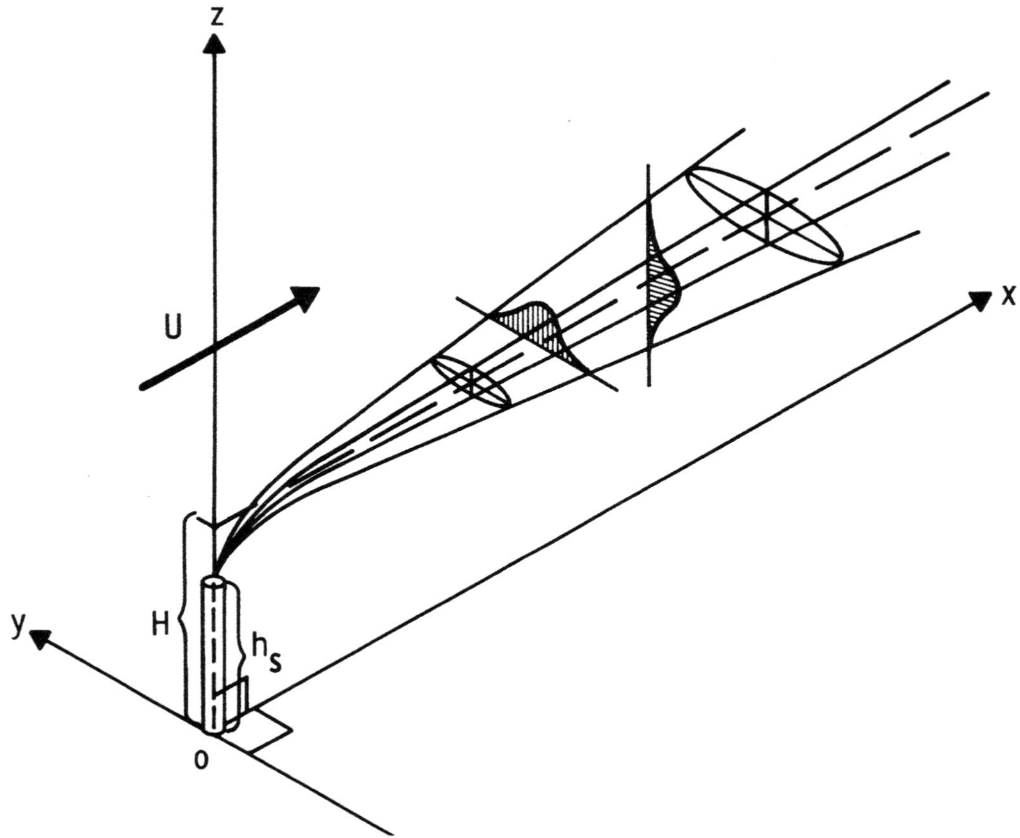

FIGURE 1: Representation of a Gaussian Plume, Showing Plume Rise and Vertical and Crosswind Diffusion (Reproduced by Kind Permission of Electric Power Research Institute)

$$E \;=\; \frac{1}{(2\ )^{\frac{1}{2}} s} \quad \exp \quad - \quad \frac{(\bar{x} - x')^2}{2s^2} \qquad \text{(I)}$$

where E is the probability density function, x' is a fluctuation about $\bar{x}$, and s^2 is the variance of the fluctuations. The factor $(2\ )^{\frac{1}{2}}$ is included so that the total area under the curve from x = + will be equal to unity.

In two dimensions, when the distribution is <u>binormal</u>, the equation becomes:

$$E \;=\; \frac{1}{2\ s_x s_y} \quad \exp \quad - \frac{1}{2} \quad \frac{(\bar{x} - x')^2}{s_z^2} + \frac{(\bar{y} - y')^2}{s_y^2} \qquad \text{(II)}$$

This is a basic working tool in diffusion theory.

Let there be a continuous point source Q of gas or weightless particles at a height h. Let the concentration at any point (x,y,z) be C. It is assumed that diffusion in the forward direction may be neglected in comparison with the wind, and that the particles diffuse in a random fashion in the y- and z-directions. Thus, any vertical cross-section through the plume will have a binormal distribution of particle positions, with variances s_y^2 and s_z^2. An equation similar to (II) will then apply.

$$\bar{u}C/Q \;=\; \frac{1}{2\ s_y s_z} \quad \exp \quad - \frac{1}{2} \quad \frac{y^2}{s_y^2} + \frac{(z - h)^2}{s_z^2} \qquad \text{(III)}$$

It is noted that the probability density function E of equation (II) has been replaced by $\bar{u}C/Q$ to preserve dimensions and to take account of the fact that the mean wind stretches the plume forward with resulting dilution. This equation is not yet complete because vertical diffusion is bounded by the surface of the earth. It is assumed that the ground acts as a perfect reflector, and that there is a <u>mirror image</u> source at z = -h as well as one at z = +h, as indicated in Figure 2.

The final form of the equation is then: (IV)

$$\bar{u}C/Q \;=\; \frac{1}{2\ s_y s_z} \ \exp(-y^2/2s_y^2) \ \exp - \frac{(z - h)^2}{2s_z^2} + \exp - \frac{(z + h)^2}{2s_z^2}$$

For ground-level centre-line concentrations (z = 0), (y = 0), equation (IV) can be simplified somewhat.

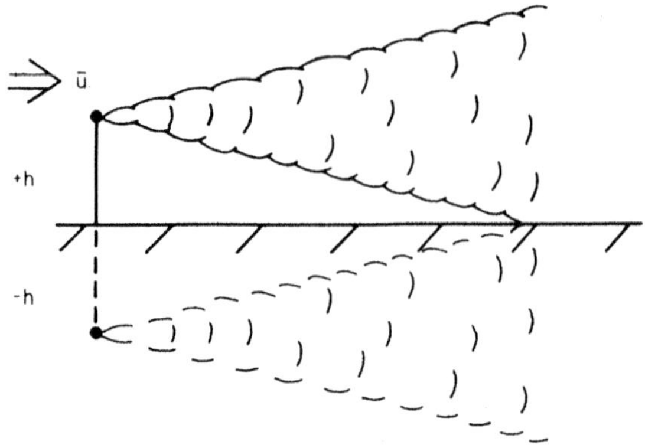

FIGURE 2: Sutton's Mirror Image Model for Diffusion from a Continuous Point Source (Reproduced by Kind Permission of the Royal Society of London)

This model contains the following assumptions:

a) steady-state conditions (the wind field does not change with time),
b) horizontal homogeneity (the wind field does not change in space),
c) wind constant with height,
d) weightless pollutant,
e) chemically passive pollutant,
f) perfect reflection at the surface of the earth,
g) diffusion in the forward direction is negligible in comparison with transport by the mean flow,
h) no exit velocity or heat to cause buoyant plume rise,
i) a non-zero wind speed,
j) no precipitation.

There are a number of uncertainties when the model is used in EIA calculations. First, the wind climatology of the area must be estimated for the projected life of the development. Secondly, the standard deviations s_y, s_z must be estimated, also for the projected life of the development, and for various downwind distances and meteorological conditions. The values of s_y and s_z are functions of the turbulent structure of the flow, that is, for more turbulence, there will be more dispersion. However, the relations are complex: the theory applies to a Lagrangian moving reference frame whereas the downwind pollutant concentration measurements are made in a fixed Eulerian co-ordinate system. The relations are therefore parameterized in the form of simple nomograms. However, the nomograms have been derived from data which show considerable scatter and which are obtained under rather ideal conditions (uniform surfaces, steady winds, no precipitation, etc.)

which rarely occur in many practical applications. Furthermore, the relations have been derived from relatively few data, for example, the numerical value of the Hay-Pasquill (9) Lagrangian-Eulerian factor, the so-called β, was determined from only eight diffusion trials. Finally, there is a need to estimate buoyant plume rise, and to perform all of these calculations for each hour for a climatologically representative period of time, which would include cases of precipitation, light winds and wind shifts.

Despite these many problems, a code of engineering practice has developed, so that the EIA diffusion calculations can be judged as acceptable or non-acceptable by regulatory bodies. No criticism of this procedure is implied, provided that an air pollution meteorologist has been involved in the calculations.

3 A CONCEPTUAL FRAMEWORK FOR EIA

3.1 The Framework

An EIA conceptualization should:

a) Identify the important elements,
b) Identify the important processes,
c) List possible interactions, non-linearities and feedbacks,
d) Establish the boundaries if the system to be conceptualized,
e) Estimate relevant times, for example, cycling times, turnover times, residence times, response time-lags, etc. for the processes being considered,
f) Fit the pieces together,
g) List the uncertainties.

In a) and b) above, the word important needs clarification. Issues perceived as such by the public and by special interest groups should be included in the conceptualization, no matter how irrational they may seem. This is in addition to issues identified as being important by the assessor and technical advisers. For example, some people may visualize a mushroom cloud over an uranium smelter even though a nuclear explosion is impossible.

The establishment of a conceptual framework can be undertaken at a 3-5 day workshop, as recommended by Holling (2) and others at the University of British Columbia. Alternatively, a series of meetings by the assessment team could be held over a period of several weeks. Some of the points to be noted with respect to EIA conceptualization are given in the following paragraphs.

3.2 Interdisciplinary Studies

As mentioned by Regier (5), an assessment is truly interdisciplinary only if "the contributions of different workers cannot readily be recognized as characteristic of the separate disciplines". This is difficult to achieve but implies, for example, that a demand-pull rather than a supply-push approach has been used, that is, the discipline-based specialist should be asking other team members to specify the information that they need from him. Usually, the specialist decides the information that he will provide to others. The former approach will quickly identify problems and should lead to real collaboration after several iterations.

It is also useful to emphasize processes rather than elements during conceptualization of an EIA. Rather than listing every conceivable item to be included in the EIA, for example, bird census, fish census and soil characteristics, one should discuss processes such as energy flows, mass flows, heat flows, biological food chains, predator-prey relations and life cycles of important or representative species.

3.3 Changing Values

With respect to socioeconomic factors, it should be recognized that society's values change with time. In retrospect, for example, suburban development in the 1950s and 1960s was based on the premise that commuter transportation costs would remain low. A desirable objective in 1981 might therefore be unacceptable in 2001.

3.4 Uncertainty

The predictions of an EIA are clearly uncertain. Even the initial conceptualization is a simplified and probably a partially incorrect view of the environment. What has been left out? What interlinkages based on conventional wisdom may in fact be incorrectly specified? It is important for the assessment team to retain a sense of scepticism with respect to predictions, although recognizing that the team is better informed than if other techniques had been used. Unexpected externalities such as a major shift in world prices could trigger irreversible socioeconomic impacts. For example, one of the most difficult aspects of an EIA is the question of future secondary impacts; if the proponent for an airport runway extension stated categorically that this would be the final development at the airport in question, is he to be believed? There are many such "foot-in-the-door" proposals, often prepared in good faith; however, the promises may be forgotten twenty years later.

For reasons such as this, the adaptive EIA strategy expounded by Holling (2) is strongly supported. Successful EIA should be a continuous process in which options are never foreclosed, at least unknowingly.

Some methods used to prepare for the unexpected include:

a) Diversification; for example, avoid single-industry towns whose economies are more fragile with respect to externalities,

b) Small is beautiful; "An action restricted to part of a system is less risky than the same action applied to the entire system" (5),

c) Self-sufficiency/ecodevelopment; that is, encourage the use of locally available materials in the construction and operation of an enterprise,

d) Development of early warning indicators of surprise,

e) Development of adaptive strategies, keeping options open.

With respect to d) some attention should be given to the question of deciding whether warnings would be heeded, even if the predictions were correct. In some cases, the capital investments and infrastructures are so great that there may be no easy way of changing course.

It is not possible to protect a system from all eventualities. Such a strategy would in fact reduce the resilience of the system. Some tradeoffs will always be required but the above checklist should be helpful to an assessment team. Items d) and e) are of course of most importance (2).

3.5 Data

The initial conceptualization should be undertaken without regard to the data available. Later, compromises will be necessary, of course, particularly when trying to match data-rich geophysical factors with data-poor socioeconomic ones. It is, for example, ridiculous to calculate future ground-water quality to three significant factors, when industrial, commercial and residential development is uncertain.

It is also important, although understandably difficult, to avoid allowing available data banks to control the nature of the assessment. Data can be used to provide a shopping-list description of the environment, but this has little relevance in an EIA. More profitably, data can be used to explore relationships (for example, amongst predators and prey; or amongst nesting, feeding and mating territories of particular species), or to validate models. As an example of the latter application, models of lake eutrophication require very specific hydrological, chemical and biological data, and it is not easy to use observations collected for other purposes.

3.6 An Illustrative Example: The CO_2 Issue

In this subsection the CO_2 climate-warming issue will be discussed as a case study illustrating the elements of a strategy for managing uncertainty. Early warnings of possible climatic effects have already been given by atmospheric scientists and the effects could be irreversible over a time span of a century. The following notes are based on a draft UNEP discussion document (4).

a) Present perception of the issue

There is agreement that atmospheric CO_2 concentrations will double before the middle of the next century, and that the concentrations of other greenhouse-type gases (for example, chlorofluorocarbons, methane, ammonia) will also rise. Current atmospheric simulation models predict that this will cause a 1-2 degree rise in world mean temperature, but the models are not capable of predicting disaggregated regional patterns or the associated changes in cloudiness, precipitation and wind. Although a few countries or economic sectors might benefit, there is concern about possible disbenefits to many countries. The issue is global.

Because the natural variability of climate from year to year is great, an upward temperature trend, or an associated change in weather patterns, will be very difficult to detect. There is therefore the possibility that by the time a climatic trend has been positively identified, it will already be too late to reverse. There is the added difficulty that if industry were asked to reduce its CO_2 emissions, its response time would be of the order of 30 years or so. An appropriate environmental management strategy to deal with this difficult issue can be conceptualized as follows:

b) Space and time scales

The issue is global, and the time scales of interest range from 10 to 100 years.

c) Keeping options open

(i) With respect to the rising concentrations of atmospheric

CO_2: develop energy alternatives that do not produce CO_2; develop an engineering capability to switch from fossil fuels if the need arises within a time interval of not more than 20 years.

(ii) With respect to the impacts of climate change: (a) maintain genetic diversity; (b) support both dry and irrigated agriculture development; (c) using historical analogies and models, undertake detailed studies of the range of impacts of climate change on climate-sensitive sectors of national and international economies, for example, agriculture, fisheries, energy and transportation.

d) Practise conservation

(i) Reduce the consumption of fossil fuels,
(ii) Reduce deforestation and encourage reforestation,
(iii) Reduce the emissions of other greenhouse-type gases through product substitution and the development of emission control equipment.

e) Development of early warning indicators

(i) Expand monitoring networks forthe world carbon cycle,
(ii) Seek to establish methods for early detection of increased primary CO_2 production in the biosphere,
(iii) Seek to establish methods for early detection of climatic effects.

f) Encouragement of research

(i) Improve current understanding of the world carbon cycle,
(ii) Improve current understanding of the resulting climatic effects,
(iii) Improve current understanding of the resulting socio-economic impacts,
(iv) Develop engineering technological capabilities to slow down or stop the present rise in CO_2 concentrations.

4 CONCLUSION

Models are widely used in the preparation of component parts of EIAs. However, it is fair to say that a big multidisciplinary model has never contributed to an EIA project decision, although it may demonstrate the virtuosity of its creators. There is, however, a real need for initial conceptualizations in which physical scientists, ecologists, social scientists, engineers and economists collectively produce overall frameworks for the assessment and data collection programs. Finally, it is emphasized that these frameworks should not be "carved in stone" but should be reviewed and updated periodically, not only during the assessment but also in the years ahead.

ACKNOWLEDGEMENT

I acknowledge some illuminating discussions with Richard Fleming of Cornell University.

REFERENCES

1. Holcomb Research Institute. *Environmental Modelling and Decision Making : The United States Experience* (New York, New York : Praeger, 1976).

2. Holling, C.S.,ed. *Adaptive Environmental Assessment and Management* (Chichester, England : John Wiley, 1978).

3. Munn, R.E., ed. *Environmental Impact Assessment : Principles and Procedures,* SCOPE Report 5, 2nd edition (Chichester, England : John Wiley, 1979).

4. United Nations Environment Programme. *Common Perceptions of Environmental Issues,* Draft Discussion Paper (Nairobi : United Nations Environment Programme, 1981).

5. Regier, H.A. *A Balanced Science of Renewable Resources, with Particular Reference to Fisheries* (Seattle, Washington : Division of Marine Resources, University of Washington, 1978).

6. Sutton, O.G. "A Theory of Eddy Diffusion in the Atmosphere". *Proceedings of the Royal Society of London Series A* 135 (1932) 143-165.

7. Drake, R.L. *et al. Mathematical Models for Atmospheric Pollutants* (Palo Alto, California : Electric Power Research Institute, 1979).

8. Munn, R.E. *Descriptive Micrometeorology* (New York, New York : Academic Press, 1966).

9. Hay, J.S. and F. Pasquill. "Diffusion from a continuous source in relation to the spectrum and scale of turbulence". *Advanced Geophysics* 6 (1959) 345-365.

ADAPTIVE ENVIRONMENTAL ASSESSMENT AND MANAGEMENT: SOME CURRENT APPLICATIONS

Robert R. Everitt

ESSA Environmental and Social Systems Analysts Ltd.
Vancouver, Canada

1 INTRODUCTION

The process of Adaptive Environmental Assessment and Management (AEAM) has been adapted to over sixty different applications. Some have been failures; some have been successes. In all cases something has been learned.

The process has been nurtured by institutions like the United Nations Environment Program, the International Institute of Applied Systems Analysis, and the Institute of Resource Ecology at the University of British Columbia.

Its practitioners are well-respected academics, scientists within Canada and the United States government, private consultants, and recently a team of resource scientists in Thailand.

Many of its central ideas are synthesized in Holling's book (1978) (1), which began with an attempt by a group of individuals with a wide range of experience and with different concerns and styles to address three major questions:

a) What, if anything, does our understanding of the nature and behaviour of ecological systems have to say about the issues, limitations, and potential of environmental assessment?
b) What can be done to bridge the abyss presently separating impact assessment studies from actual environmental planning and decision making?
c) To what extent, and under what circumstances, do present methods provide useful predictions of impact?

Many of the early applications were designed to develop and test the techniques used in AEAM. Their primary product was an increase in understanding. As the process matured, a greater concern for serving the client's specific needs developed and was combined with a greater emphasis on the policy implications of the work. The results of these efforts are well documented in the book (1) and the scientific literature, (2), (3), (4), and (5).

The case histories below describe current applications of work in progress as AEAM moves into its second decade of practice. The results of these applications and others will be the successes and failures of AEAM in the 1980s.

A diverse set of three applications has been selected to illustrate a range both in purpose and problem. One is chosen from Canada concerned with the planning and approval of lakeshore development. Two water resource problems in different parts of the globe are selected; one of water use conflicts in California, and the other an environmental management research program in Thailand. Before discussing the case histories a brief description of AEAM is presented to provide background for the case histories. There is no intention of attempting to provide a detailed representation of the philosophy, theory, and techniques of AEAM; these are well documented in the book (1).

2 AEAM IN BRIEF

Through its range of applications AEAM has developed a number of distinctive features that take account of the many problems often encountered in performing environmental impact assessment.

It brings together key scientists, managers, policy makers, and members of public interest groups in interdisciplinary workshops at an early stage of the planning process. Representation from all of these groups ensures a more balanced consideration of the relevant issues, and often identifies potential conflicts early in the process, thus increasing opportunities for their resolution well before deadlines for important decisions.

AEAM uses computer simulation modelling and other techniques of system analysis to limit the scope of the environmental assessment to relevant factors affecting decisions. These techniques help identify important gaps in data and understanding that must be filled before an important decision can be taken. Simulation is particularly useful in predicting potential consequences of various alternative strategies.

The use of the simulation modelling techniques with the workshop format promotes interdisciplinary communication among experts from different backgrounds and enhances the contribution from each of the individual participants.

AEAM is organized around a series of modelling workshops convened by a group of facilitators, or workshop staff, trained in the techniques of systems analysis, computer modelling, policy analysis, and group dynamics. During the course of a workshop the staff has four primary functions: to facilitate the workshop; to assist participants in formulating and constructing a computerized simulation model of the resource system under consideration; to assist participants in the interpretation of model output; and to aid in the integration of workshop results into relevant management activities and research.

An initial modelling workshop generally consists of five days of meetings. It begins with a scoping and bounding exercise during which considerable effort is devoted to identifying quantities (indicators) that describe the resource system, and current and proposed management actions. The resource system is then divided into a number of logical components for detailed consideration by subgroups. Before breaking into subgroups, the participants go through an exercise (looking outward) which defines how the components of the resource system interact with each other. This interdisciplinary exercise helps broaden participants' understanding of the resource system and lays the foundation for building the simulation model and for developing alternative management scenarios later in the workshop.

Participants then break into smaller working groups according to areas of expertise and interest. Each subgroup meeting is convened by a member of the workshop staff. Utilizing the indicators and management actions identified in the bounding exercise and the interactions between components of the resource system identified in the looking outward exercise, each subgroup develops the functional relationship that describes the behaviour of its component of the resource system.

Once the submodels have been programmed by the staff, workshop participants begin a process of model testing. This process is accomplished through experimentation with the model by examining courses of action (scenarios) that management agencies might take and the responses of model variables to these scenarios. Initially, the model may exhibit unrealistic behaviour. Participants may suggest appropriate changes in data or functional relationships which the workshop staff attempts to incorporate into the model. The consequences of the scenarios are re-examined in the light of the interests of the groups represented. Through this process, the participants begin to develop confidence that the model conforms to their perceptions of reasonable system behaviour. The model that emerges thus represents a group consensus concerning assumptions and behaviour. The model, while only a caricature of the real world, often has a group perspective vastly superior to that of any individual. While the model serves as a focus for the workshop, its real purpose is to develop links of communication and to identify clearly the key areas of missing information, research priorities, ways to get needed information, and the responsibilities assigned for research efforts.

The first workshop is followed by a period of independent work by the collaborating individuals which will lead to a second workshop and possibly subsequent ones in a phased sequence. Early in the sequence workshops concentrate on technical issues, but later they focus more and more on communication to policy advisers and the constituencies. The emphasis on communication enables an effective and logical move to implementation, either in a pilot project or a full-scale program.

In summary, AEAM's key features are:

a) Ecological and environmental knowledge is incorporated with economic and social concerns at the beginning of a strategic analysis rather than at the end of a design process;
b) Since linked resource/social systems are dynamic rather than static and linear, techniques of simulation modelling, qualitative modelling, and policy design and evaluation are used to reflect these features;
c) Scientists, managers, and policy people are involved and interact from the beginning and throughout the process of synthesis, analysis, and design so that learning becomes as much of a product as does problem solving;
d) Direction, design, and understanding are in the hands of those from the region who analyze, select and endure policies rather than in the hands of a separate group of analysts who lack the knowledge of needs, the responsibility and the accountability;

and e) Although prediction can be improved, AEAM recognizes that the uncertain and unexpected lie in the future of every design. Hence policies are designed both to explore opportunities and pitfalls as well as to fulfil immediate social needs.

3 CASE HISTORIES

AEAM was forged out of the practice of its many techniques over the last ten years. More can be learned about it by looking at it in action than can be gleaned from a discussion of how to do it. This section focuses on three case histories, different in most respects but similar in that they represent current applications of AEAM. The case histories focus on the institutional aspects of the project, the scientific details and results being well documented elsewhere. In each a short description of the problem is given, an examination of the institutional environment attempted, a recapitulation of what was done is given, and the institutional response is discussed.

One case history is from Canada and concerns the planning and approval of lakeshore development. The other two are water resource problems in different parts of the globe; one of water use conflicts in California, the other an environmental management research program in Thailand.

It needs to be emphasized that these case histories are only a small part of the current activities of AEAM.

3.1 Nam Pong Environmental Management Research Project

3.1.1 Setting. The Nam Pong basin, located northwest of the city of Khon Kaen in northeast Thailand, has a catchment area of some 15,000 km^2. Since 1966 when the Ubolratana dam and the Nam Pong reservoir were completed, the Nam Pong basin has undergone dramatic changes in the quality of natural resources, the productivity of resource use, and the human population and its welfare. The Nam Pong water resource development project, designed primarily for hydro-electric power generation, irrigation, and flood control, has had impacts on power supply, fisheries, agriculture, flood patterns, transportation, growth and spread of settlement patterns, income levels, health, and other environmental considerations.

During 1977-1979 the Nam Pong Environmental Management Research Project began investigation of the environmental impacts of the Nam Pong water resource developments project for the 15 year period 1964-1978. The specific objectives of the environmental management research project were:

a) To define the changes related to water resources development that have occurred and are occurring in the Nam Pong basin system;
b) To develop knowledge about the causative factors and the effects of these changes:
c) To establish preliminary guidelines for sound resource and environmental management in the basin;
and d) To develop a methodology of general applicability for evaluating and incorporating environmental dimensions into the planning and management of development projects centred on intensive use of water and related resources.

Most of the work of the thirteen basic studies undertaken during 1977-1979 focused on the first two objectives. The latter two objectives were left for Phase III of the Nam Pong project, the integration of the Nam Pong studies. The Mekong Secretariat of the United Nations Interim Committee for Co-ordination of the Lower Mekong Basin chose Adaptive Environmental Assessment and Management as the methodology appropriate for Phase III of the Nam Pong Environmental Management Research Project.

3.1.2 Institutional environment. Thailand is one of the most bureaucratized countries in the world. It is thus extremely difficult to describe the institutional structure that influences environmental problems in Thailand. This is further complicated by the role of the United Nations and international aid agencies like the Ford Foundation. For example, for the Nam Pong environmental management research project, the Thai national Mekong Committee acted as *sponsor*, the Mekong Secretariat acted as the *executing agency*, university, agency and Secretariat scientists acted as *research* groups, and the Ford Foundation acted as *donor*. In addition the participation of our AEAM group was funded by the United Nations Environment Program (UNEP).

Fortunately, an institutional analysis performed by a Thai scientist, revealed some institutional problems that would have remained obscure to a foreigner. Briefly summarized these were:

a) Most institutional decisions are not local. This promotes a top-down, centralized approach to management which is not responsive to local needs.
b) Provincial budgets, established and tightly controlled in Bangkok, are small and invariably do not account for the needs of the region. As a consequence, interest in new approaches to management is low.
c) There is a general lack of unity and co-ordination within and between resource management programs implemented in the region.
d) In general there is a shortage of qualified technical and management personnel.
e) There is an unwillingness on the part of the farmers to participate in development projects.

3.1.3 What was done. The third phase of the Nam Pong Environmental Management Research Project was executed by the Mekong Secretariat over a nine month period starting in September 1980, in collaboration with scientists from various Thai institutions. Phase III used the methodology of AEAM to achieve two major objectives:

a) To construct an integrated management-oriented simulation model of the Nam Pong basin taking into account the biological, physical, and socio-economic aspects of the system. The model was required to be responsive to potential management actions and to produce quantitative predictions of system indicators meaningful to decision makers.

b) To train a team of Thai scientists in the procedures of model building, testing, evaluation, and communication so they could take a major responsibility for development of the Nam Pong model.

The Nam Pong model was initially conceptualized and built at a four week modelling workshop at the Mekong Secretariat Offices in Bangkok during October 1980. Present at this workshop were representatives from various Thai government agencies and universities whose involvement was crucial to ensure the modelling effort was relevant to the perceived needs of the Nam Pong region.

Over the six month period following the workshop the group of Thai scientists trained at the workshop in the procedures of AEAM, and model building refined the model. The group also facilitated meetings between managers and policy people concerned with the Nam Pong basin.

The resulting refined version of the model was presented to the original workshop participants at a second two-day workshop in May 1981 and subjected to extensive evaluation and testing. Following this, a range of management scenarios was structured and applied to the model. The resulting predictions provided insight into the implications of alternative management strategies in the Nam Pong basin.

3.1.4 Institutional response. The institutional response must be considered on at least two levels, namely the international institutions (United Nations Mekong Secretariat and the Ford Foundation) and the Thai national institutions.

The international institutions responded favourably. The Ford Foundation was a valuable ally throughout the project and their staff based in Thailand provided positive guidance and insight at a number of key stages in the project. The Mekong Secretariat (the client) remained guarded throughout the project, but as the project approached fruition they became more optimistic and at the end considered the work highly successful. The final report will be circulated throughout the world to various agencies concerned with the environmental aspects of resource development.

Within Thailand there appears to be a greater understanding of the elements needed for an integrated basin management plan for the Nam Pong. The key actors (Electricity Generating Authority, Royal Irrigation Department, Royal Forestry Office, Department of Fisheries, and the Governor of Khon Kaen) all provided manpower, positive energy, and time for the project. In a subsequent presentation of the model to the Department of Fisheries, senior officials were enthusiastic about the fisheries submodel. The Governor of Khon Kaen and many of the local agencies were co-operative and pleased that they had been consulted about the model and saw its application as a useful management tool. The simulation model itself is to be transferred to the Water Resource Information Centre at Khon Kaen University where it can become a tool for management of the Nam Pong basin.

The training objectives of the project were met and Thailand now has a group of accomplished practioners of AEAM. However, at this time it is not clear whether the institutional environment is such that they can remain together as a team.

3.2 Integration of the Lakeshore Capacity Study

3.2.1 Setting. The southern half of the Precambrian Shield in Ontario, Canada contains approximately 50,000 lakes 90 hectares or greater in size. Since the 1930s cottaging and associated water recreation activities have formed the economic base of this region. Cottage subdivisions were routinely approved until deterioration of the environment became evident, bringing a halt to development, and the initiation of remedial measures. The increasing demand for cottages, coupled with an alarming decline in the health of many lake ecosystems, indicated the need for a management strategy based on an understanding of the impact of cottage development, rather than one designed to take remedial measures. The need prompted the Ontario government to undertake a Lakeshore Capacity Study (LCS) which was designed to provide a way to forecast the probable effects of the development of subdivisions.

The ultimate goal of the LCS is to implement an integrated model that will aid the Ontario Ministry of Municipal Affairs and Housing in understanding the cottage subdivision approval process and improve the means for evaluating the impact of lakeshore developments on wildlife, fish and water quality.

In 1974 five components were identified for research:

a) Land use - the source of the impact, the cottagers and how much they use their cottages;
b) Trophic status - the man-made nutrient input to a lake, (primarily phosphorous) and the resultant changes in primary and secondary production;
c) Microbiology - health effects of enrichment;
d) Fisheries - sport fish harvest and littoral zone degradation;

and e) Wildlife - changes in bird, mammal, and reptile populations as a result of habitat degradation.

It was clear from the outset that the Ontario Ministry of Municipal Affairs and Housing (MMAH) could either retain the component findings as independent studies and intuitively fit the pieces together or they could link the components and construct an integrated, albeit simplified, dynamic model of the physical/biological system. MMAH chose the latter, as it was felt that this would provide the most useful and realistic management tool. The objective of such an integration activity is explicitly to account for the interactions between the various components to ensure that the direct and indirect impact of cottage development can be fully realized throughout the system. The methodology chosen by MMAH to integrate the component study findings was Adaptive Environmental Assessment and Management.

3.2.2 Institutional environment. MMAH is responsible for subdivision approval. Upon receipt of an application for subdivision, MMAH circulates the subdivision plan to a number of departments in the Ministries of Environment and Natural Resources for comment. The current process is cumbersome and time-consuming and a major objective of the LCS study is to streamline this process.

The LCS is co-ordinated through a Steering Committee that has representatives from the three Ministries, Natural Resources, Housing and Environment. Most of the studies undertaken were done by the Ministries within the Ontario Government. The component studies varied considerably in terms of scientific quality and in terms of the degree to which they addressed the question of the impact of cottage development.

The integration component of LCS is being co-ordinated by an employee of MMAH.

3.2.2 What was done. We worked closely with the co-ordinator for the integration component of the LCS as well as with the chairman of the Steering Committee to plan the integration component, design and develop the integrated model, and bring about the preliminary stage of implementation.

In December 1980, a five-day AEAM workshop was held with about 30 selected participants. These participants included the managers of the five component studies, members of the Steering Committee, planners from the Ministry of Municipal Affairs and Housing, scientists and managers from the Ministries of Environment and Natural Resources.

The results of the workshops were a preliminary model and a detailed report of the assumptions and functional relationships contained in the model. The report was circulated and the model received considerable comment and crticism. In April 1981 a series of technical meetings with the scientists were held to make additions and refinements to the preliminary model. These technical meetings were a forerunner of an integration workshop held in May to bring the original participants of the first (December) workshop back together to evaluate and game with the refined version of the model.

Throughout early 1981, the model and the computer software and technology needed to use it were transferred to scientists within the MMAH.

3.2.4 Institutional response. Anyone who has tried to integrate a number of research studies knows the difficulties involved. In this case the task was compounded by the territorial response of various ministries. Cries of "our model should be separate from the integrated model", and "the integrated model is the consultant's model, not ours" greeted the preliminary version of the model.

Fortunately, the tenacious efforts of the co-ordinator of the integration component and the chairman of the Steering Committee overcame these objections and by the end of the May workshop the initial resistance turned to enthusiasm to get the model out in the field and working. This activity will have to wait, for the current goal is officially to end the research phase of the LCS and present the results to the Management Board of the Ontario Cabinet.

The project is currently in a consolidation phase with emphasis on report writing, careful editing, and making the computer program easy to use and error free.

The next steps will involve testing the model's applicability to the subdivision approvals process and training people how to use the model and interpret its results.

3.3 Sacramento - San Joaquin Water Management

3.3.1 Setting. The Central Valley of California extends 400 miles from the city of Red Bluff in the north to Bakersfield in the south. The Valley is 60 miles in width at its widest point, with an average width of about 40 miles. The Valley can be divided into the Sacramento Valley, the Delta-Suisun area, the San Joaquin Basin and the Tulare Basin. Each of these areas historically contained vast wetland areas consisting of marshlands, meandering sloughs, streams and wet grasslands. The Central Valley, having a climate suitable for a variety of crops, has been converted to agricultural lands. The agricultural economy of the area is developed on a wide range of truck crops, pasture, small grains and fruits. Supported by a reliable water supply it has resulted in a multi-million dollar agri-business. In less than 100 years some 5 million acres of the Central Valley were changed from grasslands, marshlands, and meandering waterways to high-value farm lands, towns, and cities crossed by various paths of communications that serve regional and national markets.

This transformation has brought about the destruction and alteration of most of the area's salmon and steelhead spawning habitats and natural wetlands used by migratory birds. In addition, 65% of the delta outflow of the Sacramento - San Joaquin water system has been diverted.

In 1979, the California Water Policy Center of the U.S. Fish and Wildlife Service began a project designed to help decision-makers evaluate possible water management options that affect fish and wildlife in the Central Valley. The California Water Policy Center chose AEAM and held the first workshop in April 1979.

3.3.2 Institutional environment. Both the management of water and the management of fish and wildlife are shared between state and federal agencies. The major actors with respect to fish and wildlife are U.S. Fish and Wildlife Regional Office and the California Department of Fish and Game; the major actors with respect to water are the State Water Resources Control Board, the California Department of Water Resources, the federal Soil Conservation Service, and the Water and Power Resources Service.

The California Water Policy Center was established by the U.S. Fish and Wildlife Service to deal with fish and wildlife problems connected with water use in California. The water policy center focuses on system-wide problems and emphasizes policy-level guidelines. Its work is designed to complement rather than duplicate the work of the state and federal agencies.

3.3.3 What was done. In April 1979 a five day workshop was held to initiate the application of AEAM to water management problems in the Central Valley. Participants at the workshop included both U.S. Fish and Wildlife Service and California Department of Fish and Game personnel. This workshop produced a preliminary simulation model with water, fish, waterfowl, and land use submodels.

In January 1980 agency participation was expanded to include the State Water Resources Central Board, the California Department of Water Resources, the federal Soil Conservation Service and the Water and Power Resources Service through a series of technical meetings.

Between the technical meetings and an Integration Workshop in March 1980 extensive reprogramming of the submodels took place as a result of the technical meetings and as new data became available. During this period a staff person of the California Water Policy Center kept up informal day-to-day contact between participating state and federal agencies. This activity, combined with the gathering of background information, resulted in a dialogue that continuously developed ideas and concepts used in the modelling activity.

During the Integration Workshop the submodels were evaluated, some refinements were made, and they were then linked together. Participants examined scenarios and suggested alternative scenarios as well as additional improvements needed.

Some improvements were incorporated into the model during the next few months and, at the end of June, a one day Technical Evaluation Workshop was held for all participants who had been involved in the process. An entire day was devoted to evaluating the model, developing scenarios, and making suggestions for a forthcoming presentation to be given to policy level personnel.

In July, a Policy Evaluation Workshop was convened. Because many of the individuals in attendance had not participated in the model building process the day was devoted to presentation of the model and the gaming with scenarios.

Since the U.S. Fish and Wildlife Service had decided not to continue the California Water Policy Center it was decided to transfer the model to the U.S. Fish and Wildlife AEAM team at Fort Collins, and rely on further initiative for use of the model to come from the regional office of U.S. Fish and Wildlife Service in California.

3.3.4 Institutional response. Throughout most of the model development stage the agencies concerned with fish and wildlife adopted a "wait and see" attitude, and some individuals were openly hostile. The water management agencies were more enthusiastic and seemed to grasp the potential of the model for explaining different policy options. The difference may be in part due to the considerable difference in level of effort that went into the water submodel as opposed to the other submodels.

As the model was refined reservations remained, but hostility was replaced by healthy scepticism, and indifference by enthusiasm.

In general, the institutions have responded positively but slowly. The design of new data collection programs was the first concrete evidence of the project's effect on water, and fish and wildlife management in the Central Valley. Recently enthusiasm and support for the model brought about two one-day meetings sponsored by the regional office of the U.S. Fish and Wildlife Service. The purpose of these meetings was:

a) To establish and encourage communication between participants;
b) To alert potential users to the use of AEAM;
c) To refine and expand the current models (or determine the refinements required);

d) To increase the participation of interest groups;

and e) To elicit support from interest groups and agencies for furthering the development of the model as a policy evaluator's tool.

4 CURRENT PRACTICE

The choice of case histories presented here was determined by the experience and active participation of the author and his colleagues. There are a number of other exciting projects either just completed or still in the midst of analysis.

The work of Dr. Carl Walters on fisheries problems continues at the Institute of Animal Resource Ecology at the University of British Columbia. Dr. Walters recently completed work for the Pearse Commission on Pacific Fisheries Policy and is in the midst of training a group of scientists in the procedures of AEAM for the Great Lakes Fisheries Commission.

The Adaptive Environmental Assessment Group, a team trained by Dr. Holling and Dr. Walters, is within the U.S. Fish and Wildlife Service at Fort Collins, Colorado. Their recent and current projects include the Truckee-Carson River Quality Assessment for the U.S. Geological Survey in Nevada, the North Dakota Wetlands study for the Special Assistant to the Secretary of the Interior and the North Dakota Natural Resources Co-ordinator, the Acid Precipitation Project for the National Power Plant Team of the U.S. Fish and Wildlife Service, the Jackson Hole Project for the Wyoming Game and Fish Department, and the Drilling Fluids Workshop for the U.S. Environmental Protection Agency.

Any of these projects would make an illuminating study of the AEAM process. The important point, however, is that AEAM is being used in many contexts, differing both in type of problem and in institutional environment. This fact underlines its essential claim to be a flexible, adaptive approach to environmental planning, assessment, and management.

REFERENCES

1. Holling, C.S. ed. *Adaptive Environmental Assessment and Management* (Chichester, England: John Wiley, 1978).

2. Clark, W.C. *et al.* "Lessons for ecological policy design: a case study of ecosystem management". *Ecological Modelling* 7(1979)1-53.

3. Walters, C.J. "An interdisciplinary approach to development of watershed simulation models". *Technological Forecasting and Social Change* 6(1974)299-323.

4. Holling, C.S. and A.D. Chambers. "Resource Science: the nurture of an infant". *Bioscience* 23(1)(1979)13-20.

5. Walters, C.J. *et al.* "Computer simulation of barren-ground caribou dynamics". *Ecological Modelling* 1(1975)303-315.

ASSESSMENT OF SPECIFIC IMPACTS

THE ENVIRONMENTAL IMPACT ASSESSMENT OF HIGHWAYS

William V. Kennedy

Office of Federal Activities
U.S. Environmental Protection Agency
Washington, D.C.

1 INTRODUCTION

Although there is general agreement on the nature of the environmental impacts associated with highway design, construction and maintenance, the way and the extent to which those impacts are assessed vary in differing national contexts. The following is based on a comparison of EIA procedures for highway planning in the United States and the Federal Republic of Germany.

Both the USA and the Federal Republic of Germany have a commitment to a federally supported highway system. Both, as well, have a body of legislation and regulations devoted to environmental protection and, specifically, to the incorporation of environmental considerations in the highway planning process. The way in which those considerations are taken into account is, however, different in the two countries.

The United States, through the National Environmental Policy Act (NEPA) has what might be called a formal-explicit process in which environmental impact assessment requirements are specifically codified in legislation and lead to the preparation of statements or documents in which the environmental effects of development projects and/or plans and proposals are addressed.

The Federal Republic of Germany has an informal-implicit process in which EIA requirements are modified or adapted to the needs of individual situations and proposals and form a part of other planning laws and regulations. Under this system an environmental impact statement as such is not prepared 1).

As an attempt to assess the advantages and disadvantages of these two approaches, research was undertaken at the International Institute for Environment and Society in West Berlin comparing the planning of actual highway projects (1). Six pairs of similar highway projects were studied. Based on interviews with highway planners and affected citizens and a comparison of planning documents the projects were analyzed in three areas of comparison:

(a) The number and type of alternatives to the proposed highway project and the way they were examined;
(b) The number of environmental impacts and the way they were considered in highway project planning; and
(c) The amount and type of inter-agency coordination and public participation.

2 HIGHWAY PROJECT PLANNING PROCEDURES

Both West Germany and the USA, through special legal procedures, regulate the development stages of highway planning from the approximate corridor selection to the exact highway alignment including the detailed planning involving the road bed, bridges, tunnels, planting schemes, drainage systems and traffic signs. In both countries there is also a division of labor between the federal and state governments. The federal governments control and allocate funds for highway planning and construction. The detailed technical planning takes place, primarily, at the state level.

A further common characteristic of the planning procedures in both countries is the gradual incorporation of environmental protection considerations in the planning decisions of various levels. Highway planning procedures are older than explicit environmental policy. In previous procedures special guidelines were developed for resolving conflicts between highway and other public or private interest concerned with land use. Initially they were directed to noise protection - later expanded to include nature, landscape and air quality considerations. Technical guidelines related to water problems which were first formulated to protect roads from flooding and provide drainage systems for rain water, streams and rivers, were later expanded to include environmental protection perspectives.

The following review of the present status of planning procedures in both countries serves to assess the fundamental comparability of the procedures.

2.1 The United States

Except for those projects built entirely on federal lands, such as National Parks and Forests, the planning and construction of U.S. federal highways is the responsibility of the states. If federal funds are used, however, the Federal Highway Administration (FHWA) reviews and approves the project's development at key stages.

Highway projects are developed through studies conducted at both the state and metropolitan area levels. These studies include inventories of existing facilities and travel, and prediction of future traffic volumes.

Once the need for a project is determined, location studies are made to determine which location, or corridor, will provide the maximum amount of service. After the location determination, the project is advanced through the basic design stage and, finally, into the Plans, Specification and Estimate (P, S & E) stage.

After the design work is completed, the state highway agency awards the construction contract and supervises the construction. Following completion, the state highway agency maintains the facility, primarily at state expense. Formal federal approvals are made at the location, design and P, S & E stages, and the project is monitored during the construction.

In 1969, the FHWA issued a Policy and Procedure Memorandum to insure full opportunity for effective public participation in the cosideration of highway location and design proposals by highway departments before submission to the Federal Highway Administration for approval (2). Specifically, it required that public hearings be held at location and design approval stages (previous requirements had not specified the exact stages of the planning process to be subjected to public hearings). The Memorandum also set down a list of social, economic and environmental effects to be considered in highway project planning. It did not, however, provide instructions as to how these effects should be considered and/or balanced with other (i.e. transportation) goals. It was not until the passage of the National Environmental Policy Act of 1969 (NEPA) that the consideration of environmental factors became a legal requirement in the U.S. With NEPA the requirements for environmental impact assessment became woven into the planning of federal highways.

The National Environmental Policy Act requires that a detailed statement of environmental impact, known as an environmental impact statement (EIS), be prepared for all major federal actions significantly affecting the quality of the human environment.

Specifically, the EIS must contain information on:

(a) the environmental impact of the proposed action;

(b) any adverse environmental effect which cannot be avoided should the proposal be implemented;

(c) alternatives to the proposed action;

(d) the relationship between local short-term uses of man's environment and the maintenance and enhancement of long-term productivity; and

(e) irreversible and irretrievable commitments of resources which would be involved in the proposed action should it be implemented.

Guidance for the preparation of environmental impact statements for highway projects is provided in FHWA's Federal-Aid Highway Program Manual (3). In addition to addressing procedural matters, the manual requires that the EIS include information on impacts to natural, cultural, and ecological systems, relocation of individuals and families, air quality, noise, water quality, wetlands and coastal zones, stream modification or impoundment and flood hazard as well as construction and socio-economic impacts.

Although EISs are the ultimate responsibility of the Federal Highway Administration, they are generally prepared by state or local agencies. Under this arrangement, a draft EIS is prepared early in consultation with the FHWA. After the draft EIS is signed by the FHWA Division Administrator to indicate FHWA acceptance for review purposes, it is then circulated to concerned agencies and the public for review comment. A waiting period for review and comments of at least 45 days is required. Public hearings are held at least 30 days after circulation of the draft. The public hearing on the draft EIS is incorporated into the location approval hearing.

The final EIS, also prepared by the state or local agency on behalf of the FHWA, incorporates comments, indicates their reaction to them, and shows which of the alternatives discussed in the draft EIS has been selected. Final EIS's are approved by FHWA Regional Administrators, except that in certain cases (e.g., new freeway projects) the prior concurrence of the FHWA Washington headquarters and the Office of the U.S. Secretary of Transportation is needed. After the final EIS is approved, it is filled with the Environmental Protection Agency and made available to the public and other federal agencies for a 30-day review period prior to any construction. The presentation of the final EIS coincides with the design approval hearing.

2.2 The Federal Republic of Germany

The West German highway planning procedure is a complicated one. It is made up of a combination of several planning levels and decision levels which not only increase in coverage as highway planning proceeds, but also partly overlap in their implementation and which vary in terms of the obligations of their outcomes. The traditional planning procedure has three stages - pre-investigation, pre-design and construction design - and involves two determinations: the Initial Route Determination Procedure and the Plan Approval Procedure.

During the pre-investigation, alternative routes are examined and compared. In addition to the technical and engineering factors which are weighed, regional planning principles and the federal ministry's long range plan for highways must be taken into consideration in selecting the route.

After the pre-investigation, the determination of the route takes place in accordance with Section 16 of the Federal Highway Act (4) which states that the procedures should incorporate the participation of the federal ministries concerned with regional planning and state agencies as well.

The actual procedure varies from state to state. For example, in those states which have separate regional planning procedures (Bavaria, Hesse, Lower Saxony, the Rhineland Palatinate, and Schleswig-Holstein) the route determination process involves more 'outside' comment and participation from other agencies. Otherwise the procedures is 'internalized' by state highway-planning agencies. In any event, the initial route determination must be approved by the federal Ministry of Transportation. The federal Ministry reviews not only the notification and the grounds for the preferred route but data related to the rejected alternatives as well. All of this information is included in an explanatory report or initial route determination document, which is submitted to Bonn.

A general highway construction circular (5) issued by the federal Ministry of Transportation spells out the content of the explanatory report. For each route considered the state agency must present data related to seven points including length of road, technical description, relationship of the route to the overall highway system, interchanges, etc. The report must also include information on environmental effects, specifically impacts on (a) water protection areas; (b) landscape and nature preserves; (c) built-up areas; and (d) single objects particularly worthy of protection. Noise protection measures, however, need not be decided upon until the final plan approval procedure.

In a later circular, the federal Ministry pointed out the need for a stronger consideration of environmental factors. The circular states that in addition to the consideration of effects on protected areas, ecological effects should also be considered. Specifically, such questions as the destruction of valuable biospheres, ecosystem overload, and endangered species should be covered, to the extent possible, at the initial route determination phase of planning (6).

After the initial route determination has been approved by the federal government, planning proceeds to the pre-design stage, in which the selected route is outlined in more precise detail, and continues to the construction stage at which point it takes on its final, ready-to-be-built, configuration.

When the construction design for the chosen route has been determined by the highway-planning agency, the plan approval procedure takes place. The essentials of this procedure are that other agencies (including those responsible for nature protection and land use planning) and the public (through public hearings) are informed of the final construction design and given an opportunity to voice their comments and/or objections. The process varies from state to state but always has two stages: a public hearing and the preparation of the plan approval report. These stages are implemented by the same or different agencies depending on the state. At the conclusion of the procedure the final report is issued which, having weighed all comments and objections, represents the final decision on the road.

3 SUMMARY OF RESEARCH RESULTS

3.1 Alternatives

In neither the American nor the West German cases did the consideration of corridor alternatives lead to the abandonment of the project. In terms of the actual number of construction alternatives considered the American cases were more thorough, having examined, on the average, twice as many potential routes as the West German ones. The number of alternatives considered alone, however, does not provide much insight into the thoroughness or outcome of the planning procedure. Indeed, given the lower population and urbanization density of the United States compared to the Federal Republic of Germany one would rightly assume that for any given highway project more alternatives would be considered in the U.S. than in Germany. The more interesting question, therefore, is not the <u>number</u> of alternatives, but (a) the <u>way</u> in which they were examined and compared and (b) the influence of environmental factors in making a corridor selection.

Regarding (a) the American EIS procedure must be judged to be more thorough. In each EIS all alternatives are described in detail in terms of their engineering characteristics and environmental impacts. Tables and maps summarizing the comparison are included which provide the "outside reader" with an overview of the project and the decision parameters.

The West German documents, on the other hand, give no evidence of such a comparison. A recommended alternative is presented and described. Rejected alternatives are only referred to in terms of those characteristics (engineering, environmental, etc.) which, in the opinion of the highway planners, are inferior to the recommended route. One can assume that a thorough comparison of alternatives for the German projects also took place. That comparison, however, and the way in which it was carried out are matters of "internal" agency decision making which remain closed to the "outside" observer.

In terms of (b) - the influence of environmental factors in selecting a route alternative - the situation is not as clear-cut. In both countries there are examples in which the evidence indicates that a recommended alternative was rejected in favour of another alternative because of fewer or less damaging environmental impacts. Often, however, those changes came about not because of an objective rethinking of the route on the part of highway planners but because of massive public opposition. At the same time, there are examples of projects in both countries where an environmentally superior alternative was rejected on cost and engineering grounds.

The "no-build" alternative, which in the U.S. belongs to the obligatory assessment schedule, is not examined in West Germany. Although one occasionally finds statements in West German planning documents that it would not be possible to abandon the project, a specific investigation of the "no-build" alternative does not take place. On the other hand, that alternative in the American statements, though always described in some detail, usually has the character of an "alibi function" for a construction alternative. Standard arguments related to higher traffic accidents, detrimental air quality and increases in noise are presented for rejecting the "no-build" alternative without examining the unique characteristics of the project in question.

The situation regarding mitigation measures for the selected alternative is also similar in both countries. Examples of needed protection measures as a result of the assessment of negative environmental impacts are generally more frequent than changes in the route configuration. In the area of noise protection, the planning of mitigation measures is a standard part of highway planning in both countries. The construction of noise walls and embankments as well as cut and fill, and planting schemes belong to the standard mitigation repertoire in the USA and West Germany. Other measures related to bridge and dam construction, drainage systems and underpasses for wild animal migration are also carried out.

Regarding the dimension of mitigation measures, standards play an important role. Noise protection measures, for example, are generally so designed and constructed that standards are barely adhered to. In both countries the construction of a new highway in a previously quiet, rural area, bringing a dramatic increase (though still under the standard) in noise level is viewed as a major impact to deal with and, if possible, avoid. This situation, however, is not as seriously addressed as when standards are exceeded regardless of the preconstruction noise situation. The main difference regarding mitigation measures results from the number of environmental impacts assessed. Once a problem is identified and recognized as serious, the mitigation measures taken in both countries are, in principle, the same.

Im summary, it is difficult to determine whether the selection of a highway alternative reflects a consideration of environmental impacts to a greater degree in the USA with an EIA procedure or in West Germany without it. Given the unique individual characteristics of highway projects, the landscapes and communities they impinge upon and the growing controversy regarding their construction in both countries, highway corridors seem to follow the geographical path of least political resistance regardless of the physical environment in which they are built.

3.2 Environmental Impacts Considered

The consideration of environmental impacts in the American EISs appears to be substantially more thorough and far-reaching than in the West German planning documents. American highway environmental impact statements are generally several hundred pages in length and often thousand page documents divided into several volumes. The West German Initial Route Determination and Plan Approval Documents, on the other hand, are short volumes and devote, on the average, 2 to 5 pages to environmental effects. In addition, environmentally related considerations are occasionally presented under other document headings such as "alternatives". In both countries, there are additional planning documents which address environmental questions. These are generally specific individual studies related to noise, air quality, general ecology and climatology or other special problems associated with individual cases. Some of these additional documents relate environmental questions to engineering or economic ones such as traffic safety, geology/topography or construction costs. Although the number of additional documents is greater for the West German cases than the American ones the total number and type of impacts considered is still greater for the American projects.

The function of the additional documents or studies is somewhat different in the two countries. In the USA they provide the scientific basis for the EIS. As such they contain numerous data and computation descriptions which are not necessary for inclusion in the EIS itself. In the impact statement the results of the studies are summarized and evaluated. In West Germany, on the other hand, the studies are often the only source of assessed environmental impacts. They are not, however, part of materials which must be presented at public hearings. Occasionally, individuals studies are made public but that is the exception to the rule.

The EIS in the USA provides a complete and wide-ranging overview of all environmental impacts. Indeed, EISs are often criticized for providing too much information, often including for example, superfluous data on plant and animal species or ecosystems that is not necessary for making an adequate assessment of potential environmental damage. In West Germany, on the other hand, the assessment of environmental impacts is problem-oriented and includes only those effects which have been identified through the concerns of outside or specialized agencies (*Fachbehoerden*) or the public.

A further obvious difference is the limitation of environmental impact assessment in West Germany to the recommended route. In the USA impacts are assessed for all alternatives. Usually, only route variations are assessed in West German studies and even then, not as a matter of course.

3.3 Coordination with Other Agencies and the Public

No other area of comparison reveals so dramatically the differences between West German and American highway planning procedures as this one. Although in both countries numerous other federal, state and local agencies are involved in highway project planning, the extent of that coordination, the point in time at which it takes place and the influence it has on planning outcomes are vastly different.

In West Germany the extent of cooperation with other agencies or Fachbehoerden depends upon the type of planning procedure within the individual state. In states where highway projects are subjected to a regional planning procedure the coordination with other agencies, even at the pre-investigation phase of planning, is relatively broad-based. The effectiveness of the cooperation is limited, however, as the concluding State Planning Assessment (*landesplanerische Beurteilung*) is not legally binding. It often provides suggestions for further planning but it need not be heeded by the highway planning agency and it is not made public.

In states where no regional planning procedure is required, planning is totally the domain of the responsible highway agency and outside agencies are involved at its discretion. Particularly, nature protection agencies often complain that they are either involved too late in the planning process or have too little influence on planning outcomes. At the same time, however, there are indications that nature protection agencies and other *Fachbehoerden* are under-staffed and under-financed and cannot take advantage of all the possibilities for cooperation. When the *Fachbehoerden* have serious reservations about a particular project, the highway agency officials generally attempt to clear them up before the public hearing in order to ensure a unified position of agencies *vis à vis* the citizens.

In the USA, through the coordination of other agencies in the EIS process, a uniform procedure is guaranteed. A conciliation process with outside agencies also takes place where, however, the positions and objections of the other agencies are made public. For particular problems, the Department of Transportation EIS Guidelines provide for cooperation between highway planning agencies and other agencies such as those responsible for nature protection or regional planning. Environmental protection sections within state highway planning agencies ensure coordination with other agencies. The documents and correspondence from reviewing agencies attached to the final EIS indicate that in many cases considerable disagreement and controversy took place. Such controversies often continue at public hearings where agencies present their views.

Formal citizen participation in highway planning in West Germany is limited to the Plan Determination phase, although communities are often informed of the planned action, individual citizens or citizens groups are not involved. In the USA citizens are incorporated into the planning procedures at a much earlier stage. The first public hearing often takes place even before the preparation of a draft EIS and always after its completion. A second hearing is conducted when the final EIS is completed. Nor is public participation limited to hearings. The formation of citizen advisory groups, the distribution of questionnaires and polls,and other participation strategies often accompany the formal public hearing process. In addition, American citizens have the right of access to all the planning documents. In West Germany, during the Plan Determination Procedure, certain documents are presented and distributed; however, agencies are not required to make other documents available including specially commissioned environmental studies.

Lastly, in the USA citizen action groups often, together with public interest law firms, have the advantage of being able to sue federal agencies if it appears that the EIS was inadequately prepared. Such class action suits and public interest law firms are unknown in West Germany and only individuals whose rights (i.e. to health or property) are threatened can take legal action.

4 IMPLICATIONS OF THE RESEARCH

The goal of this study was to compare two different environmental planning procedures in order to determine their advantages and disadvantages in assuring that environmental impacts are adequately considered in federal highway planning. The results indicated that in terms of consideration of alternatives, environmental impacts and inter-agency coordination/public participation the American formal-explicit procedure was more comprehensive and effective than the German informal-implicit procedure. The question remains, however, can the difference be explained by the procedure alone? Or, in other words, if West Germany had a law similar to NEPA requiring the preparation of environmental impact statements, would the planning outcomes be improved to give a broader consideration of alternatives, a greater number and type of environmental impacts considered and an improvement in inter-agency coordination and public participation? Unfortunately, that question cannot be answered. Simply because a policy or policy instrument such as EIA is successful in one country, it cannot be assumed that it will be successful in another country with a different political, legal, and administrative culture. Neither the American nor the West German EIA procedures are operating in a vacuum. Both are affected by other legislation and procedures, are operating in different administrative and legal systems and are being implemented by different types of civil servants with different backgrounds, training, and attitudes. Although these parameters of policy implementation were not the focus of this research their relevance remains undisputed and they should definitely be addressed in future research comparing EIA procedures in various national settings.

FOOTNOTE

1. Full accounts of EIA procedures in the USA and in West Germany are contained in the paper by C.P. Wolf (this volume) and in W.V. Kennedy, "The West German Experience", pp. 155-185 in T. O'Riordan and W.R.D. Sewell, eds. *Project Appraisal and Policy Review* (Chichester, England: John Wiley, 1981) respectively.

REFERENCES

1. Kennedy, W.V. and R. Lummert. "Umweltvertaeglichkeitspruefrang in der Fernstrassenplannung". *Zeitschrift für Umweltpolitik* 4 (in press).

2. U.S. Federal Highway Administration. *Public Hearings and Location Approval,* Policy and Procedure Memorandum 20-8 (Washington, D.C.: U.S. Federal Highway Administration, 1969).

3. U.S. Federal Highway Administration. *Federal-Aid Highway Program Manual,* vol. 7 (Washington, D.C.: U.S. Federal Highway Administration, 1976).

4. "Bundesfernstrassengesetz rom 6 August 1953 (BGB1 IS 903)", in M. Kloepfer. *Deutsches Umweltschutzrecht* (Percha am Starnberger) See: Verlag R.S. Schulz, 1979).

5. Bundesministerium für Verkehr. *Allgemeines Rundschreiben Strassenbau,* No. 1 (Bonn: Bundesministerium für Verkehr, 1974).

6. Bundesministerium für Verkehr. *Allgemeines Rundschreiben Strassenbau,* No. 8 (Bonn: Bundesministerium für Verkehr, 1976).

CURRENT TRENDS IN LANDSCAPE EVALUATION

Xaver Monbailliu

Land Use Consultant, Paris

INTRODUCTION

This paper provides a description of quantitative techniques for visual assessment which are of importance in the selection of sites for least impact for development proposals. Landscape evaluation methods can be grouped into three categories, namely:

(a) Intuitive field based appreciation of landscape scenery
(b) Grading or scaling methods
(c) Statistical techniques such as multiple regression analysis as introduced by Warwickshire County Council in 1971. (1)

This paper will briefly examine techniques within each of these categories.

INTUITIVE APPRECIATION

The first group of techniques used in the assessment of visual resources are based on the allocation of qualitative scores to landscape units using descriptions such as "spectacular or beautiful", "unsightly" etc. Fines (2) used this technique for an assessment of Eastern Sussex, in England, with the assistance of some 45 trained observers, from the design professions. These observers graded twenty pictures showing different scenes of the East Sussex landscape. These views of the landscape were then evaluated in the field and the results subjected to a standard deviation in order to allow the ranking system of each landscape tract to be refined. Landscape tracts can be defined as:

(a) Large areas of land with uniform physiognomic characteristics
(b) Visual units as seen from the road, limited by vegetation cover or by relief (3)
(c) Grid squares (for example 1 km^2 grids) or a hexagonal grid.

SCALING TECHNIQUES

This simplified technique consists of attributing a quality rating to landscape scenery depending on topography, vegetation and land use. Terminology of landscape appreciation is replaced by a nominal value system (for example, from -10 to +10). Appraisal is undertaken by one observer. This technique has been used by several landscape architects, for example Tandy (4) and Hebblethwaite (5). Both consultants used a grid square unit in their evaluation.

In a study to find possible sites for industry in the Clyde Estuary, Prof. E. Weddle (6) of Sheffield University, tried to assess the landscape under the assumption that a view or scenery is of value only when it is of some use to the community. He evaluated what he termed the "inherent values of a landscape" such as views, background scenery and foreground vegetation, as well as the "acquired value" by aggregating the importance of the landscape to the residential and visiting populations. Each land unit was given a 1, 2 or 3 star rating. This study is of particular interest, since it integrates the frequency that a landscape is viewed by its inhabitants with the more intrinsic values of the landscape scenery.

STATISTICAL TECHNIQUES

Statistical techniques may approach a higher level of objectivity in landscape evaluation, but are both time consuming and expensive. However, they are relatively reliable techniques since they establish a relationship between the overall attractiveness of a site and the value of the elements that compose the landscape scenery. Regression analysis, known by statisticians as "stepwise multiple linear regression analysis", is often used in statistical landscape evaluation techniques which are based on the following:

(a) A random sample of grid squares (1 km^2 grid) in the study area.
(b) A team of assessors touring around in a vehicle evaluating the views of each grid square, while

attributing a value score to each landscape unit, for example a value of +10 for outstanding beauty.

(c) The selection and measurement of a number of the most representative landscape components. These variables can be expressed as the percentage of moorland cover, percentage of different woodland types, percentage of grazing land, length of shelterbelt and hedgerows, dykes, canals, streams etc. The variables are measured for each grid square from air photographs or from topographical maps (1:50,000 or 1:25,000).

(d) Stepwise multiple regression analysis is applied using the mean score of the landscape view as the dependent variable.

(e) The regression equations are subsequently used to evaluate the attractiveness of the landscape in other sectors of the study area.

Refinement of this technique can be attained firstly by enlarging the number of the assessing team. Needless to say a representative sociological sample will be required if one is to achieve an objective evaluation. Secondly, by introducing the observing team to several scenes of the study area prior to the evaluation. A session in which transparencies representing different views are shown can be a very useful tool in this context. A scoring range, for example -10, can be attributed to the most "unsightly" scenes with +10 being applied to the most attractive areas. A landscape team from the University of Manchester (UK) calls this reference point system a "heaven and hell system" (7), where "heaven" relates to a paradise-on-earth scenery. "Hell", as a result of a French study north of Paris (8), referred to abandoned waste lands with power lines, and other industrial equipment situated in the urban fringe.

In a study for South Yorkshire, Briggs and France (9) formulated two different landscape scores. The first score referred to the scenery within a grid square and was termed the "internal score". The second score related to the same scene, but taking into account the surrounding views outside the grid square in question and was termed the "total score". Their study demonstrated a close correlation between the internal score and the total score. However, a disparity of such grading systems will occur in mountainous areas, where the valley floor is being used for sand extraction, hydro-electric power or ski resorts (all having negative scores), but the views of the mountain slopes contribute significantly to the overall attractiveness of the valley landscape.

DRAWBACKS OF LANDSCAPE EVALUATION METHODS

Landscape evaluation methods can be a useful tool leading to the classification or protection of the most attractive sceneries of a region. They can also be particularly useful in the integration of linear infrastructure networks, such as roads, railways, canals, powerlines and pipelines into the landscape. Usually, environmental impact studies for these types of development projects are to be produced in a limited time period and the existence of regional landscape evaluation studies undertaken by local authorities (county or region), can often be of assistance in the definition of the corridor of least impact for the proposed project.

Government officials frequently consider that landscape assessment studies are limited in what they measure. It is true that visual analysis excludes several elements in landscape assessment. For example:

(a) The historical significance of a landscape (for example, a battlefield)
(b) A scenery once described by a popular author
(c) The tactile and olfactory quality of the Mediterranean maquis
(d) The fertility of land for an agriculturalist
(e) The accessibility and visibility of the terrain
(f) A landscape of childhood memories
(g) A scenery depicted by a famous painter

According to the nature of the project and the importance of the study area, some of the above elements may be considered alongside the purely visual assessment. This also means that landscape evaluation is somehow a subjective phenomenon, since it depends on the judgements of people for its existence. This implies that most assessors try to appraise the intrinsic qualities of the landscape experience as objectively as possible. This quest for objectivity is certainly justified, since landscape studies can, in some cases, lead to the classification of areas of outstanding natural beauty. The lack of objectivity is certainly one reason for politicians and decision-makers to justify not basing their final judgements of a development on visual criteria. The route of a road might be preceded by a visual assessment, but other parameters such as the needs of agriculture, nocturnal migration corridors of deer, or the presence of a military camp often carry more weight in the final decision.

Other drawbacks associated with landscape evaluation relate to the use of grid squares as landscape units. Grid squares exclude the views outside the square or limit themselves to the view from one particular direction. This is especially true when the landscape assessment is undertaken through the use of pictures or slides of set landscape views. Evaluation of video panoramic views (360°) could eliminate this problem. Another drawback is the time consuming nature of visual analysis. The techniques indicated above are designed to provide a basis for planning measures and, therefore, guide landscape change on a regional scale. Evaluation techniques are not, as yet, adjustable to the often restricted financial resources of a proponent of a development, or to the urgent nature of a planning application.

In environmental impact assessment, simple techniques will often be used to justify the choice of the site of least impact for certain types of development (for example, power stations, gas plants, quarries etc). These techniques are often limited to a more comparative description of a few potential construction sites. Perspective sketches, photomontages of "before-and-after" scenes will often accompany this type of visual appraisal. In the worst cases, environmental impact studies are only accompanied by a simple visual analysis of the "before-and-after" situation without any justification of the geographical choice of the construction site. Linear infrastructure projects, highways for example, will mostly require a regional assessment, which includes landscape criteria, in order to justify the corridors and their variants of least impact. In such situations landscape assessment techniques should be designed in accordance with the particular needs of the infrastructure project that is proposed. For example, the visual impact of a high voltage powerline will be completely different from that of a pipeline.

CONCLUSIONS

It is clear that there are various approaches to landscape assessment and that the subjective and objective evaluation techniques both require greater research in order to improve the role of landscape assessment in appraisal of development projects. Such landscape assessment studies ought to be prepared by local authorities and retained for repeated use during individual project EIAs thereby avoiding unnecessary repetition and delay.

REFERENCES

1. Warwickshire County Council. *Coventry-Solihull-Warwickshire, a Strategy for the Sub-region*, Supplementary Report No. 5 (Warwick, England: Warwickshire County Council, 1971).

2. Fines, K.D. "Landscape Evaluation: A Research Project in East Sussex". *Regional Studies* 2 (1) (1968) 41-55.

3. Countryside Commission for Scotland. *A Planning Classification of Scottish Landscape Resources*, CCS Occasional Paper No. 1 (Perth, Scotland: Countryside Commission for Scotland, 1971).

4. Tandy, C. *Landscape Evaluation Techniques* (Croydon, England: Land Use Consultants, 1971).

5. Hebblethwaite, R.L. *East Hampshire AONB: A Study in Countryside Conservation* (Winchester, England: Hampshire County Council, 1968).

6. Weddle, A.E. "Landscape Evaluation: Case Study in the Clyde Estuary". *Journal Royal Town Planning Institute* 55 (9) (1969) 387-389.

7. Robinson, D. *et al*. *Landscape Evaluation: Report of the Landscape Evaluation Research Project* (Manchester, England: University of Manchester, 1976).

8. Institut d'Aménagement et d'Urbanisme de la Région d'Ile-de-France. *Méthodes d'Evaluation des Caractères Physiques et Humains du Paysage* (Paris: I.A.U.R.I.F., 1978).

9. Briggs, D.J. and J. France. *Environmental Mapping of the European Community: South Yorkshire Case Study* (Barnsley, England: South Yorkshire County Council, 1979).

THE ATMOSPHERIC COMPONENT OF ENVIRONMENTAL IMPACT ASSESSMENT

R. E. Munn

Institute for Environmental Studies, University of Toronto, Canada

1 INTRODUCTION

Most Environmental Impact Assessments (EIAs) contain a chapter on the climate of the area, or at least on air pollution. Included are graphs or tables giving monthly mean, maximum and minimum temperatures, and precipitation, as well as a wind rose from the nearest weather observing station. It is customary, also, to present a description of the large-scale climatological features of the region. This information is usually given with only a cursory attempt to show its relevance to the development under consideration.

If air pollution is a consideration, the EIA may contain computer-calculated ground-level concentration field; if a breakwater is being built, the EIA may include information on strong winds; or if an airport is to be constructed, a fog frequency table may be inserted. Various horror tales could be mentioned here; for example, chimney-height calculations for a site in the Crows Nest Pass (an east-west canyon in the Canadian Rockies) based on winds measured at Cowley, Alberta (just east of the Rockies and having a strong tendency for winds to be either north or south); a breakwater orientation at Sept Isles, Quebec, based on airport winds a few kilometers inland; and an airport siting at Torbay, Newfoundland, based on fog frequencies observed at a climatological station in the city of St. John's, Newfoundland.

In this paper a plea is made for more relevance in the atmospheric part of EIAs, both in terms of the representativeness of the climatological data with respect to the site being considered, and in terms of the relevance of the climatological elements selected for analysis relative to the problem being considered. In this latter connection, the word climatology is rarely used alone except in a very general sense. Instead one hears expressions such as air pollution climatology, building climatology, agro-climatology, hydroclimatology, forest climatology and road construction climatology.

The main climatological data banks contain information from first-order weather observing stations (mostly at airports, thus showing an aviation climatological bias), upper-air rawinsonde stations and voluntary observer second-order stations (daily precipitation and maximum and minimum temperatures). When skilfully used, these data banks can shed light on the problem at hand. However, it is necessary:

a) to understand the spatial variations of the relevant meteorological elements, that is, the mesoclimatology, in order to know how to use the available climatological data banks in an intelligent way. It is necessary to produce "synthetic" climatologies for the site being considered or to demonstrate by a sensitivity analysis that the likely error induced by using the wind rose from an adjacent station does not matter;

b) to realize that a large development will change the microclimate and may even affect the mesoclimate.

c) to recognize that climatological data banks may have limitations with respect to the development being considered, thus necessitating the establishment of additional observing stations of a special field programme.

2 TYPES OF ATMOSPHERIC IMPACTS

Although the atmospheric component of an EIA often relates mainly to air pollution, there are nevertheless several kinds of atmospheric impacts that should be considered. In the case of an hydroelectric/irrigation/flood control system, for example, air pollution is not a factor except for the problem of blowing dust during construction; the main interest is in monthly evapotranspiration rates and the return periods of large rainfalls. A checklist of atmospheric impacts that should be considered at the initial EIA screening stage is as follows:

a) Air pollution, including impacts on health and the biosphere, materials (such as buildings), and socio-economic and climate effects (See Section 3 below).

b) Noise, particularly during construction.

c) Impairment of the use of the atmosphere as a resource. Microscale impacts of development can be exceedingly troublesome for people who live in the immediate vicinity. For example, the erection of a tall building may shade the property of a neighbour, reducing the efficiency of solar energy devices, or compromising future uses. The building may also spoil a scenic view or may cause wind channelling and gustiness on public sidewalks. As a final example, a cooling tower may occasionally cause fog patches to drift over an adjacent highway, creating a hazard for motorists.

d) Climatic impacts. Often mesoscale atmospheric impacts of development are an issue. For example, would the creation of a large man-made lake affect the climate of the surrounding area? If an ice channel were kept open for navigation in the arctic, what might be the climatic impacts?

e) Water quantity and quality. In the case of irrigation and other hydrologic works, estimates of evapotranspiration from the drainage basin are required. As another example, if a forest were cut down, what would be the effect on the hydrologic cycle, particularly runoff, and what would be the effect on frost frequencies and forest regeneration?

f) Safety and related considerations. Engineers do not design fail-safe systems. For every tall tower, dam and bridge, there is a calculated risk (admittedly extremely small) of structural failure due to atmospheric extremes. For flood-control systems, the concept of "return period" is used, based on the statistical theory of extreme values (1). For this category of impact, there is a wide body of accepted engineering practice, as well as government codes and standards.

Finally, it should be mentioned that not all atmospheric changes caused by development are harmful. In fact the atmosphere can sometimes be exploited in clever ways. Soviet architects refer to children's microclimates and design outdoor winter playgrounds that minimize the wind chill. Similarly, shelter belts are used extensively in the countryside, and careful planning of orchards in hilly country can enhance drainage and reduce fruit losses by hard frosts.

3 AIR POLLUTION IMPACTS

A schema of the air pollution problem is given in Figure 1. Some measure of source emission control is practised but complete removal of pollutants is technologically impossible. (Yet the remaining 1% to 3% that escapes into the atmosphere may amount to hundreds of tonnes per day in the case of SO_2 from a smelter.) The problem is to ensure that emissions are kept down to an acceptable level, where "acceptable" is sometimes a controversial term.

The main impacts of air pollution are:

a) health effects
b) vegetation damage
c) damage to materials
d) socioeconomic impacts
e) climatic effects

Air quality standards are designed to protect society from the impacts listed above. The standards are of course not absolutes but change as new research information is received and as society's values are modified. Examples of changing values are to be found in the United States where there are conflicting objectives for regulating the automobile with respect to energy conservation and clean air (2), and where asthmatics in California have recently been found to be much more sensitive to SO_2 than previously thought, leading to the following question being posed: "Should the entire populace assume the burden of preventing aggravation of a disease in a relatively small group of people who unfortunately live in large cities?" (3)

Despite these contentious issues, it is generally accepted that EIAs should adopt the numerical values contained in air quality standards currently in force within the jurisdiction in question. The air pollution section of an EIA should then demonstrate that the proposed development will not violate these standards. This demonstration can only be done in an engineering design sense, of course, because of the uncertainties in air pollutant diffusion calculations (see below). This means that project design should have an "after-construction" fall-back position in case of violations, in case air quality standards become more stringent, or in case the assimilative capacity of the atmosphere must subsequently be shared with other industrial developments.

Pollutant diffusion calculations are based on methods recommended by the responsible air pollution control authority, or permitted by their approvals branch. Several handbooks have been produced and used, for example, the Turner Workbook (4), the Handbook on Atmospheric Diffusion (5) and the PADC Manual (6). To obtain an air pollution climatology, several approaches are used:

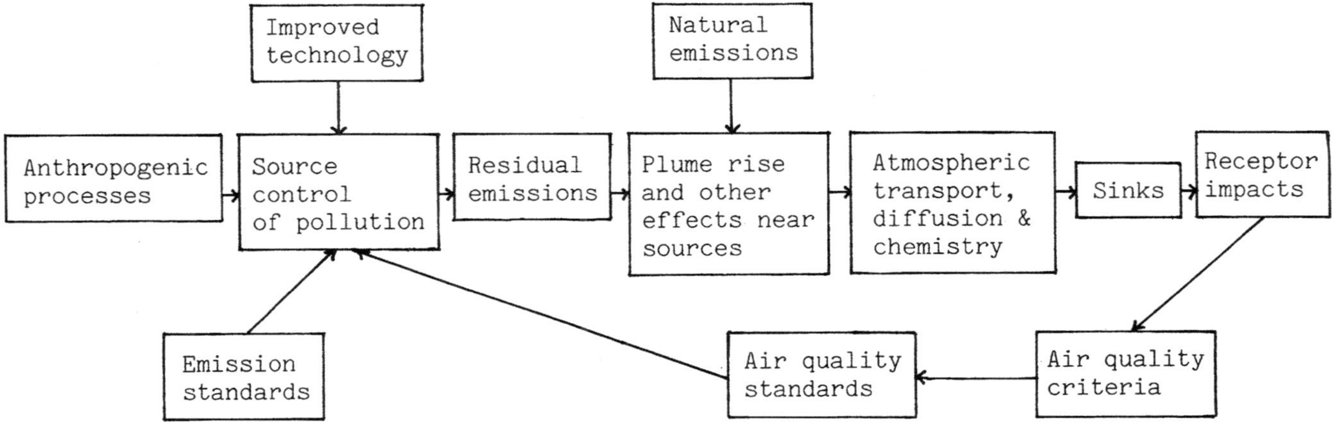

Fig. 1: Schema of the air pollution problem

a) One or several design cases are presented in detail (for example, a moderate inversion with steady winds). This so-called maximum credible accident approach is not recommended unless the frequency of occurrence is also specified.

b) The Pasquill-Gifford climatology, in which a 30-year weather record is used to subdivide the diffusion regime for the area into subsets. To adjust for meso-climatological differences between the observing station and the construction site, a number of adjustments may be made, the net result being a so-called synthetic diffusion climatology.

c) Numerical 3-dimensional models. This approach is often used in complex terrain.

Some of the assumptions buried in the glossy computer outputs may include the following:

i) no vertical wind shear.
ii) horizontal homogeneity.
iii) steady state
iv) perfect reflection by the surface of the earth.
v) chemically passive and weightless pollutants.
vi) varying sets of assumptions to accommodate the plume-rise formula used.
vii) a non-zero wind speed. (If the wind is calm, the diffusion equation "blows up" in the mathematical sense.)

Thus, for example, the Pasquill-Gifford and other diffusion climatology schemes fail in the most important cases: stagnation episodes and aerodynamic downwash around apartment blocks and other tall buildings.

4 MESOCLIMATE

Returning to the question of the representativeness of climatic data for a particular site under investigation, three further points should be made.

a) In the first place, the climatic statistics of the last 30 years may not be the same as those for the next 30 years. Climatological means obtained from time series rarely converge.

b) Secondly, the proposed development will certainly affect the microclimate and could affect the mesoclimate around the site.

c) Thirdly, mesoclimatic variations are considerable, particularly during clear skies and light regional winds. Even over relatively flat terrain, fine weather wind roses from stations 30 km apart can be quite different. Furthermore, many developments are sited in complex terrain, for example, in river valleys, on shorelines, or in urban complexes.

5 CONCLUDING REMARKS

The words of caution given above apply equally with respect to other kinds ot atmospheric impacts. The lesson to be learned by those who prepare or review EIAs is to seek the advice of a meteorological specialist rather than to rely on engineering handbooks.

REFERENCES

1. Munn, R.E. *Biometeorological Methods* (London: Academic Press, 1970).

2. Lave, L.B. "Conflicting objectives in regulating the automobile". *Science* 212 (1981) 893-899.

3. Smith, R.J. "Utilities choke on asthma research". *Science* 212 (1981) 1251-1254.

4. Turner, D.B. *Workbook on Atmospheric Dispersion Estimates* (Research Triangle Park, North Carolina: U.S. Environmental Protection Agency, 1967).

5. Hanna, S.R. *et al. Handbook on Atmospheric Diffusion* (Oak Ridge, Tennessee: National Oceanic and Atmospheric Administration, in press).

6. Clark, B.D. *et al* *A Manual for the Assessment of Major Developments* (London: Her Majesty's Stationery Office, 1981).

WATER POLLUTION IMPACTS

L. W. Canter

Professor of Civil Engineering and Environmental Science and Co-Director, National Center for Ground Water Research, University of Oklahoma, Norman, Oklahoma, United States of America

Water pollution impacts occur from many types of projects, including the construction and operation of multi-purpose reservoirs, industrial complexes, and housing developments. Water pollution impacts may be visible in terms of floating debris and the occurrence of dead fish. The purpose of this paper is to present a systematic approach for identifying and evaluating water quality and quantity-related impacts of projects. Background information on water pollution and quality considerations will be presented along with a conceptual framework for addressing water pollution impacts. Specific information associated with each of the framework elements will be summarized. Examples of scientific approaches for accomplishing the various elements of the conceptual framework are also included.

BASIC INFORMATION ON WATER POLLUTION AND QUALITY

Water pollution can be defined in a number of ways; however, the basic features of most definitions address excessive concentrations of particular substances for sufficient periods of time to cause identifiable effects. Water quality represents a term associated with the composite analysis of physical, chemical, and bacteriological parameters. Physical parameters include color, odor, temperature, solids (residues), turbidity, oil and grease. Sub-parameters for solids include suspended and dissolved solids as well as organic (volatile) and inorganic (fixed) fractions. Chemical parameters associated with the organic content of water include biochemical oxygen demand (BOD), chemical oxygen demand (COD), total organic carbon (TOC), and total oxygen demand (TOD).

Inorganic chemical parameters include salinity, hardness, pH, acidity, alkalinity, iron, manganese, chlorides, sulfates, sulfides, heavy metals (mercury, lead, chromium, copper and zinc), nitrogen (organic, ammonia, nitrite, and nitrate), and phosphorus. Bacteriological parameters include coliforms, fecal coliforms, specific pathogens, and viruses.

In evaluating water pollution impacts associated with the construction and operation of a potential project, two main sources of water pollutants should be considered. Non-point pollutants refer to those substances which can be introduced into receiving waters as a result of urban or rural runoff. Point sources are related to specific discharges from municipalities or industrial complexes.

The effects of pollution sources on receiving water quality are manifold and dependent upon the type and concentration of pollutants. Soluble organics, as represented by high BOD wastes, cause depletion of oxygen. Trace quantities of certain organics cause undesirable taste and odors, and some may be biomagnified in the aquatic food chain. Suspended solids decrease water clarity and hinder photosynthetic processes; if solids settle and form sludge deposits, changes in benthic ecosystems result. Color, turbidity, oils, and floating materials are of concern due to their aesthetic undesirability and possible influence on water clarity and photosynthetic processes. Excessive nitrogen and phosphorus can lead to algal overgrowth with concomitant water treatment problems resulting from algae decay and interference with treatment processes. Chlorides cause a salty taste to be imparted to water, and in sufficient concentration limitations on water usage can occur. Acids, alkalis, and toxic substances have the potential for causing fish kills and creating other imbalances in stream ecosystems. Thermal discharges also cause imbalance as well as reductions in stream waste assimilative capacity. Stratified flows from thermal discharges minimize normal mixing patterns in receiving streams and reservoirs (1).

CONCEPTUAL FRAMEWORK

A conceptual framework for addressing the water pollution impacts of a potential project is shown in Figure 1. Each of the boxes represents elements associated with prediction and assessment of water pollution impacts. The first element involves assemblage of information on impact-causing factors associated with the construction and operation of a proposed project. Information is needed on project type as well as associated specific pollutants. The second element involves the description of existing water quality and quantity conditions for the subject resource. It may be necessary to conduct field studies as a part of gathering base-

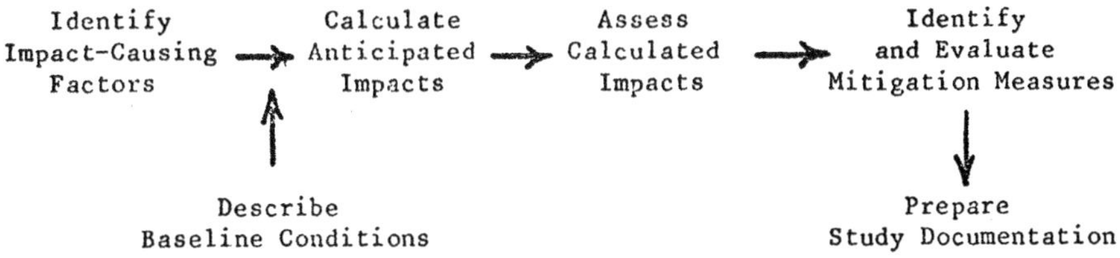

Figure 1: Conceptual Framework for Water Pollution Impact Study

line information.

The most important technical element is associated with calculations related to project impacts. Calculations may include usage of mass balance approaches or specific numerical models. Qualitative projections can also be utilized. The key focus is to delineate anticipated changes in baseline conditions that would result from the construction and operation of the proposed project or alternatives. The fourth element involves the assessment or interpretation of the significance of the predicted impacts. Water quality standards or criteria can be utilized as the technical basis for assessment; however, it is also necessary to utilize professional judgment. Another work element is associated with identifying and evaluating potential mitigation measures for minimizing undesirable water pollution impacts. Finally, it is necessary to document the study in order to provide a written record which can be utilized as a part of an environmental impact assessment report or impact statement.

IDENTIFICATION OF IMPACT-CAUSING FACTORS

It is necessary to identify impact-causing factors associated with the construction and operation of a proposed project. Information should be based on specific features of the project or alternative under consideration, review of pertinent literature, and professional knowledge and judgment. For some projects, it may be possible to utilize unit waste generation factors, with this term representing the rate at which a pollutant is released to a drainage area or watercourse as a result of some activity, such as land clearing or production by industry, divided by that activity (1). Information on the transport and fate of various pollutant materials within the aquatic environment is also necessary.

An initial activity should be the undertaking of literature reviews on similar projects to the one under consideration. Since there is over ten years of experience in the conducting of environmental impact studies within the United States as well as numerous other countries, it is appropriate to review published literature and other reports related to the water pollution impact concerns of specific project types. To serve as an illustration, Table 1 contains example references on the impacts of impoundment and channelization projects. Table 2 contains a similar listing of example reference materials associated with the impacts of dredging and development proiects. These references were identified through the process of computer-based literature searches based on selected descriptor words. Specific data bases which were searched included National Technical Information Service, Pollution Abstracts, Compendex (Engineering Index), and Biosis

Table 1: Example References on Impacts of Impoundment and Channelization Projects

Project	Reference and Comments
Impoundment	(2) -- Diurnal stratification in deep impoundments.
	(3) -- Prediction of water quality impacts from outboard engines.
	(4) -- Impact prediction for proposed impoundment in southwestern Wisconsin.
	(5) -- Effects of hydraulic structures.
	(6) -- Summary of impacts on fisheries.
	(7) -- Effects on fish migration.
	(8) -- Study of suspended sediment and phosphorus in Missouri flood detention reservoir.
	(9) -- Prediction of mixing effects from boating activities.
Channelization	(10) -- Study of impacts on invertebrates in Georgia.
	(11) -- Study of ecological changes in six Pennsylvania streams.
	(12) -- Effects on fish populations in Wisconsin.
	(13) -- Model presented for calculating flooding decrease.
	(14) -- Hydrological effects in Louisiana.

Table 2: Example References on Impacts of Dredging and Development Projects

Project	Reference and Comments
Dredging	(15) -- Investigation of pollution properties of dredged material.
	(16) -- Evaluation of impacts of dredged material disposal on ground water and surface water.
	(17) -- Study of physical effects of aquatic disposal.
	(18) -- Study of impacts on fresh water benthos.
	(19) -- Prediction of fate of dredged material placed in oceans, estuaries, lakes, and rivers.
	(20) -- Summary of dredging technology.
	(21) -- Guide for reviewing environmental impact statements on dredging in estuaries.
	(22) -- Chemical and biological effects of dredging and spoil disposal.
	(23) -- Study of physical and biological impacts of dredging.
Development projects in coastal zone	(24) -- Impact of construction activities in wetlands.
	(25) -- Effects on wetland ecosystems.
Artificial islands	(26) -- Effects of sand dredging for island construction material.
Gabions	(27) -- Effects of gabion installation and stream reconstruction.
Bank stabilization	(28) -- Effects on game fish habitat from revetments, retards, and jetties.

(Biological Abstracts).

If the project under consideration involves point source discharges, unit waste generation factor information should be assembled. There is an extensive literature base on unit waste generation factors for municipal and industrial wastewater discharges (29,30). To serve as an example, brief information will be presented on the pollutant discharges from petroleum refineries. The total amount of water used in a petroleum refinery has been estimated to be 770 gallons per barrel of crude oil (30). Approximately 80 to 90 percent of the water usage is for cooling purposes only, and is not contaminated except by leaks in the lines. Process wastewaters, comprising 10 to 20 percent of the total, may include free and emulsified oil from leaks, spills, tank draw-off, and other sources; waste caustic, caustic sludges, and alkaline waters; acid sludges and acid waters; emulsions incident to chemical treatment; condensate waters from distillate separators and tank draw-off; tank-bottom sludges; coke from equipment tubes, towers, and other locations; acid gases; waste catalyst and filtering clays; and special chemicals from by-product chemical manufacture.

Table 3 contains a listing of conventional and toxic pollutants in the wastewaters from petroleum refineries. Conventional pollutants refer to those which have received historical attention, while toxic pollutants relate to parameters being given increasing attention due to their potential environmental toxicity (31). In delineating information on wastewater characteristics five refinery subcategories, based on throughputs and process capacities, can be used as defined in Table 4. Table 5 presents ranges and median loadings in raw wastewater of conventional pollutants for the petroleum refining industry subcategories. Raw wastewater has been defined as the effluent from the oil separator, which is an integral part of refinery process operations for product/raw material recovery prior to wastewater treatment.

Non-point sources of water pollution have been recognized as potential major contributors to the total waste load within the aquatic environment, and it is vitally important in environmental impact studies to consider non-point sources along with point sources of water pollution. Information on unit waste generation factors for non-point sources is increasing, with Table 6 presenting information on the rates of erosion from various land uses (32). More extensive information could be developed by using the universal soil loss equation; however, these rates represent factors useful for estimating total pollutant loadings. Table 7 contains some example references on non-point sources of water pollution, including urban and rural runoff, erosion and sediment, and other sources such as acid rainfall and industrial areas.

Table 3: Conventional and Toxic Water Pollutants from Petroleum Refineries (Reproduced by kind permission of U.S. Environmental Protection Agency)

Conventional Pollutants:

- BOD
- COD
- TOC
- Phenol
- Sulfide
- Oil and Grease
- Emulsified Oil
- pH
- Temperature
- Ammonia
- Chlorides
- Acidity
- Alkalinity
- Suspended Solids

Toxic Pollutants:

Metals and inorganics
- Antimony
- Arsenic
- Asbestos
- Beryllium
- Cadmium
- Chromium
- Copper
- Cyanide
- Lead
- Mercury
- Nickel
- Selenium
- Silver
- Thallium
- Zinc

Phthalates
- Bis(2-ethylhexyl) phthalate
- Di-n-butyl phthalate
- Diethyl phthalate
- Dimethyl phthalate

Phenols
- 2-Chlorophenol
- 2,4-Dichlorophenol
- 2,4-Dinitrophenol
- 2,4-Dimethylphenol
- 2-Nitrophenol
- 4-Nitrophenol
- Pentachlorophenol
- Phenol
- 4,6-Dinitro-*o*-cresol
- Parachlorometa cresol

Aromatics
- Benzene
- 1,2-Dichlorobenzene
- 1,4-Dichlorobenzene
- Ethylbenzene
- Toluene

Polycrylic aromatic hydrocarbons
- Acenaphthene
- Acenaphthylene
- Anthracene
- Benzo(a)pyrene
- Chrysene
- Fluoranthene
- Flourene
- Naphthalene
- Phenanthrene
- Pyrene

Polychlorinate biphenyls and related compounds
- Aroclor 1016
- Aroclor 1221
- Aroclor 1232
- Aroclor 1242
- Aroclor 1248
- Aroclor 1254
- Aroclor 1260

Halogenated aliphatics
- Carbon tetrachloride
- Chloroform
- Dichlorobromomethane
- 1,2-Dichloroethane
- 1,2-*Trans*-dichloroethylene
- Methylene chloride
- 1,1,2,2-Tetrachloroethane
- Tetrachloroethylene
- 1,1,1-Trichloroethane
- Trichloroethylene

Pesticides and metabolites
- Aldrin
- α-BHC
- β-BHC
- δ-BHC
- γ-BHC
- Chlordane
- 4,4'-DDE
- 4,4'-DDD
- α-Endosulfan
- β-Endosulfan
- Endosulfan sulfate
- Heptachlor
- Isophorone

Table 4: Subcategorization of the Petroleum Refining Industry Reflecting Significant Differences in Wastewater Characteristics (Reproduced by kind permission of U.S. Environmental Protection Agency)

Subcategory	Basic Refinery Operations Included
Topping	Topping and catalytic reforming whether or not the facility includes any other process in addition to topping and catalytic process. This subcategory is not applicable to facilities which include thermal processes (coking, visbreaking, etc.) or catalytic cracking.
Cracking	Topping and cracking, whether or not the facility includes any processes in addition to topping and cracking, unless specified in one of the subcategories listed below.
Petrochemical	Topping, cracking, and petrochemical operations, whether or not the facility includes any process in addition to topping, cracking, and petrochemical operations,[a] except lube oil manufacturing operations.
Lube	Topping, cracking, and lube oil manufacturing processes, whether or not the facility includes any process in addition to topping, cracking, and lube oil manufacturing processes, except petrochemical operations.[a]
Integrated	Topping, cracking, lube oil manufacturing processes, and petrochemical operations, whether or not the facility includes any processes in addition to topping, cracking, lube oil manufacturing processes, and petrochemical operations.[a]

[a] The term "petrochemical operations" shall mean the production of second generation petrochemicals (i.e., alcohols, ketones, cumene, styrene, etc.) or first generation petrochemical and isomerization products (i.e., BTX, olefins, cyclohexane, etc.) when 15% or more of refinery production is as first generation petrochemicals and isomerization products.

Table 5: Raw Wastewater[a] Loadings in Net Kilograms/1000m³ of Feedstock Throughput by Subcategory in Petroleum Refining
(Reproduced by kind permission of U.S. Environmental Protection Agency)

	Topping subcategory		Cracking subcategory		Petrochemical subcategory	
Characteristics	Range[b]	Median	Range[b]	Median	Range[b]	Median
Flow[c]	8.00 - 558	66.6	3.29 - 2,750	93.0	26.6 - 443	109
BOD_5	1.29 - 217	3.43	14.3 - 466	72.9	40.9 - 715	172
COD	3.43 - 486	37.2	27.7 - 2,520	217	200 - 1,090	463
TOC	1.09 - 65.8	8.01	5.43 - 320	41.5	48.6 - 458	149
TSS	0.74 - 286	11.7	0.94 - 360	18.2	6.29 - 372	48.6
Sulfides	0.002 - 1.52	0.054	0.01 - 39.5[d]	0.94[d]	0.009 - 91.5	0.86
Oil and grease	1.03 - 88.7	8.29	2.86 - 365	31.2	12.0 - 235	52.9
Phenols	0.001 - 1.06	0.034	0.19 - 80.1	4.00	2.55 - 23.7	7.72
Ammonia	0.077 - 19.5	1.20	2.35 - 174	28.3	5.43 - 206	34.3
Chromium	0.0002 - 0.29	0.007	0.0008 - 4.15	0.25	0.014 - 3.86	0.234

	Lube subcategory		Integrated subcategory	
Flow[c]	68.6 - 772	117	40.0 - 1.370	235
BOD_5	62.9 - 758	217	63.5 - 615	197
COD	166 - 2.290	543	72.9 - 1,490	329
TOC	31.5 - 306	109	28.6 - 678	139
TSS	17.2 - 312	71.5	15.2 - 226	59.1
Ammonia	6.5 - 96.2	24.1		
Phenols	4.58 - 52.9	8.29	0.61 - 22.6	3.78
Sulfides	0.00001 - 20.0	0.014	0.52 - 7.87[d]	2.00[d]
Oil and grease	23.7 - 601	120	20.9 - 269	74.9
Chromium	0.002 - 1.23	0.046	0.12 - 1.92	0.49

[a]After refinery oil separator.

[b]Probability of occurrence less than or equal to 10% or 90% respectively.

[c]1,000 m³/1,000 m³ of feedstock throughput.

[d]Sulfur.

Table 6: Representative Rates of Erosion from Various Land Uses (Reproduced by kind permission of U.S. Environmental Protection Agency)

Land use	Erosion Rate MT/km^2-yr	Erosion Rate Tons/mi^2-yr	Relative to forest=1
Forest	8.5	24	1
Grassland	85	240	10
Abandoned surface mines	850	2,400	100
Cropland	1,700	4,800	200
Harvested forest	4,250	12,000	500
Active surface mines	17,000	48,000	2,000
Construction	17,000	48,000	2,000

Table 7: Example References on Non-Point Sources of Water Pollution

Topic	Reference and Comments
Urban runoff	(33) -- Development of model for direct runoff from urban watersheds.
	(34) -- Effect on water quality in a reservoir.
	(35) -- Computer simulation model for estimating pollutant loading.
	(36) -- Models for effects on receiving stream dissolved oxygen and hydraulic characteristics.
Rural runoff	(37) -- Ecological effects from agriculture and silviculture activities.
	(38) -- Summary of non-point models for wildland areas.
Erosion and sedimentation	(39) -- Literature review on the physical and biological effects of sedimentation in streams.
	(40) -- Sediment transport model for natural channels.
	(41) -- Effects of grazing practices on runoff and sediment yield.
	(42) -- Relationship of land management practices to erosion and sedimentation.
	(43) -- Influence of urban land use on erosion and sedimentation.
Other sources	(44) -- Non-point pollution from an industrial site.
	(45) -- Effects of acid rainfall on plant communities in lakes.

Table 7 (continued)

Topic	Reference and Comments
	(46) -- Non-point pollution from unconfined animal production areas.
	(47) -- Model for estimating atmospheric contaminant rainout.

In addition to information on pollutant types and quantities, it may also be necessary to assemble information on the transport and fate of specific pollutant materials. Table 8 contains some example references on the fate of petroleum products, other organics, nutrients, metals, and radioactivity in the water environment. Of particular concern is the potential reconcentration of pollutant materials into aquatic organisms and their subsequent harvesting and consumption by man. Multiple questions could be raised relative to the unique water chemistry of metals. For example, due to its density and insolubility, mercury discharged from industrial or municipal point sources typically ends up in the sediments of small bays, rivers, lakes and estuaries. However, bacteria have the capacity to transform inorganic mercury into organic forms which are readily absorbed by aquatic faunal species. Therefore, a key concern is the microbial methylation process and the resultant uptake of the solubilized mercury by fish and seafood.

DESCRIPTION OF BASELINE CONDITIONS

The second element involves assembling information on existing surface and ground water quantity and quality levels in the area of the project, particularly focusing on quality parameters related to anticipated water pollutants to be emitted from the construction and operational phases of the project or alternatives under consideration. Information gathering can be achieved through the usage of existing data collected by governmental agencies and/or the planning and conducting of specific baseline studies. Within the United States the typical agencies which would have pertinent data include city, county, and state water resources agencies as well as federal agencies such as the Environmental Protection Agency, Geological Survey, Bureau of Reclamation, and Army Corps of Engineers.

Planning of baseline studies involves the development of a sampling network design, including selection of water quality and biological parameters for monitoring. Table 9 summarizes twelve steps which can be used for the design of a sampling network (60). The twelve steps listed should be viewed as general guidance as opposed to specific rules for design in every environmental impact study. Steps 1 through 3 are related to defining the objectives of the sampling program. Steps 4 through 7 are self-explanatory, while step 8 involves revisions of sampling station locations and sampling frequencies based on information gathered to that point. Step 9 on the development of operational plans includes consideration of sampling equipment, field or *in situ* monitoring, as well as laboratory analyses. Step 10 focuses on data formats and the associated data reporting. Step 11 relates to adjustment of the first ten steps for study compatability. Finally, step 12 is associated with the preparation of a network design report.

Table 8: Example References on Fate of Pollutants in the Water Environment

Pollutant	Reference and Comments
Petroleum products	(48) -- Fate and effects of petroleum hydrocarbons in the marine environment.
	(49) -- Book on the effects of petroleum on marine environments and organisms.
	(50) -- Methodology for determining effects of petroleum in intertidal ecosystems.
	(51) -- Prediction of environmental impacts of oil spills.
Other organics	(52) -- Literature review on effects of phenolics in aquatic ecosystems.
Nutrients	(53) -- Phosphorus loadings to 15 New York lakes and ecological responses.
	(54) -- Phosphorus movement in channelized and meandering streams.
	(55) -- Study of effects of development on lake nutrient budgets.
Metals	(56) -- Study of heavy metals in fresh-water wetlands.
	(57) -- Environmental pathways of human exposure to lead.
	(58) -- Literature review on bioaccumulation and toxicity of heavy metals and related trace elements in aquatic ecosystems.
Radioactivity	(59) -- Model for calculating radionuclide transport between receiving waters and bottom sediments.

Table 9: Steps Related to Sampling Network Design (Reproduced by kind permission of Colorado State University)

1. Determine monitoring objectives and relative importance of each.
2. Express objectives in statistical terms.
3. Determine budget available for monitoring and amount to be allocated for each objective.
4. Define the characteristics of the area in which the monitoring is to take place.
5. Determine water quality variables to be monitored.
6. Determine sampling station locations.
7. Determine sampling frequency.
8. Compromise previous objective design results with subjective considerations.
9. Develop operating plans and procedures to implement the network design.
10. Develop data and information reporting formats and procedures.
11. Develop feedback mechanisms to fine tune the network design.
12. Prepare a network design report.

A critical component in the conducting of baseline studies is associated with the parameters which should be monitored for determining baseline water quality characteristics. There are hundreds of water constituents which could be selected for monitoring, and it will be necessary to compromise between the number of parameters measured and the marginal difference that information on each parameter makes to subsequent interpretation and impact assessment. One approach which can be used in the selection of water quality parameters is to consider general recommendations for conduction of water quality surveys. Table 10 contains a reference list of parameters used for river water quality surveys (61). Another approach is to take information from the first element on the types of pollutants anticipated from the construction and operation of the proposed project and include those in the baseline monitoring programme.

Selection of biological parameters is complicated due to system relationships in terms of material and energy flows. Figure 2 is a schematic diagram indicating material and energy flows in an aquatic ecosystem along with system inputs and outputs (62). Aquatic organisms include primary producers, plant eaters, meat eaters, and decomposers. Primary producers include algae, while plant eaters encompass zooplankton, fish, and benthic organisms. Meat eaters include zooplankton and fish as well as benthic organisms. A monitoring program related to water pollution impacts should include consideration of sampling of various planktonic and benthic forms, as well as fish. Table 11 identifies some useful references related to biological indicators and the planning of sampling programs.

An important aspect of describing the baseline setting is the interpretation of the extant or collected data. Interpretation can be based on water quality criteria, water quality standards which have been adopted for a given stream setting, or the quality and quantity requirements of existing or potential water users in the area. To serve as an example, Table 12 summarizes the stream quality requirements for public and private drinking water supplies utilized by the State of Oklahoma in the United States (72). Professional judgment will have to be exercised in the data interpretation phase, particularly as related to biological parameters and characteristics.

Due to the large number of water quality parameters which might be associated with describing the baseline setting for a given project, presentation of summary information related to these parameters may be difficult. One approach which could be used is an empirical index which combines data from several parameters into one numeric indicator of water quality. The usage of water quality indices has been increasing in the United States and other countries (73). The most frequently used index in the Unites States

Table 10: Selection of Parameters for River Water Quality Surveys (Reproduced by kind permission of UNESCO and WHO)

Type of survey	Physical parameters	Chemical parameters			Biological parameters	
		Inorganic	Organic	Nutrients	Microbiological	Hydrobiological
Proposed for inclusion in all surveys	Colour pH Specific conductance Suspended solids Total solids		Chemical Oxygen Demand (COD) Total Organic Carbon (TOC)		Coliforms, total and faecal	
Recommended for collection of baseline data	Odour	Acidity Alkalinity Calcium, Ca Chlorides, Cl Dissolved oxygen Hardness Iron, Fe Magnesium, Mg Manganese, Mn Potassium, K Selenium, Se Silver, Ag Sodium, Na	Biochemical Oxygen Demand (BOD); immediate, 5-day, ultimate	Nitrate nitrogen, NO_3	Total plate count	
Recommended additional parameters where municipal and/or industrial pollution are expected	Floating solids	Arsenic, As Barium, Ba Beryllium, Be Boron, B Cadmium, Cd Chromium, Cr Copper, Cu Dissolved Carbon Dioxide, CO_2 Fluorides, F Hydrogen sulphide, H_2S Lead, Pb Mercury, Hg Nickel, Ni Vanadium, V Zinc, Zn	Cyanide, CN Dissolved organic carbon Methylene Blue Active Substances (MBAS) Oil and grease Pesticides Phenolics	Ammonia nitrogen, NH_3 Nitrite nitrogen, NO_2 Organic nitrogen Soluble phosphorus Total phosphorus	Faecal streptococci Salmonella	Benthos Plankton counts
Optional parameters for surveys of special purpose	Bed load Light penetration Particle size Sediment concentration Settleable solids	Aluminium, Al Sulphates	Carbon Alcohol Extract (CAE) Carbon Chloroform Extract (CCE) Chlorine demand	Organic phosphorus Orthophosphates Polyphosphates Reactive silica	Shigella Viruses: —Coxsackie A&B —Polio —Adenoviruses —Echoviruses	Chlorophylls Fish Periphyton Taxanomic composition

MATERIAL AND ENERGY FLOWS IN AN AQUATIC ECOSYSTEM

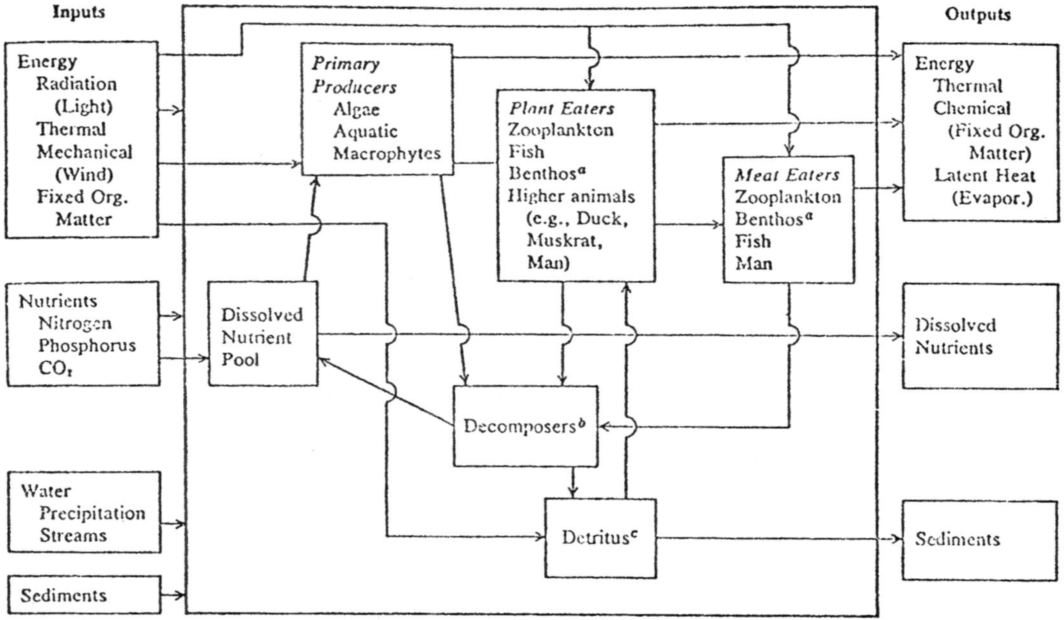

a. Organisms living at or on the bottom of bodies of water.
b. Fungi and bacteria.
c. Small particles of organic matter.

Figure 2: Material and Energy Flows in an Aquatic Ecosystem
(Reproduced by kind permission of University of Oklahoma)

Table 11: Example References on Biological Monitoring

Topic	Reference and Comments
Biological indicators	(63) -- Study of indicators for the marine environment.
	(64) -- Use of community diversity index of benthic macroinvertebrates and fish.
	(65) -- Use of zooplankters as bioindicators of water quality.
Biological sampling	(66) -- Systematic approach for inventorying aquatic habitat.
	(67) -- Quantitative studies of marine intertidal benthic environments.
	(68) -- Handbook for the biological monitoring of rivers.
	(69) -- Methods of zooplankton sampling and analysis for quantitative surveys.
	(70) -- Study of efficiency of three macrobenthic sampling methods.
	(71) -- Methods of phytoplankton sampling and analysis for quantitative surveys.

Table 12: Stream Quality Requirements for Public and Private Drinking Water Supplies (Reproduced by kind permission of Oklahoma Water Resources Board)

Characteristics	Standards
Physical	
Color	Color producing substances from other than natural sources, shall be limited to concentrations equivalent to 75 color units (CU).
Odor	Taste and odor producing substances from other than natural origin shall be limited to concentrations that will not interfere with the production of a potable water supply by modern treatment methods.
Temperature	At no time shall heat be added to any stream in excess of the amount that will raise the temperature of the receiving water more than 5°F. In streams, temperature determinations shall be made by averaging representative temperature measurements of the cross sectional area of streams at the end of the mixing zone. The normal daily and seasonal variations that were present before the addition of heat from other than natural sources shall be maintained. The maximum temperature due to man-made causes shall not exceed 68°F in trout streams, 84°F in smallmouth bass streams, or 90°F in all other streams and lakes.
Inorganic Elements (mg/l)	
Arsenic	0.05
Barium	1.0
Cadmium	0.01
Chromium	0.05
Copper	1.0
Fluoride (at 90°F)	1.6
Lead	0.05
Mercury	0.002
Nitrates	10.0
pH	The pH value shall be between 6.5 and 8.5, pH values less than 6.5 or greater than 8.5 must not be due to water discharge(s).

Table 12 (continued)

Characteristics	Standards
Selenium	0.01
Silver	0.05
Zinc	5.0
Organic Chemicals (mg/l)	
Cyanide	0.2
Detergents (total)	0.2
Methylene blue active substances	0.5
Oil and grease	All waters shall be maintained free of oil and grease to prevent a visible film of oil or globules of oil or grease on or in the water. Oil and grease shall not be present in quantities that adhere to stream banks and coat bottoms of water courses or which cause deleterious effects to the biota. For public and private water supplies, the water shall be maintained free from oil and grease and taste and odors that emanate from petroleum products.
Phthalate esters	0.003
Microbiological	
Coliform organisms	The bacteria of the fecal coliform group shall not exceed a monthly geometric mean of 200/100 ml, as determined by multiple tube fermentation or membrane filter procedures based on a minimum of not less than five (5) samples taken over not more than a thirty (30) day period. Further, in no more than 10% of the total samples during any thirty (30) day period shall the bacteria of the fecal coliform group exceed 400/100 ml.

is the National Sanitation Foundation Water Quality Index (WQI). The WQI was developed in 1970 using a formal procedure based on the Delphi technique, and it is composed of nine parameters -- dissolved oxygen (DO), fecal coliforms, pH, BOD, nitrates (NO_3), phosphates (PO_4), temperature, turbidity, and total solids. Information on the nine parameters is aggregated into one index through the use of a geometric formulation as follows:

$$WQI = \prod_{i=1}^{n} I_i^{W_i}$$

where I_i = sub-index value for ith parameter.

W_i = importance weight for ith parameter.

Importance weights for the nine parameters are listed in Table 13. Interpretation of parameter measurements is based on the usage of functional relationships, with the example for dissolved oxygen shown in Figure 3. Table 13 includes example calculations for the water quality index, with the resultant index being 39. The suggested stream classification system based on the water quality index is as follows: 0-25, very bad; 26-50, bad; 51-70, medium; 71-90, good; and 91-100, excellent (73).

CALCULATIONS FOR ANTICIPATED IMPACTS

The most important technical element in addressing water pollution impacts is the scientific prediction of the effects of various activities, with these predictions being based on appropriate calculations. Calculations can range from the use of mass balance approaches to sophisticated computer models. Qualitative projections can also be used in the absence of specific information on pollutants or modelling techniques. Water pollutants can be categorized into conservative, non-conservative, bacterial, and thermal groups. Different technical approaches are necessary depending upon pollutant group. Conservative pollutants refer to those materials not biologically degraded in the aquatic environment nor lost from the water phase due to precipitation, sedimentation, or volatilization. Non-conservative pollutants include organic materials, nutrients, and chemical substances that may undergo precipitation or volatilization or sedimentation. The key characteristic of non-conservative pollutants is that these materials are not conserved in the aquatic environment in their original state. Organic materials can be biologically decomposed by bacteria, while nutrient materials can be incorporated in aquatic biomass.

Table 13: Calculations for Water Quality Index

Variable	Measurement	I_i(a)	W_i(b)	$I_i^{W_i}$
DO (mg/l)	60%	60	0.17	2.01
Fecal Coliforms (no./ml)	10^3	20	0.15	1.57
pH	7	90	0.12	1.72
BOD_5 (mg/l)	10	30	0.10	1.41
NO_3 (mg/l)	10	50	0.10	1.48
PO_4 (mg/l)	5	10	0.10	1.26
Temp. Deviation (^oF)	5	40	0.10	1.45
Turbidity (JTU)	40	44	0.08	1.35
Total Solids (mg/l)	300	60	0.08	1.39
			WQI =	38.8

Notes: (a) subindex values are from functional relationships, see Figure 3 for one example.

(b) importance weights are from the Delphi study to solicit opinions from over 100 water quality professionals in the United States.

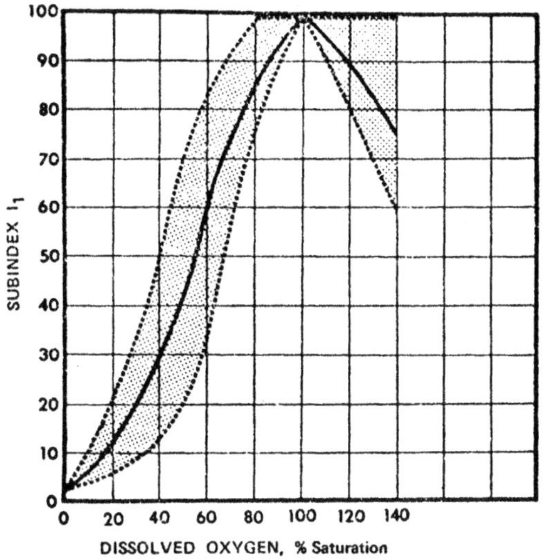

Figure 3: Functional Relationship for Dissolved Oxygen (Reproduced by kind permission of Ann Arbor Science Publishers, Inc.)

The basic modeling approach for conservative pollutants such as chlorides, are mass-balances considering the influence of hydraulics on system flows and dilution factors. Modeling approaches for organic materials must include consideration of the changes in DO resulting from bacterial demand for oxygen in the decomposition process. A classical DO model developed by Streeter and Phelps is as follows (1):

$$D_t = \frac{K_1 L_a}{K_2 - K_1} \left(10^{-K_1 t} - 10^{-K_2 t}\right) + D_a 10^{-K_2 t}$$

where D_t = dissolved oxygen deficit at any flow time t downstream, the flow time t expressed in days

= saturation dissolved oxygen concentration - actual dissolved oxygen concentration

K_1 = coefficient of deoxygenation, day^{-1}

K_2 = coefficient of reaeration, day^{-1}

L_a = ultimate BOD in the stream following mixing, mg/1

D_a = dissolved oxygen deficit upstream of waste discharge, mg/1

The saturation dissolved oxygen concentration is a function of temperature, pressure, and salt content. It should be noted that K_1, K_2, and L_a are influenced by temperature. Specific mathematical relationships describing the temperature influence are as follows:

$$K_{1(T)} = K_{1(20)} (1.047)^{T-20}$$

$$K_{2(T)} = K_{2(20)} (1.016)^{T-20}$$

$$L_{a(T)} = L_{a(20)} (0.02T + 0.6)$$

where T = water temperature, $^{\circ}C$

$K_{1(20)}$, $K_{2(20)}$, $L_{a(20)}$ = values at $20^{\circ}C$

In addition to the Streeter-Phelps model, numerous other techniques have been developed based on the classical Streeter-Phelps approach. Table 14 provides a summary of several water quality models for both non-conservative and conservative pollutants (74). Table 15 contains references on additional models for water quality and quantity impact prediction. Included are lake

Table 14: Summary of Some Water Quality Models
(Reproduced by kind permission of the Urban Institute)

Name	Type of Water Body	Pollutants Modeled	Input	Output
Streeter-Phelps	Streams and reservoirs	DO-BOD	Effluent content of point sources, velocity of flow, and ambient DO concentrations	DO concentrations as a function of distance downstream from a source
EPA's Simplified Model	Streams and estuaries	Chloride, dissolved solids, total P, bacteria suspended solids, DO-BOD	Effluent content of point sources, average and low flow velocity, general type of flow, channel geometry and slope, ambient water quality	Pollutant and indicator values as a function of location in stream/estuary
Auto-Qual	Streams and elongated estuaries	DO-BOD based on both carbon and N compounds, suspended solids, temperature, chloride	Flow and velocity, channel configuration at each junction, effluent content of point sources, water temperature, rates of DO uptake by sediments, and production by photosynthesis	Pollutant and indicator values as a function of location in stream/estuary and of time
Hydrocomp Simulation Program (HSP)	Streams, lakes, impoundments	Dissolved solids, temperature, bacteria, DO-BOD based on both carbon and N compounds, phytoplankton, zooplankton, benthal organisms, various forms of N and P	Effluent content of point sources	Pollutant and indicator values as a function of stream estuary location and of time

Table 15: Example References on Models for Water Cuality/Quantity Impact Prediction

Topic	Reference and Comments
Lake and reservoir models	(75) -- Multi-segment deep reservoir water quality simulation model for prediction of algal and DO-BOD dynamics.
	(76) -- Mixed-layer model for seasonal temperature predictions in lakes.
	(77) -- Literature review of mathematical modeling of surface water impoundments.
	(78) -- One-dimensional temperature-oxygen model for reservoirs.
Stream and estuary models	(79) -- Methodology for determining assimilative capacity of small streams.
	(80) -- Modeling of dissolved oxygen in estuaries.
Ground water	(81) -- Summary of ground water modeling techniques.
	(82) -- Review of several types of analytical models for ground water evaluation.
Thermal pollution	(83) -- Vertical mixing in statified impoundment from submerged discharge of heated water.
	(84) -- Model for prediction of temperature distribution in lakes resulting from thermal discharges.
Non-point pollution	(85) -- Methodology for assessing impact of urban stormloads on receiving water quality.
	(86) -- Modeling of effects of land use changes on streamflow quantity and quality.

Table 15 (continued)

Topic	Reference and Comments
	(87) -- General summary of procedures for predicting impacts of urban stormwater.
	(88) -- Methodology for assessment of point and non-point pollution on rivers, impoundments, and estuaries.
Irrigation return flow	(89) -- Conjunctive use model for predicting the mineral quality of irrigation return flow.
	(90) -- Literature review of irrigation return flow models.

and reservoir models which address both chemical constituents as well as temperature fluctuations, stream and estuary models related to dissolved oxygen, ground water models, and models relating to specific types or sources of water pollution such as thermal pollution, and irrigation return flows.

Bacterial pollution can also be addressed through the application of a specific model relating the aquatic environment characteristics and the bacterial death rate. This approach involves the prediction of bacterial self-purification, with this defined as the decrease of bacteria of all types, and especially those of fecal origin, as a function of flow distance or flow time in a river (91). The mathematical relationship that describes bacterial self-purification is as follows (1):

$$B_t = B_0 10^{-Kt}$$

where B_t = bacterial residual after any time t, days

B_0 = initial number of bacteria in stream

K = bacterial death rate, day^{-1}

Impact calculation approaches are also available for addressing the biological characteristics of the aquatic ecosystem. Table 16 summarizes some references on models for biological impact prediction. The first two examples represent aquatic habitat approaches in which a series of identifier parameters are used to develop an aquatic habitat index similar in concept to the WQI. Numerous ecosystem models have been developed for the aquatic environment, with several listed in Table 16. The models allow for calculation of chemical cycling as well as bioconcentration in aquatic organisms. Qualitative impact projections can be made in the absence of specific models. For example, Hazen, *et al.* (99) presented summary information on 47 case studies of California water projects that altered streamflows and causally affected fish and wildlife. This information would be useful for both impact identification and projection.

In summary, relative to impact calculations, there is considerable technology available for determining anticipated changes in water quality and aquatic biology that might occur as a result of the construction and operation of a given project type. However, it should be noted that the considerable exercise of professional judgment is involved in the selection of an appropriate model or models, and the interpretation of the results of the model applications.

Table 16: Example References on Models for Biological Impact Prediction

Topic	Reference and Comments
Habitat approach	(92) -- Systematic methodology for aquatic and terrestrial ecosystem evaluations.
	(93) -- Procedural manual for estimating and comparing development project impacts on fish and wildlife resources.
Ecosystem models	(94) -- Model for the Chesapeake Bay ecosystem with submodels on wetlands, plankton, seagrasses, other benthos, and fish trophic levels.
	(95) -- Model of nitrogen-cycle dynamics in an estuarine system.
	(96) -- Simulation models for determining pollutant impacts on marine biota.
	(97) -- Model for predicting chemical pollutant toxicity in fish.
	(98) -- Model for estimating bioconcentration of organic chemicals in fish.

ASSESSMENT OF CALCULATED IMPACTS

Assessment or interpretation of the significance of identified impacts represents a vital element in the conceptual framework for addressing water pollution impacts. This element requires the considerable exercise of professional judgment along with water quality criteria or standards and other scientific information. Typical criteria or standards give consideration to multiple uses of water resources and the quality requirements associated with the uses. Within the United States water quality standards have been established for stream segments, river basins, lakes, estuaries, and coastal areas. In addition, there is growing usage of ground water quality standards. A sound approach for impact assessment is to evaluate the calculated impacts relative to existing and resultant water quality if the project is implemented.

One of the difficult areas is related to interpretation of anticipated changes on the aquatic ecosystem. There are some laws and executive orders within the United States which address aquatic biological features, including Executive Order 11990 (Protection of Wetlands); Coastal Zone Management Act of 1972; Deep Water Port Act of 1974; Endangered Species Act Amendments of 1978; Fish and Wildlife Co-ordination Act of 1966; Marine Mammal Protection Act of 1972; Marine Protection, Research and Sanctuaries Act of 1972; and the Clean Water Act of 1977. Most of these regulatory documents provide general guidance for protection of the aquatic environment; however, specific standards for aquatic species or species diversity are not included. The most appropriate technical approach to utilize in aquatic ecosystem impact assessment involves the application of specific biological principles and recommended criteria. Table 17 contains some example references on these bases for assessment of aquatic ecosystem impacts.

IDENTIFICATION AND EVALUATION OF MITIGATION MEASURES

One of the most valuable results of a study of water pollution impacts can be the identification of mitigation measures which could be utilized in the planning, construction, or operation of a given project so as to minimize undesirable effects on the water environment. Mitigation may include avoiding the impact altogether by not taking a certain action or parts of an action; minimizing impacts by limiting the degree or magnitude of the action and its implementation; rectifying the impact by repairing, rehabilitating, or restoring the affected environment; reducing or eliminating the impact over time by preservation and maintenance operations during the life of the actions; and/or compensating for the impact by replacing or providing substitute resources or environments (104).

Appropriate mitigation measures must be identified on a

Table 17: Example References on Assessment of Biological Impacts

Topic	Reference and Comments
General approaches	(100) -- Methodology using weighted criteria to assess impacts on stream habitats.
	(101) -- Methodologies for determining instream flow requirements for fish, terrestrial wildlife, and water quality.
Specific biological changes	(102) -- Methodology for recommendation of minimum discharges for a warm water fishery.
	(103) -- Summary of temperature criteria for 34 fresh-water fish species.

project specific basis. However, some general examples can be cited for minimizing water pollution impacts. One appropriate approach is to attempt to minimize the non-point source pollution that would occur during the construction phase of a project. Table 18 identifies several references associated with minimizing impacts from construction activities as well as other non-point sources of pollution. For those project types which involve discharge of wastewaters, appropriate treatment could be included as mitigation. It is beyond the scope of this paper to review wastewater treatment processes and their cost-effectiveness; however, excellent references are available on this topical area (29, 30).

As illustrated in Figure 1, it is necessary to evaluate the effectiveness of proposed mitigation measures by making appropriate impact calculations for the project with mitigation, and then assessing the significance of the resultant changes on the baseline environmental setting. There may be several iterations required in evaluating the effectiveness of mitigation measures.

PREPARATION OF STUDY DOCUMENTATION

The end product of a study of water pollution impacts should be a report which addresses the findings and provides appropriate interpretation. This report should be incorporated in the overall environmental impact study report for a given project. The water pollution impacts report should summarize the impact-causing factors, baseline environmental setting, impact calculations and assessment, and mitigation measures. The report should incorporate technical writing principles. Examples include clear delineation of study objectives; inclusion of visual display materials such as maps, tables, figures, and photographs; and appropriate referencing of data sources and utilized scientific methodologies. Information should be provided on the bases for impact assessment and evaluation of appropriate mitigation measures.

SUMMARY

Water pollution impact prediction and assessment can be achieved by the application of existing technology and professional knowledge and judgment. However, many environmental impact studies have not used available scientific methods for impact prediction and assessment. Some possible reasons for this lack of extensive use include lack of knowledge about available approaches on the part of many practitioners, non-existence of current technology during the early years following the initiation of environmental impact studies, and general reluctance to use approaches perceived to be difficult and time- and cost-consuming.

Table 18: Example References on Mitigation of Impacts

Topic	Reference and Comments
Construction	(105) -- Guidelines for minimizing erosion from construction projects.
	(24) -- Summarizes mitigation of construction project impacts in wetland areas.
	(106) -- Organizing planning efforts to mitigate adverse impacts of construction.
Non-point pollution	(107) -- Linear programming model to allocate land use so as to minimize undesirable impacts.
	(108) -- Impacts of land clean-up and restoration following chemical spills.
	(109) -- Effects of soil and water conservation practices on minimizing non-point source pollution.
	(110) -- Guidance for controlling non-point pollution.
	(111) -- Summarizes agricultural management practices in terms of minimizing runoff and sediment production.

Usage of scientific methods and techniques is expected to increase as a result of the expanding knowledge base for conducting of studies of water pollution impacts, and the emphasis being given to public justification and accountability in project planning and decision-making. Opportunities for increased scientific emphasis in the analysis of water pollution impacts include:

(a) Use of computer-based literature searches to identify water pollutant and flow-related impact factors.

(b) Development and conducting of cost-effective baseline monitoring programs which have been planned based on the type of project and anticipated impacts.

(c) Use of water quality indices and indicators of biological quality and aquatic habitat.

(d) More extensive analyses of hydraulic/quantitative impacts and the development of information on cumulative impacts associated with project plans.

(e) Conducting of worst case analyses such as those related to accidental releases or spills of toxic materials.

(f) Systematic analyses of the cost-effectiveness of mitigation measures.

(g) More extensive analyses of potential water use limitations resulting from water quality changes.

(h) Development of better information relating water quality or quantity changes and resultant aquatic ecosystem changes, including emphasis on material cycling and toxic effects.

While there exists a strong technical basis for studies of water pollution impacts, research needs also exist. Examples of needed research include:

(a) Techniques for identifying and predicting short-term secondary impacts resulting from project construction and operation, with secondary impacts referring to population growth and associated infrastructure changes that potentially result from the construction and operation of large projects.

(b) Techniques for predicting long-term future impacts resulting from both primary and secondary impacts. Currently, most impact calculations focus on anticipated changes in the short term as opposed to several decades in the future.

(c) Techniques for identifying and predicting impacts from acid rain (precipitation washout) on both soil chemistry and surface and ground water resources.

In summary, water pollution impacts can be addressed by using a systematic approach involving identification of impact-causing factors, describing the baseline setting, and calculating anticipated changes in the water environment. Also included would be an assessment of the significance of calculated changes and the identification and evaluation of appropriate mitigation measures. Throughout the process it is important that recognition be given to system characteristics of the water environment, including interrelationships between water quality parameters and aquatic ecosystem features.

REFERENCES

1. Canter, L.W. *Environmental Impact Assessment* (New York, New York: McGraw-Hill Book Company, 1977).

2. Armaly, B.F. and S.P. Lepper. *Diurnal Stratification of Deep Water Impoundments,* Report No. 75-HT-35 (New York, New York: AMerican Society of Mechanical Engineers, 1975).

3. Environmental Control Technology Corporation. *Analysis of Pollution from Marine Engines and Effects on the Environment,* EPA-670/2-75-062 (Ann Arbor, Michigan: Environmental Control Technology Corporation, 1975).

4. Keeney, D.R. "A Prediction of the Quality of Water in a Proposed Impoundment in Southwestern Wisconsin, USA". *Environmental Geology* 2(6)(1978)341-349.

5. King, D.L. "Environmental Effects of Hydraulic Structures". *Journal of Hydraulics Division, American Society of Civil Engineers* 104(2)(1978)203-221.

6. Nelson, W. *et al* *Assessment of Effects of Stored Stream Flow Characteristics on Fish and Wildlife, Part A,* FWS/OBS-76/28 (Rockyville, Maryland: Environmental Control, Inc., 1976).

7. Raymond, H.L. "Effects of Dams and Impoundments on Migrations of Juvenile Chinook Salmon and Steelhead from the Snake River, 1966 to 1975". *Transactions of the American Fisheries Society* 108(1979)505-529.

8. Schreiber, J.D. and D.L. Rausch. "Suspended Sediment - Phosphorus Relationships for the Inflow and Outflow of a Flood Detention Reservoir". *Journal of Environmental Quality* 8(4)(1979) 510-514.

9. Yousef, Y.A. *et al*. *Mixing Effects Due to Boating Activities in Shallow Lakes,* Technical Report No. 78-10 (Orlando, Florida: Environmental Systems Engineering Institute, Florida Technological University, 1978).

10. Benke, A.C. *et al*. *Biological Basis for Assessing Impacts of Channel Modification: Invertebrate Production, Drift, and Fish Feeding in a Southwestern Blackwater River,* Report No. ERC 06-79 (Atlanta, Georgia: Environmental Resources Center, Georgia Institute of Technology, 1979).

11. Duvel, W.A. *et al*. "Environmental Impact of Stream Channelization". *Water Resources Bulletin* 12(4)(1976)799-812.

12. Headrick, M.R. *Effects of Stream Channelization on Fish Populations in the Buena Vista Marsh, Portage County, Wisconsin* (Stevens Point, Wisconsin: U.S. Fish and Wildlife Service, 1976).

13. Huang, Y.H. and R.K. Gaynor. *Effects of Stream Channel Improvements on Downstream Floods,* Research Report No. 102 (Lexington, Kentucky: Kentucky Water Resources Research Institute, University of Kentucky, 1977).

14. Stone, J.H. and G.F. McHugh. *Simulated Hydrologic Effects of Canals in Barataria Basin: A Preliminary Study of Cumulative Impacts*, Final Report (Baton Rouge, Louisiana: Louisiana State University Center for Wetland Resources, 1977).

15. Brannon, J.M. *Evaluation of Dredged Material Pollution Potential,* Technical Report No. DS-78-6 (Vicksburg, Mississippi: U.S. Army Engineers Waterways Experiment Station, 1978).

16. Chen, K.Y. *et al. Confined Disposal Area Effluent and Leachate Control Laboratory and Field Investigations,* Technical Report No. DS-78-7 (Vicksburg, Mississippi: U.S. Army Engineers Waterways Experiment Station, 1978).

17. Engler, R.M. "Impacts Associated with the Discharge of Dredged Material Into Open Water", pp. 213-223 in *Proceedings of the Third U.S. - Japan Experts' Meeting on Management of Bottom Sediments Containing Toxic Substances,* Report No. EPA-600/3-78-084 (Washington, D.C.: U.S. Environmental Protection Agency, 1978).

18. Flint, R.W. "Responses of Freshwater Benthos to Open-Lake Dredged Spoils Disposal in Lake Erie". *Journal of Great Lakes Research* 5(3-4)(1979)264-275.

19. Holliday, B.W. *Processes Affecting the Fate of Dredged Material*, Technical Report No. DS-78-2 (Vicksburg, Mississippi: U.S. Army Engineers Waterways Experiment Station, 1978).

20. JBF Scientific Corporation. *Dredge Disposal Study, San Francisco Bay and Estuary, Appendix M - Dredging Technology*, (San Francisco, California: U.S. Army Corps of Engineers, 1975).

21. Massoglia, M.F. *Dredging in Estuaries - A Guide for Review of Environmental Impact Statements*, Report No. NSF/RA-770284 (Corvallis, Oregon: Oregon State University, 1977).

22. National Marine Fisheries Service. *Physical, Chemical and Biological Effects of Dredging in the Thames River (CT) and Spoil Disposal at the New London (CT) Dumping Ground,* Final Report (Highlands, New Jersey: Division of Environmental Assessment, 1977).

23. Slotta, L.S. *et al. An Examination of Some Physical and Biological Impacts of Dredging in Estuaries* (Corvallis, Oregon: School of Oceanography, Oregon State University, 1977).

24. Darnell, R.M. "Minimization of Construction Impacts on Wetlands: Dredge and Fill, Dams, Dikes, and Channelization", pp. 29-36 in *Proceedings of the National Wetland Protection Symposium, June 6-8, 1977, Reston, Virginia* (College Station,Texas: Texas A and M University, 1978).

25. Elkington, J.B. "The Impact of Development Projects on Estuarine and Other Wetland Ecosystems". *Environmental Conservation* 4(2)(1977)135-144.

26. de Groot, S.J. "An Assessment of Potential Environmental Impact of Large-Scale Sand-Dredging for the Building of Artificial Islands in the North Sea". *Ocean Management* 5(3)(1979)211-232.

27. Bradt, P.T. and G.E. Wieland. *The Impact of Stream Reconstruction and a Gabion Installation on the Biology and Chemistry of a Trout Stream,* Completion Report (Bethlehem, Pennsylvania: Department of Biology, Lehigh University, 1978).

28. Witten, A.L. and R.V. Bulkley. *A Study of the Effects of Stream Channelization and Bank Stabilization on Warmwater Sport Fish in Iowa,* Report No. 76/2 (Ames, Iowa: Office of Biological Services, U.S. Fish and Wildlife Service, 1975).

29. Council on Environmental Quality. *Evaluation of Municipal Sewage Treatment Alternatives* (Washington, D.C.: Council on Environmental Quality, 1974).

30. Nemerow, N.L. *Industrial Water Pollution: Origin, Characteristics and Treatment,* (Reading, Massachusetts: Addison-Wesley Publishing Company, 1978).

31. U.S. Environmental Protection Agency. *Treatability Manual, Vol. II, Industrial Descriptions,* EPA-600/8-80-042b (Washington, D.C.: U.S. Environmental Protection Agency, 1980).

32. U.S. Environmental Protection Agency. *Methods for Identifying and Evaluating the Nature and Extent of Non-Point Sources of Pollutants,* EPA-430/9-73-014 (Washington, D.C.: U.S. Environmental Protection Agency, 1973).

33. Hossain, A. *et al.* "Estimation of Direct Runoff from Urban Watersheds". *Journal of the Hydraulics Division - American Society of Civil Engineers* 104(1978)169-188.

34. Larson, F.C. *The Impact of Urban Stormwater on the Water Quality Standards of a Regulated Reservoir,* Research Report No. 62 (Knoxville, Tennessee: Water Resources Research Center, 1978).

35. McCuen, R.H. *et al. Estimates of Nonpoint Source Pollution by Mathematical Modeling,* Technical Report No. 43 (College Park, Maryland: Maryland Water Resources Research Center, University of Maryland, 1978).

36. Smith, R. and R.G. Eilers. *Stream Models for Calculating Pollution Effects of Stormwater Runoff,* EPA-600/2-78-148 (Cincinnati, Ohio: U.S. Environmental Protection Agency, 1978).

37. Unger, S.G. *Environmental Implications of Trends in Agriculture and Silviculture. Volume II: Environmental Effects of Trends,* EPA-600/3-78-102 (Manhattan, Kansas: Development Planning and Research Associates, Inc., 1978).

38. U.S. Environmental Protection Agency. *Non-point Water Quality Modeling in Wildland Management: A State-of-the-Art Assessment,* EPA-600/3-77-036 (Athens, Georgia: U.S. Environmental Protection Agency, 1977).

39. Iwamoto, R. *et al. Sediment and Water Quality: A Review of the Literature Including a Suggested Approach for Water Quality Criteria,* EPA-910/9-78-048 (Washington, D.C.: U.S. Environmental Protection Agency, 1978).

40. Johanson, R.C. and K.C. Leytham. "Modeling Sediment Transport in Natural Channels", pp. 861-885 in *Watershed Research in Eastern North America, A Workshop to Compare Results, Volume II, February 28-March 3, 1977,* Report No. NSF/RA-770255 (Edgewater, Maryland: Chesapeake Bay Center for Environmental Studies, 1977).

41. Lusby, G.C. *Effects of Grazing on Runoff and Sediment Yield from Desert Rangeland at Badger Wash in Western Colorado, 1953-73,* Water Supply Paper 1532-I (Washington, D.C.: U.S. Geological Survey, 1979).

42. Rickert, D.A. and G.L. Beach. "Assessing Impacts of Land Management Activities on Erosion-Related Nonpoint Source Problems". *Journal of Water Pollution Control Federation* 50(11)(1978)2439-2445.

43. Whipple, W., Jr. *et al. Erosional Aspects of Managing Urban Streams,* Completion Report (New Brunswick, New Jersey: Water Resources Research Institute, Rutgers University, 1980).

44. Brookman, G.T. *et al. Technical Manual for the Measurement and Modeling of Non-Point Sources at an Industrial Site on a River,*

EPA-600/7-79-049 (Research Triangle Park, North Carolina: Research Corporation of New England, 1979).

45. Hendrey, G.R. and F.W. Barvenik. *Impacts of Acid Precipitation on Decomposition and Plant Communities in Lakes,* CONF-7805164-1 (Upton, New York: Brookhaven National Laboratory, 1978).

46. Robbins, J.W.D. *Environmental Impact Resulting from Unconfined Animal Production,* EPA-600/2-78-046 (Washington, D.C.: U.S. Environmental Protection Agency, 1978).

47. Watson, C.W. *et al. Rainout Assessment: The ACRA System and Summaries of Simulation Results,* LA-6763 (Los Alamos, New Mexico: Los Alamos Scientific Laboratory, 1977).

48. Anderson, J.W. "An Assessment of Knowledge Concerning the Fate and Effects of Petroleum Hydrocarbons in the Marine Environment", pp. 3-21 in W.D. Vernberg, *et al.,* eds. *Marine Pollution, Functional Responses* (New York, New York: Academic Press, 1979).

49. Malins, D.C. *Effects of Petroleum on Arctic and Subarctic Environments and Organisms* (New York, New York: Academic Press, 1977).

50. Moore, S.F. and D.S. McLaughlin. *Design of Field Experiments to Determine the Ecological Effects of Petroleum in Intertidal Ecosystems,* EPA-600/7-78-231 (Lafayette, California: Resource Management Associates, 1978).

51. Straughan, D. "Coast Environmental Impact Assessment: Lessons from Oil Spills". pp. 116-123 in *The Present and Future of Coasts, Proceedings of the First Annual Conference of the Coastal Society, held at Arlington, Virginia, November 1975.* Los Angeles, California: University of Southern California, 1975).

52. Buikema, A.L., Jr. *et al.* "Phenolics in Aquatic Ecosystems: A Selected Review of Recent Literature". *Marine Environmental Research* 2(2)(1979)87-181.

53. Oglesby, R.T. and W.R. Schaffner. "Phosophorus Loadings to Lakes and Some of Their Responses". *Limnology and Oceanography* 23(1)(1978)135-145.

54. Rosendahl, P.C. and T.D. Waite. "Transport Characteristics of Phosphorus in Channelized and Meandering Streams", *Water Resources Bulletin* 14(5)(1978)1227-1238.

55. Watson, V.J. *et al.* "Impact of Development on Watershed Hydrologic and Nutrient Budgets". *Journal of the Water Pollution Control Federation* 51(12)(1979)2875-2885.

56. Damman, W.H. *Mobilization and Accumulation of Heavy Metals in Freshwater Wetlands* (Storrs, Connecticut: Institute of Water Resources, Connecticut University, 1979).

57. Drill, S. *et al. The Environmental Lead Problem: An Assessment of Lead in Drinking Water from a Multi-Media Perspective,* EPA-570/9-79-003 (McLean, Virginia: Mitre Corporation, 1979).

58. Leland, H.V. *et al.* "Bioaccumulation and Toxicity of Heavy Metals and Related Trace Elements". *Journal of the Water Pollution Control Federation* 51(6)(1979)1592-1616.

59. Booth, R.S. "A Systems Analysis Model for Calculating Radionuclide Transport Between Receiving Waters and Bottom Sediments" in *Proceedings of the 18th Rochester International Conference on Environmental Toxicity* (Oak Ridge, Tennessee: Oak Ridge National Laboratory, 1975).

60. Sanders, T.G., ed. *Principles of Network Design for Water Quality Monitoring* (Fort Collins, Colorado: Colorado State University, 1980).

61. IHD-WHO Working Group on the Quality of Water. *Water Quality Surveys* (Paris and Geneva: UNESCO and WHO, 1978).

62. Canter, L.W. *Supplement to Environmental Impact Assessment* (Norman, Oklahoma: University of Oklahoma, 1979).

63. Iverson, R.L. "Use of Biological Indicators for Monitoring Effect of Pollutants on the Marine Environment", pp. 95-106 in *Proceedings, First American-Soviet Symposium on the Biological Effects of Pollution on Marine Organisms,* EPA-600/9-78-007 (Gulf Breeze, Florida: U.S. Environmental Protection Agency, 1978).

64. Lee, C.D. *et al.* "Benthic Macroinvertebrates and Fish as Biological Indicators, With Reference to Community Diversity Index", pp. 233-238 in *Water Pollution Control in Developing Countries. Proceedings of the International Conference Held at Bangkok, Thailand, February 1978* (New York, New York: Pergamon Press, Inc. 1978).

65. Polivannaya, M.F. and O.A. Sergeyeva. "Zooplankters as Bioindicators of Water Quality". *Hydrobiological Journal* 14(3) (1978)39-43.

66. Collotzi, A.W. and D.K. Dunham. "Inventory and Display of Aquatic Habitat", pp. 533-542 in *CLassification, Inventory and Analysis of Fish and Wildlife Habitat - the Proceedings of a Natural Symposium, January 24-27, 1977, Phoenix, Arizona,* FWS/OBS-78/76 (Washington, D.C.: U.S. Forest Service, 1978).

67. Gonor, J.J. and P.F. Kemp. *Procedures for Quantitative Ecological Assessments in Intertidal Environments,* EPA-600/3-78-087 (Corvallis, Oregon: School of Oceanography, Oregon State University, 1978).

68. Hellawell, J.M. *Biological Surveillance of Rivers: A Biological Monitoring Handbook* (Stevenage, England: Water Research Centre, 1978).

69. Jacobs, F. and G.C. Grant. *Guidelines for Zooplankton Sampling in Quantitative Baseline and Monitoring Programs,* EPA-600/3-78-026 (Gloucester Point, Virginia: Virginia Institute of Marine Science, 1978).

70. Pollard, J.E. and W.L. Kinney. *Assessment of Macroinvertebrate Monitoring Techniques in an Energy Development Area,* EPA-600/7-79-163 (Las Vegas, Nevada: U.S. Environmental Protection Agency, 1979).

71. Stofan, P.E. and G.C. Grant. *Phytoplankton Sampling in Quantitative Baseline and Monitoring Programs,* EPA-600/3-78-025 (Gloucester Point, Virginia: Virginia Institute of Marine Science, 1978).

72. Oklahoma Water Resources Board. *Oklahoma's Water Quality Standards - 1979,* Publication No. 101 (Oklahoma City, Oklahoma: Oklahoma Water Resources Board, 1979).

73. Ott, W.R. *Environmental Indices - Theory and Practice* (Ann Arbor, Michigan: Ann Arbor Science Publishers, Inc., 1978).

74. Keyes, D.L. *Land Development and the Natural Environment: Estimating Impacts,* URI 13500 (Washington, D.C.:The Urban Institute, 1976).

75. Baca, R.G. *et al. Water Quality Models for Municipal Water Supply Reservoirs. Part 3. User's Manual* (Richland, Washington: Battelle Pacific Northwest Laboratory, 1977).

76. Ford, D.E. and Stefan, H.G. "Thermal Predictions Using Integral Energy Model". *Journal of the Hydraulics Division, American Society of Civil Engineers* 106(1980)39-55.

77. Orlob, G.T. *Mathematical Modeling of Surface Water Impoundments, Volumes I and II* (Lafayette, California: Resource Management Associates, 1977).

78. Snodgrass, W.J. and M.F. Holloran. "Utilization of Oxygen Models in Environmental Impact Analysis", p. 1 in *Proceedings of the 12th Canadian Symposium on Water Pollution Research* (Hamilton,

Ontario: McMaster University, 1977).

79. Bathala, C.T. *et al.* "Assimilative Capacity of Small Streams". *Journal of the Environmental Engineering Division, American Society of Civil Engineers* 105(1979)1049-1061.

80. Ozturk, Y.F. "Mathematical Modeling of Dissolved Oxygen in Mixed Estuaries". *Journal of the Environmental Engineering Division, American Society of Civil Engineers* 105(1979)883-904.

81. Prickett, T.A. "State-of-the-Art of Ground Water Modeling". *Water Supply and Management* 3(2)(1979)134-141.

82. Walton, W.C. "Progress in Analytical Ground Water Modeling". *Journal of Hydrology* 43 (1979)149-159.

83. Hoopes, J.A. *et al. Selective Withdrawal and Heated Water Discharge: Influence on the Water Quality of Lakes and Reservoirs, Part II - Induced Mixing with Submerged, Heated Water Discharge,* WRS SRC 79-04 (Madison, Wisconsin: Water Resources Center, University of Wisconsin, 1979).

84. Uzzell, J.C., Jr. and M.N. Ozisik. "Three-Dimensional Temperature Model for Shallow Lakes". *Journal of the Hydraulics Division, American Society of Civil Engineers* 104(1978)1635-1645.

85. Driscoll, E.D. *et al. A Statistical Method for Assessment of Urban Stormwater,* Final Report (Washington, D.C.: U.S. Environmental Protection Agency, 1979).

86. Overton, D. "Stormwater Modeling", pp. 267-274 in *Proceedings of International Symposium on Urban Hydrology, Hydraulics and Sediment Control, July 18-21, 1977, Lexington, Kentucky* (Lexington, Kentucky: Office of Research and Engineering, University of Kentucky, 1977).

87. U.S. Environmental Protection Agency. *Areawide Assessment Procedures Manual, Volumes I-III,* EPA-600/9-76-014-1,2,3 (Cincinnati, Ohio: Municipal Environmental Research Laboratory, 1976).

88. Zison, S.W. *et al. Water Quality Assessment: A Screening Method for Nondesignated 208 Areas,* EPA-600/9-77-023 (Lafayette, California: Tetra Tech Inc., 1977).

89. U.S, Bureau of Reclamation. *Prediction of Mineral Quality of Irrigation Return Flow: Volume I. Summary Report and Verification,* EPA-600/2-77-179a (Denver, Colorado: U.S. Bureau of Reclamation, 1977).

90. Walker, W.R. *Assessment of Irrigation Return Flow Models,* EPA-600/2-76-219 (Fort Collins, Colorado: Department of Agricultural and Chemical Engineering, Colorado State University, 1976).

91. Phelps, E.B. *Stream Sanitation* (New York, New York: John Wiley, 1944).

92. U.S. Army Corps of Engineers. *A Habitat Evaluation System for Water Resources Planning* (Vicksburg, Mississippi: U.S. Army Corps of Engineers, 1980).

93. U.S. Fish and Wildlife Service. *Habitat Evaluation Procedures* (Fort Collins, Colorado: U.S. Fish and Wildlife Service, 1979).

94. Green, K.A. *A Conceptual Ecological Model for Chesapeake Bay,* FWS/OBS-78/69 (Washington, D.C.: U.S. Fish and Wildlife Service).

95. Najarian, T.O. and D.R.F. Harleman. "A Real Time Model of Nitrogen-Cycle Dynamics in an Estuarine System". *Progress in Water Technology* 8(4-5)(1977)323-345.

96. Snedaker, S.C. "Design of Simulation Models to Determine Biological Impact of Pollutants on the Marine Environment", pp. 1-9 in *Proceedings, First American-Soviet Symposium on the Biological Effects of Pollution on Marine Organisms,* EPA-600/9-78-007 (Gulf Breeze, Florida: U.S. Environmental Protection Agency, 1978).

97. Taylor, M.H. *An Indicator-Prediction Model for Ecosystem Parameters of Water Quality,* Technical Completion Report (Newark, Delaware: Water Resources Center, University of Delaware, 1979).

98. Veith, G.D. *et al.* "Measuring and Estimating the Bio-Concentration Factor of Chemicals in Fish". *Journal of the Fisheries Research Board of Canada* 36(1979)1040-1048.

99. Hazen, C. *et al. Assessment of Effects of Altered Stream Flow Characteristics on Fish and Wildlife, Part B: California, Case Studies,* FWS/OBS-76/34 (Washington, D.C.: U.S. Fish and Wildlife Service, 1976).

100. Bovee, K.D. and T. Cochnauer. *Development and Evaluation of Weighted-Criteria, Probability-of-Use Curve for Instream Flow Assessments: Fisheries,* FWS/OBS-77/63, IFIP-3 (Fort Collins, Colorado: U.S. Fish and Wildlife Service, 1977).

101. Stalnaker, C.B. and J.L. Arnette. *Methodologies for the Determination of Stream Resource Flow Requirements: An Assessment,* FWS/OBS-76/03 (Washington, D.C.: U.S. Fish and Wildlife Service, 1976

102. Bovee, K.D. *The Determination, Assessment, and Design of 'In-Stream Value' Studies for the Northern Great Plains Region* (Missoula, Montana: Department of Geology, Montana University, 1974).

103. Brungs, W.A. and B.R. Jones. *Temperature Criteria for Freshwater Fish: Protocol and Procedures*, EPA-600/3-77-061 (Duluth, Minnesota: U.S. Environmental Protection Agency, 1977).

104. Council on Environmental Quality. "National Environmental Policy Act - Regulations". *Federal Register* 43(230)(1978) 55978-56007.

105. Anton, W.F. and J.L. Bunnel. "Environmental Protection Guidelines for Construction Projects". *Journal of American Water Works Association* 68(12)(1976)643-646.

106. Gordon, J.R. "Interdisciplinary Team Approach to Mitigating Adverse Environmental Impacts of Highway Construction", Transportation Research Record 551, pp. 1-11 in Federal Highway Administration. *Mitigating Adverse Environmental Effects of Highway Construction* (Sacramento, California: Federal Highway Administration, 1975).

107. Bammi, D. *et al.* "Urban Planning to Minimize Environmental Impact". *Environment and Planning Journal* 8(3)(1976)245-259.

108. U.S. Environmental Protection Agency. *The Ecological Impact of Land Restoration and Cleanup*, EPA-520/3-78-006 (Washington, D.C.: U.S. Environmental Protection Agency, 1978).

109. Walter, M.F. *et al.* "Nonpoint Source Pollution Control by Soil and Water Conservation Practices". *Transactions of the American Society of Agricultural Engineers* 22(4)(1979)834-840.

110. Whalen, N.A. *Nonpoint Source Control Guidance, Hydrologic Modifications* (Washington, D.C.: U.S. Environmental Protection Agency, 1977).

111. Whisler, F.D. *et al.* *Agricultural Management Practices to Effect Reductions in Runoff and Sediment Production* (Starkville, Mississippi: Water Resources Research Institute, Mississippi State University, 1979).

RISK AND HAZARD ASSESSMENT

Robert G.H. Turnbull

Planning Consultant, Edinburgh

INTRODUCTION

The major accidents due to industrial activities and involving dangerous substances which have occurred in recent years have justifiably caused concern among the general public and raise important issues for land use and land use planners.

Incidents such as Flixborough (1974), Beek (1975), Seveso (1976) and the more recent Three Mile Island close-down, are only a few examples among many which had dramatic consequences for the surrounding population and the environment. They have also highlighted the enormous difficulties involved in reconciling safety with economics, the very natural fears of the population at large, and the need to implement more specific control over potentially hazardous activities.

This paper provides a brief account of the position on risk and hazard control in the United Kingdom, and from the planning viewpoint highlights some of the important issues, topics, concerns, successes, failures and areas where research is still required in relation to land-use and EIA.

RESPONSIBILITY FOR HEALTH AND SAFETY

In the UK responsibility for the safety of industrial activities rests with the Health and Safety Commission answerable to the Secretary of State for Employment. The Secretary of State for Industry is also concerned because of his responsibility for relations with the manufacturing sector. Similarly, the Secretary of State for Energy may be involved because of his responsibility for energy installations.

There are many specific laws to control polluting or dangerous establishments in order to protect the environment. Under the Health and Safety at Work Act 1974, occupiers of industrial installations are required to conduct their operations with full regard for the safety of workers and the public. The Act is enforced by the Health and Safety Executive.

ADVICE TO PLANNING AUTHORITIES

The Health and Safety Executive (HSE) has no direct jurisdiction in planning matters but can advise planning authorities and developers about potential major hazards. It can provide evidence at planning enquiries or appeals related to major hazards, express an opinion on the risk associated with major hazard sites and advise on the measures to minimize risk and advise on any developments alongside or near to existing hazards. The HSE has power to enforce health and safety legislation and can issue a Prohibition Notice, if any plant or process gives rise to risk or serious bodily harm.

Most enquiries concerning risks take time because experts in various fields have to be consulted. The degree of hazard which dangerous materials actually pose depends on many factors - quantity, physical state, type of usage and where used. Moreover, if an accident occurs, the consequential risk to people will depend on the protection provided for the workers on the site; the distribution of population around the site, wind velocity, weather and the topography of the surrounding area.

In many instances a detailed and expensive hazard survey report containing information on substances, plants, accident hazard factors, prevention and control measures and emergency plan will be required to be undertaken by the developer as advised by the HSE. When a development has been authorized, HSE will wish to be satisfied that an appropriate management system with effective methods for on-going identification and evaluation of hazards together with adequate emergency procedures has in fact been provided (1).

LAND USE PLANNING ISSUES

Safety of the surrounding population has not always been taken into account when setting up a potentially hazardous industry such as an oil refinery or petrochemical production plant. Indeed, most of today's chemical industry has been sited without the slightest consideration for health or safety of anyone but the employees of the undertaking concerned. This is hardly surprising since the potential for disaster had not been fully appreciated or recognized until the time of Flixborough. If should also be appreciated that since that incident very few members of the public in the UK have been injured by sudden events occurring in the chemical industry.

Although particularly dangerous activities have been controlled for many years, after the Flixborough explosion there was an upsurge of concern by planning authorities over developments in their areas coupled with an increase in the number of requests to HSE for appraisal of risk. This resulted in the HSE setting up a Major Hazards Branch with the responsibility for developing and co-ordinating policy on major hazards between central and local government. A Risk Appraisal Group was also set up and meets weekly to consider planning applications and appraisals of risk prepared by specialist inspectors.

SCOTTISH EXPERIENCE

The discovery of North Sea oil in the late sixties has had a major impact on land issues in Scotland, not least because of concern for the hazard implications of proposed and potential developments for the transport and processing of oil and gas products. These include major oil and gas terminals at pipeline landfalls, the transport of oil, natural gas, natural gas liquids and liquid petroleum gases by ship and pipelines, the development of petrochemical plants, and synthetic natural gas plants, all of which present major locational and safety problems in relation to specific development proposals and in the forward identification of sites.

Risk appraisal has increasingly become concerned with statistical probabilities based on an historical pattern of events and I am in some doubt as to the relevance of such exercises as part of an EIA. The arithmetical precision of some studies is highly questionable. Faced with the need to become involved in a project which demands the assessment of risk, the text of Lord Rothschild's 1978 lecture on risk (2) makes interesting reading.

In the context of identifying potential large sites the appraisal of the risk posed by a potential development is only appropriate, in my view, when enough is known about plants and processes to enable a detailed audit and risk analysis to be carried out. Such information is not usually available at the pre-planning, or indeed, the outline planning application stage and thus the question that arises from this is, should there be a change in planning procedures for hazardous plants whereby permission should not be sought until most details of the plant and processes are known? In view of this hazard dimension, the whole question of planning procedures in relation to major hazardous development requires reconsideration.

Some of the problems associated with site selection for chemical plants involve consideration of the possible effects of an accident on the surroundings as well as site suitability for the development (3). Threats which are important in this context may be sub-divided into toxic, explosion and fire. It is when consideration is given to the likely consequences of a specific accident under specific circumstances that difficulties arise. It is easy to say that a hazardous plant should not be built near a hospital or school but much less easy to state at what distance from such developments the plant could reasonably be permitted. Three substantial "unknowns" enter into the equation:

(i) the assessment of probability of large escapes of toxic gases varies according to the scale of measurement used. Very small leaks happen almost daily.
(ii) the consequences of massive failure.
(iii) the acceptability of any particular level of risk in particular circumstances, even if that level of risk could be defined with some degree of precision.

It would be clearly unreasonable to say that no risk is acceptable because this would bring all new development to a halt with major economic consequences for the country and localities concerned. However, under present control arrangement, planning authorities and people near to plants which are believed or likely to be dangerous have to decide whether or not a particular risk is acceptable in all the circumstances, and this is best determined through the planning legislation. Inevitably, this gives rise to differences of opinion, controversy and confrontation between all concerned as to what is and what is not acceptable in relation to plant explosion hazards, to escape of vapour clouds and the possible consequences of fire hazards. To provide statistical evidence on the likely occurrence of a hazardous incident does little to satisfy the perceived notions and fear of what the result and consequences of such an incident would be.

SUCCESSES AND FAILURES

At national level, both oil and gas developments have made a significant contribution to domestic energy requirements by way of their impact on the balance of payments, and have provided the basis for the growth of a home-based petrochemical industry. Sub-regional economies have also benefitted through the creation of jobs and opportunities which have transformed local activity levels and provided prospects for further development.

The British Gas Corporation's large coastal gas reception facilities at St. Fergus in the North-East of Scotland provide a unique example of an integrated development process involving environmental and risk and hazard assessments spread over a decade. However desirable in terms of promoting economic growth, the build-up of the terminal and uncertainty about the number of associated pipelines and hazard implications created land use problems and tested the development plan system and development control procedures. It was instrumental in the progressive revision of development plans and the development of environmental and hazard studies for assessing the merits of particular planning applications.

An interesting example of an unforeseen hazard arose several years after the initial phase of the development had been approved. In the early days, attention had been drawn to the fact that the metal towers in the terminal might obstruct transmission from a proposed nearby radio station but the risk of gas explosions resulting from these transmissions was not raised. This risk was first brought to light with the possibility that the structures in a proposed natural gas liquid plant on the Firth of Forth, some 100 kilometres south, might extract energy from radio waves and that radio frequency sparks capable of igniting concentrations of vapour might occur. The HSE were requested to provide an assessment of this phenomenon and their report took a year to complete.

The importance attached to safety considerations and the operation of the radio station resulted in the setting up of an inter-departmental committee to consider further research on the outcome of the HSE report. In the event there were no reasons why the radio station should not start transmitting. For different reasons, the experience gained highlighted the need for much closer liaison and co-operation between government departments concerned. In future, all such applications would be notified to the HSE who would alert interested departments at an early stage and take whatever action was needed to advise planning authorities.

This concern for community safety, and the need for staged hazard and operability audits has drawn attention to the need for clarification of the roles and responsibilities of developers, independent consultants, the HSE, and the planning authority in carrying out audits and in discharging safety conditions attached to any planning application. It has highlighted the need for continuous involvement between interested parties and the need for early agreement on the form, content and coverage of each audit in terms of the nature and scale of the main hazard elements, the assessment of probabilities, and the inter-relationships of these and other relevant factors.

The commercial pressure to obtain early planning permission strained both the financial and manpower resources of local authorities. The inadequacy of information provided by developers highlighted the need for special studies which in turn led to delays. This also led to questioning of the values of outline planning applications supported by detailed environmental and risk appraisals for plan designs which may be changed at a later stage, and drew attention to the possibility that the public and local authorities might be misled as to the eventual design and layout of particular proposals. In retrospect, the number and timing of environmental and hazard studies in an uncertain situation gave rise to a degree of duplication and overlap which might have been avoided with a more co-ordinated approach. From 1972 to 1981 some 21 environmental studies, including 7 dealing with safety aspects, had been undertaken on the St. Fergus terminal complex. These did not include post-operability studies.

ONSHORE PIPELINES

Although underground pipelines are the safest method of transporting hazardous materials, events in America, the Middle East and elsewhere, served to focus attention on some of the complexities of planning, administration and consultation which has enabled the successful and timely installation of a major pipeline network in Scotland. It also raised concern about safeguarding the communities affected by pipeline routes and what might happen in the event of a major pipeline failure. The three main elements of concern were:

(i) the route adopted for the pipeline and its contents in relation to levels of population.
(ii) the standard of construction adopted for the pipeline and its installations such as pumping stations.
(iii) the monitoring system adopted to check the regularity of its functioning and the range of procedures laid down to deal with emergencies.

The second and third elements are controlled by the Department of Energy Pipelines Inspectorate which lays down pipeline construction standards related directly to the degree of hazard arising from the contents of the pipe and its location in relation to buildings. However, planning authorities were concerned that no adequate policy in planning a route had been adopted by HSE for commercial pipelines to safeguard the communities in the event of pipeline leakage. This led the planning authorities to the formulation of their own standards which resulted in HSE issuing interim advice about consultation distances on either side of the pipeline. The advice notes prepared by the Scottish Office on methane gas pipelines operated by the British Gas Corporation and commercial pipelines were mainly concerned with procedural aspects and drawing together advice available at the time (4, 5).

The British Gas Corporation has developed the most advanced and satisfactory procedure for pipeline routing, adopting adequate safeguarding distances (½ mile) and density controls (not more than 1 person per acre) within which the pipe is laid. It is significant that these pipelines have been laid with the full co-operation of the local authorities and with little public reaction. Local authorities are also involved in the preparation of emergency plans drawn up under the Pipelines Act 1962 by developers and other interested parties.

SOME RECENT CASES

Since the Risk Appraisal Group (HSE) was set up in 1974, it has dealt with over one thousand planning applications for major plants and developments in their vicinity. Each case was taken entirely on its merits and in light of all the circumstances known to the Group. Due regard was, of course, paid to precedents where the essential features were the same, but this is not very often the case, and there are usually important differences to be taken into account. It is of interest to examine some of the recent cases:

(i) chlorine tanks were considered to pose a threat such that new residential development was considered undesirable on a site some 250 metres away, but another site about 750 metres distant was thought to be suitable for development.

(ii) new housing was discouraged at a distance of 500 metres from large LPG tanks but lighly populated industrial development was considered to be appropriate up to the boundary of the installation.

(iii) a new petrochemical complex was discouraged at less than 1500 metres from substantial housing, but again, lightly populated industrial developments were not discouraged at lesser distances from the complex.

(iv) the separation distance between two large new associated petrochemical plants was recommended to be in accordance with best modern refinery practice.

(v) non-industrial developments were not discouraged at about about 300 metres from an LPG tank used for firing furnaces.

GENERAL CONSIDERATIONS

In the UK there is a pragmatic approach to the problem of siting major hazardous industries and, through the Risk Appraisal

Group's activities, a reasonably consistent and uniform policy is developing for new installations and developments in their vicinity. A major criticism from planning authorities has been the delay in receiving replies about particular proposals submitted for comment. However, there appears to be no universal law which could be expressed mathematically which is capable of taking into account the very many factors which are considered to be important. Such procedures would necessarily prohibit some developments with trivial risks whilst allowing others to be unsuitably sited. A pragmatic approach, like democracy, may not be perfect, but it is the best we have at the present time.

REFERENCES

1. Health and Safety Executive. *Canvey - An Investigation of Potential Hazards from Operations in the Canvey Island/Thurrock Area*, vol. 1 (London: Her Majesty's Stationery Office, 1978).

2. Lord Rothschild. *Risk*, Richard Dimbleby Lecture (London: BBC Publications, 1978).

3. Cremer and Warner. *Guidelines for Layout and Safety Zones in Petrochemical Developments*, Report No. C.2056 (Inverness, Scotland: Highland Regional Council, IV3 5NX, 1978).

4. Scottish Development Department. *High Pressure Methane Gas Pipelines*, Planning Advice Note 17 (Edinburgh: Scottish Development Department, 1977).

5. Scottish Development Department. *Commercial Pipelines*, Planning Advice Note 25 (Edinburgh: Scottish Development Department, 1980).

SOCIAL IMPACT ASSESSMENT: THE STATE OF THE ART

C. P. Wolf

Department of Social Sciences
Polytechnic Institute of New York
Brooklyn, New York

This is a brief "state of the art" survey of social impact assessment (SIA), including a review of topics and concerns in the field, an appraisal of successes and failures to date, and some suggestions for further improvement.

How might one go about sensing "the state of the art" in SIA -- where it has come from, where it is now, and where it should be in the future? The field has undergone rapid expansion in the past few years; in tracing its development, we are definitely tracking a "moving target". One method might be to compare early and later discussions on this subject. The first was written in 1974 (1), before there was even much of an art. The most recent (2) shows marked advances on many fronts, even though some of the earlier problems remain (3). Other indicators of field development might be the growing body of SIA literature (4, 5), its codification in primers (6) and handbooks (7), or the diverse contexts in which SIA is finding application (8, 9), or yet again the degree of methodological sophistication that has come into wide-spread use (10, 11).

A more revealing procedure would be to measure progress in SIA against internal criteria of the field itself and external criteria of society as a whole. My sense of these is given in Table 1 (12). While these objectives were addressed mainly to SIA practitioners and scholars, there is considerable overlap with the external criteria of societal concern as well, in regard to public involvement and decision-making, social goal orientation and goal attainment. Judged by internal criteria, there is appreciable movement in several strategic directions, such as impact mitigation. Although major improvements are required at every point, the progress to date has been encouraging. On

Table 1: Social Impact Assessment Objectives
(Reproduced by kind permission of National Technical Information Service)

1. Assess the impacts.

2. Increase the scientific quality and general credibility of SIA procedures.

3. Build systematic and cumulative knowledge in the SIA field.

4. Encourage methodological development, e.g. bridging macro-micro levels of assessment.

5. Build professional competence and consensus.

6. Formulate and institute workable guidelines for the full and proper use of social knowledge and social research in development planning.

7. Develop operational methodologies in operational contexts.

8. Institutionalize SIA as a continuing process in appropraite decision and planning contexts.

9. Participate effectively and fully in the larger process of comprehensive and integrated assessment.

10. Improve the planning effectiveness of SIA procedures, e.g. by alleviating time, money and data constraints; lengthening lead times by early entry into the planning process, etc.

11. Build receptive audiences and responsive constituencies, e.g. secure the active support of colleagues in relevant professions and disciplines.

12. Elicit greater public involvement in the corporate and voluntary sectors.

13. Effectively assert the governmental interest in SIA.

14. Improve the quality of decision-making by providing timely and appropriate impact information.

15. Promote constructive uses of SIA, e.g. in environmental design practice.

16. Facilitate the achievement of legitimate social goals.

17. Address and express social concerns in operational terms and contexts, e.g. by improving the conceptual understanding and operational measurement of social planning objectives ("quality of life", "community cohesion", etc.).

18. Forge the linkage between development planning and social goal attainment.

19. Enlarge the range of planning alternatives--"non-structural", community-based, etc.

20. Assist in the mitigation of unavoidable adverse impacts.

external criteria, however, the situation appears less promising. Are societal problems being solved, with the assistance of SIA efforts? Is societal guidance becoming more precise and responsive, aided by SIA utilization? Here it must be said the results thus far seem meager. How do we get these answers?

1 SIA TOPICS

Traditionally, the most common SIA topics have been in "hardware" (physical planning) fields such as transportation, water resources and energy. That is from the "sending" side; usually small communities in rural areas have been on the receiving end. In the transportation field, highway planning received the earliest and most consistent attention; some later work has focused on mass transit and airports. Rail has been largely neglected, except in a few cases of abandonment; barges are usually considered in conjunction with water resources development. Reservoir construction has been the context for much research on recreation impacts, together with studies of parks, forests, wildlands, and, occasionally, urban open spaces. In energy, nuclear development has been the most intensively assessed, followed by coal and now oil and gas, the last mostly offshore. Industrial, rural, community and even regional development may all be combined in a single impact situation such as the Tennessee-Tombigbee Waterway. Housing and urban development, including tall buildings and convention centers, have been a lesser SIA interest, though no less socially important.

The least common topics of SIA application have been in areas of social programs (in which evaluation research is the strongest, implying an unfortunate disjunction between the two)--health, human services, education, manpower, etc.--although SIA is perhaps even more appropriate to social than to physical planning. Changing this order of priorities is a needed redirection of the field. Of course, topics such as casino gambling and prison construction have physical dimensions as well. Until recently (13), thermonuclear war was an unpopular topic of utmost importance. "Coming up" SIA topics include hazardous and toxic wastes, tourism, and cultural impacts on indigenous peoples in northern, western and other resource development regions particularly. This only highlights some major (and minor and upcoming) SIA topics; there is practically no end of suitable candidates, nor should there be.

2 SIA CONCERNS

SIA concerns are as broad as the impact situations to which they apply. The open "issues list" below involves philosophical, methodological and institutional questions of general concern:

a) Scientific versus humanistic concerns in SIA.

b) Social philosophy.

c) How rational can we be?

d) Objective versus subjective.

e) The disjunction of personal and social goals.

f) Comprehensive and integrated assessment.

g) Counterintuitivity.

h) Policy relevance.

i) Predicting discontinuity.

j) Weighting preference.

k) The concept of representation.

l) The "direction of the sign".

m) "Balance".

n) "Pleasing everybody".

o) How much is enough?

p) Credibility.

The list could be extended to great length and refined in great detail; to give some of its flavor, I will briefly mention some main points of concern for a few items.

2.1 Scientific Versus Humanistic Concerns in SIA

In social science we try to engage the concerns of the humanities with the methods of the sciences. It is perhaps natural that a tension should exist between our concern for humanistic ends and our expression of that by scientific means. This dualism is evident at many points, e.g. "objectivity vs. subjectivity". The fear of subjectivity leads to a preoccupation with methods of quantitative analysis in the effort to gain credibility and policy relevance. It complicates our procedures for weighting preference, even assuming a concept of representation that allows for value change, and it confounds our attempts to achieve societal consensus, resulting in the disjunction of personal and social values.

2.2 Social Philosophy

Although many people come to the field with methodological concerns seemingly uppermost in their minds, underlying these are social concerns of a philosophical bent. We may assess distributive impacts and talk about equity, but this presupposes some notions of fairness and social justice which cannot be reduced to income transfers. We must speak to people in value terms. In many methodological arguments, what is really at issue is some value question. We would be better assessors if we were better philosophers, not just of social science but also of social (moral) philosophy.

2.3 Counterintuitivity

"Counterintuitive" is a term popularized by Jay Forrester. The French parable of the "29th day" and of exponential growth in general speak to this point. An example is the case of traffic congestion where the solution is seen to be to build a bigger road. Counterintuitively, the consequence is not a solution but a bigger problem. SIA is about problems that are yet to be, and its aim is to forestall their coming about. How then are we to convince decision-makers (policy relevance) of a condition that does not yet exist and to act so as to prevent its coming into existence?

2.4 The "Direction of the Sign"

This is a problem of impact evaluation--deciding whether an assessed (measured or estimated) value on some impact indicator should be considered positive (+) or negative (-) or neutral (0). --e.g. the number of hospital beds or lawyers per 1000 population. To give a different example, do the stringent regulations to which nuclear development is held imply that it is extremely risky or well controlled? Did the accident at Three Mile Island prove that reactors are highly dangerous or--there were no outright deaths attributable to the accident--benign? How one answers has little to do with the bare facts and much to do with interpretation.

2.5 Credibility

The "bottom line" for SIA is its credibility in the eyes of decision-makers and publics alike. What convincing demonstration of its value can we produce? Would increasing the scientific quality of our procedures help, or would their policy involvement vitiate any claims to validity? Social impacts are seen as important but imponderable--"intangibles" "difficult to quantify" and "hard to measure". Are SIA methods adequate to surmount these difficulties? After all, the same was said of basic physical measurements in the early history of science. We should uphold and practice "good science" in SIA, but it will not save us. The questions of ultimate concern are value questions; our problem is to bring scientific information to bear on them--in effect, to unify the world of facts and the realm of values. Alvin Weinberg calls problems of this order "trans-scientific" (14); they are characteristic of our leading policy questions today. To play the assessment game is to accept those stakes and to bet on their resolution. So long as action is open to choice, there can never be a final solution.

3 SIA SUCCESSES AND FAILURES

It would help our credibility if we could point to some conspicuous examples in which SIA was successful. For instance, Jim Kent calculates that the Foundation for Urban and Neighborhood Development's "Social Resource Management" approach has saved the U.S. Forest Service millions of dollars in one national forest alone. He believes that enlightened labor practices reducing turnover could save energy companies billions over the coming decade. We can also point to more modest successes, for example the arterial that was not built through a stable minority neighborhood in Lexington, Kentucky because a sociology professor and a couple of graduate students went out and talked to the people and brought a video tape back with them to the city council (15). On occasion, SIA has been used as a "tie-breaker", where economic and environmental tradeoffs were indifferent between alternatives.

The failings of SIA are chiefly those of omission from major development planning decisions being taken all over the globe. "Hearing it from the people" is not all that popular under many regimes. The "triumph of the economic ideology" in the U.S. and its technocratic counterparts conspire to demean noneconomic value to sentiment, and sentiment to sentimentality. To get a hearing, you have to shout. For their part, assessors have not always done their homework or done it well enough to make a difference. In contrast, I think of Charles Komanoff and nuclear economics: if rated capacity is estimated at 75% and performance is 50%, all cost comparisons, e.g. with fossil fuels, are grossly overstated (16). In any case, I recall the final, ironic sentence of Michael Young's *The Rise of the Meritocracy*: "The failures of sociology are as illuminating as its successes" (17).

4 FIELD DEVELOPMENT IN SOCIAL IMPACT ASSESSMENT

What can we do to enhance our successes and reduce our failures? Advances in the state of the art must be sought in four linked areas of field development; theoretical, methodological, institutional and professional.

4.1 Theoretical Development

The central theoretical question for SIA field development concerns ways and means of building and applying systematic and cumulative knowledge. While in SIA there is a wealth of practical experience, little effort has been exerted to systematize its knowledge base--partly due to the highly pragmatic and pressured circumstances under which much SIA work has been performed. The general paradigm for SIA needs to be crystallized. I have been attempting this for some time in terms of the "main pattern" of assessment steps outlined elsewhere in this volume. In clarifying and codifying this, we must incorporate major bodies of theory, such as social change theory (modernization, etc.), and assimilate nodes of part-theory, e.g. diffusion. The intellectual foundation for SIA needs strengthening across the board.

4.2 Methodological Development

As with theory development, a major task here is the retrieval and reprocessing of much social science information accumulated over many decades in numerous areas of basic and applied research, e.g. by means of secondary analysis of survey data for purposes of trend impact analysis. Certain areas of research, such as institutional analysis, are in particular need of development and deployment; often an interdisciplinary team effort is required to devise, instrument and validate these techniques. Specific methodological questions arise at every point in the main pattern --for example, how to define "neighborhood" boundaries and whether the homogeneity assumption should be accepted. Conditions of fieldwork are such that methods must be flexibly employed; the adaptation of ethnographic methods to short duration studies is one conspicuous need. The budget of methodological problems could be expanded almost indefinitely. Building research capability and achieving an effective division of labor between academic and applied researchers are also institutional problems.

4.3 Institutional Development

SIA cannot fully succeed without strong institutional support. Since it is meant to be applied to a broad range of situations and conditions, this implies the engagement and involvement of all segments of the population across all sectors of society. In some of these the state of institutional organization is itself in disarray; the U.S. Department of Energy's national laboratories are a current example, and in fact the agency itself is scheduled to be scrapped. SIA cannot treat the general turbulence in the society by itself, but it is seriously constrained by it. There are areas such as voluntary action in community development where institution-building is actively taking place, however, in contexts such as alternative energy. Community-based SIA can be greatly strengthened with the proper institutional framework, such as "model" impact ordinances and guidelines like those promulgated by Alberta Environment. Perhaps the weakest institutional development, and greatest need, occurs on the international level. Corporate involvement is also required to a far greater extent. As we perceive it, the institutional problem for SIA is that the public interest in its success is nowhere matched by the public investment. Partly this is due to our own shortcomings as professionals.

4.4 Professional Development

Self-help has been the story of professional development in SIA to date and that must continue regardless of the amount of institutional support made available. Once more, an inventory of the areas of needed development is practically coextensive with the professions themselves. The interdisciplinarity of SIA underscores the need for professional involvement in all the social sciences, and the cultivation of friendly relations with neighboring disciplines. Hybrid fields, such as policy studies and futures research, offer fertile ground for co-operation, but the extent of planner involvement so far has been slight. The disparity between "disciplines" and "professions"-- consultants and other nonacademic practitioners come in the latter category--is an institutional problem within the field. Even the rudimentary tools of assessment work, e.g. documentation and computer access, are in poor repair. Our professional responsibility is to put them in better working order. This is a condition of gaining credibility,but achieving that depends in turn on some better working conditions. On balance, however, we have done quite a lot with very little in the way of resources. One is tempted to imagine what could be done with a little more.

In summary, the state of the art in SIA is improving and improvable. We have come a long way in a short time, and if we now can see more clearly how much farther we have to go, we can also detect significant progress along the way. We will not get there alone, however; in a sense, SIA is everybody--practitioners, academics, policy makers and publics. All of us know more than any of us, and any number can play.

REFERENCES

1. Wolf, C.P. "Social impact assessment: The state of the art", pp. 1-44 in C.P. Wolf, ed. *Social Impact Assessment* (Milwaukee, Wisconsin: Environmental Design Research Association, 1974).

2. Lang, R. and A. Armour. *The Assessment and Review of Social Impacts*, Technical Report 1 (Ottawa, Ontario: Federal Environmental Review Office, 1981).

3. Friesema, H.P. and P.J. Culhane. "Social impacts, politics and the environmental impact statement process". *Natural Resources Journal* 16(1976)339-356.

4. Carley, M.J. and E.O. Derow. *Social Impact Assessment: A Cross-Disciplinary Guide to the Literature*, Research Paper 80-1 (London: Policy Studies Institute, 1980).

5. Tester, F.J. and W. Mykes, eds. *Social Impact Assessment: Theory, Method and Practice* (Calgary, Alberta: Detselig, 1981).

6. Finsterbusch, K. *Understanding Social Impacts: Assessing the Effects of Public Projects* (Beverly Hills, California: Sage, 1980).

7. McEvoy, J. and T. Dietz, eds. *Handbook for Environmental Planning: the Social Consequences of Environmental Change*(New York, New York: John Wiley, 1977).

8. Bowles, R.T. *Social Impact Assessment in Small Communities: An Integrative Review of Selected Literature* (Toronto, Ontario: Butterworths, 1981).

9. Torgerson, D. *Industrialization and Assessment: Social Impact Assessment as a Social Phenomenon* (Toronto, Ontario: York University Publications in Northern Studies, 1980).

10. Finsterbusch, K. and C.P. Wolf, eds. *Methodology of Social Impact Assessment,* 2nd edition (Stroudsburg, Pennsylvania: Dowden, Hutchinson and Ross, 1981).

11. Leistritz, F.L. and S.H. Murdock. *The Socioeconomic Impact of Resource Development: Methods for Assessment* (Boulder, Colorado: Westview, 1981).

12. Wolf, C.P. *Social Impact Assessment and Environmental Design* (Springfield, Virginia: National Technical Information Service, 1980).

13. Katz, A.M. *Life After Nuclear War: The Economic and Social Impacts of Nuclear Attacks on the U.S.* (Cambridge, Massachusetts: Balinger, 1981).

14. Weinberg, A. "Science and Trans-science". *Minerva* 10(1972) 209-222.

15. McCoy, C.B. "The impacts of an impact study: contribution of sociology to decision-making in government". *Environment and Behaviour* 7(3)(1975)358-372.

16. Komanoff, C. *et al.* *The Price of Power: Electric Utilities and the Environment* (New York, New York: Council on Electric Priorities, 1972).

17. Young, M. *The Rise of the Meritocracy, 1870-2033, An Essay on Education and Equality* (London: Thames and Hudson, 1958).

AUDITING

ENVIRONMENTAL IMPACT ASSESSMENT, MONITORING AND POST-DEVELOPMENT AUDITS

Ronald Bisset & Paul Tomlinson

University of Aberdeen,
Aberdeen

INTRODUCTION

An examination of research on environmental impact assessment (EIA) indicates a concentration on the means of incorporating EIA into procedures for authorizing development, and on the formulation of methods to guide assessment and preparation of environmental impact statements (EISs). Such research focuses on the period prior to construction and does not consider the actual effects of operational developments. On the other hand, environmental science literature abounds with studies of the impacts of specific developments such as nuclear power stations, water impoundments and highways. Unfortunately, these studies do not try to link their results with predictions made prior to project authorization in order to verify or refute the impact predictions and thereby add to knowledge of the ability of particular techniques and methods accurately to predict impacts.

Information on the accuracy of impact predictions might lead to better identification of impacts in future EIAs, and also improved application of mitigating measures and monitoring schemes. At present, actions to mitigate and/or monitor impacts are made under conditions of considerable uncertainty. Such actions can be expensive and in some cases may be unnecessary as the expected harmful impacts may not occur. Existing knowledge of the nature of most development impacts does not allow the certain identification of those impacts which will actually require mitigation and/or monitoring.

A brief description of an EIA research project currently being undertaken by the PADC Environmental Impact Assessment and

Planning Unit of Aberdeen University is presented in this paper. The objective of the research is to compare the predicted and actual impacts of selected operational projects in the United Kingdom, and thereby to assess the accuracy of the predictions. From this study it is hoped that some recommendations which can be applied to improve current EIA practice will be made.

This paper is divided into 7 main sections. First, the objectives of the research will be set out. Second, a number of definitions will be presented. These will be followed by an account of the method used in examining the case studies of UK operational projects. Fourth, the results of the case studies to date will be provided, accompanied by an analysis of the main findings. Next, some of the main problems involved in the research will be mentioned. Finally, a number of important implications, arising from the research, for the practice of EIA will be presented as a contribution to the debate on the way EIA should develop in the future.

OBJECTIVES OF THE RESEARCH

The main objectives of the research are:

a) to identify impact predictions and test their accuracy
b) to identify and test techniques used to predict impacts
c) to ascertain whether methods used in preparation of EISs have identified all impacts known to have occurred.

Basically, the objectives involve identification of impacts predicted for existing developments, by the use of EISs or similar reports, coupled with tests of their accuracy. By this means it is considered possible to test techniques, such as air pollutant dispersion models and hydrological models, which are used to formulate impact predictions. In addition, by ascertaining the coverage of impacts, that is the numbers and types of actual impacts, and comparing them with impact predictions contained in EISs, it is possible to make judgements on the ability of methods (checklists, matrices and networks) 1) to identify, in advance, the range of impacts which will occur. The objectives of the research are to be achieved by examining a number of case studies.

DEFINITION OF TERMS

There are three terms which require definition: "monitoring", "post development" and "audit". In terms of this research, "monitoring" is concerned with identification and measurement of impacts from developments. This specific type of monitoring has been defined succinctly as "... the process of repetitive observation of one or more elements or indicators of the environment according to pre-arranged schedules in time and space in order to test postulates about man's impact on the environment". (1).

The second term, "post-development", refers to the time period extending from the initiation of the first specific activities associated with a proposed project. It does not only refer to the development phase which begins when a project becomes operational but encompasses all phases of development, as impacts occur at all periods during the life of a project. For example, the planning or pre-construction stage can result in land use impacts such as blight and economic effects such as land speculations. The term "post-development" is used because of the need to direct thinking away from a sole concern with the operational phase of development.

Finally, the term "audit" is used to describe the activities involved in comparing predicted impacts of development with those impacts which appear to have occurred. The aim of the comparison is to test the accuracy and coverage of predictions made in EISs. By auditing the predictions in a number of EISs it should be possible to investigate the utility of the techniques and methods used to identify, predict and assess impacts. In addition, should these audits indicate that current EIA practice is less than satisfactory in terms of impact prediction, it would be possible to formulate recommendations to improve EIA in the future.

RESEARCH METHOD

The first task was to select appropriate case studies. Initially, 130 EISs and similar reports were studied. It was considered necessary for projects selected to have been operational for at least three years and to have been associated with monitoring schemes which might aid auditing. The three-year period was chosen, as it was thought that it might be difficult to ascertain actual impacts if development had only been underway for a short time. The existence of the baseline data was also a criterion.

After this survey the Sullom Voe and Flotta oil terminals were selected for auditing. These projects are similar in nature and are located in marine environments in the Shetland and Orkney Islands off the North coast of Scotland. The main impacts for these terminals are associated with both low-level and large-scale oil spills.

To overcome this similarity the number of case studies was increased by adding the Cow Green reservoir and Redcar steelworks, both in North-east England, and the Anglesey aluminium smelter in North Wales. These developments affect terrestrial systems in a number of different ways, for example through air pollutants and climatic changes, and were considered a necessary contrast to the Flotta and Sullom Voe case studies. At the time of writing, the Flotta audit is the most advanced with the Sullom Voe and Cow Green cases in their final stages. The Redcar and Anglesey studies are still in the preliminary stages. As this work is still in progress it should be borne in mind that the results presented in this paper are preliminary and conclusions reached may be modified in the light of new data.

Once the case studies had been selected the task of identifying impact predictions began. For four of the cases this involved examination of EISs; however, the Cow Green and Anglesey projects were not preceded by EISs. They were accompanied by a variety of reports which were considered to fulfil the function of EISs. By examining two non-EIS projects it was considered possible to make a limited comparison between the accuracy of EIS impact predictions and impact predictions contained in other types of documents.

The activity of impact identification was very time-consuming and took longer than expected. For example, there were six separate EISs, of differing scope, prepared for the Flotta oil terminal. Consequently, impact predictions were repeated, often with slightly different emphases and it was necessary to attempt to distil a common prediction from a variety of different formats. Also, gaining access to information on monitoring data was not always easy because of the variety of individuals and organizations involved. Access to information was rarely withheld for reasons of confidentiality or because of the sensitive nature of the data. Most organizations were very willing to co-operate. It should be noted that analysis of raw monitoring data has not been a major component of the research because of time and manpower constraints. It has been necessary to rely on the interpretations of those responsible for particular monitoring schemes.

Once impact predictions had been identified for each case study it was necessary to collect relevant baseline and monitoring data. In some cases specific monitoring data could be used directly to audit a prediction or a number of related predictions. In many cases, however, there were not enough data to allow an audit to take place. Similar problems existed for the baseline data needed to identify whether an impact had occurred and whether it might have been caused by the project under audit.

RESULTS TO DATE

In this section an overview of results for the Flotta, Sullom Voe and Cow Green case studies will be presented. It should be noted that although the Flotta case study is complete, the Sullom Voe and Cow Green results have not yet been finalized. At this stage it is not possible to provide preliminary results for the Redcar and Anglesey case studies.

The section is divided into two parts. First, the results of auditing specific impact predictions will be presented. Then there will be an account of the results of that part of the study concerned with identifying whether impacts not mentioned or discussed in the appropriate EISs have occurred.

Auditing of Specific Impacts

Table 1 shows the total number of impacts identified for Sullom Voe, Flotta and Cow Green. A considerable disparity is apparent between the 500 impacts identified for Flotta and the 8 for Cow Green. The 500 Flotta impacts were abstracted from 6 EISs as against the usual single EIS. They contained many single predictions concerning oil slick movement and times of arrival, under different weather conditions, at various shores. The Sullom Voe EIS, containing 44 impacts, also dealt with oil impacts, but not in nearly as much detail as the Flotta EISs. The small number of impacts identified for Cow Green is thought to be due to: the non-preparation of an EIS, the nature of the impacts likely from a reservoir (less varied than from a major industrial plant), and the current status of the audit. It is likely that more impact predictions will be identified as work progresses, but the total is unlikely to approach the numbers for Sullom Voe and Flotta.

The Flotta case study will be discussed below, with reference when appropriate to the other case studies, because it exemplifies aspects common to all the case studies. Table 2 provides a breakdown of the Flotta impact predictions into categories relevant for auditing. It can be seen that 367 predictions refer to oil slick movements, a large proportion of the total number of impacts identified. It was necessary to divide all impacts into three categories; untestable, testable and tested. 25 impacts were found to be untestable. The reasons for this deserve attention as they apply to the other case studies, and also because the implications are important for the practice of EIA in the future.

There are two main reasons why impact predictions have proved to be untestable. First, there is the problem of the wording/phrasing of impact predictions. Some impact predictions were written in vague, imprecise and "woolly" language. These predictions had unclear meanings, were open to more than one interpretation and were non-specific in their reference to the environmental parameters or systems concerned. For example, one EIS contains the following predictions, "... a major oil spill would

TABLE 1: Breakdown of Audited Impacts for Sullom Voe, Flotta and Cow Green (Interim Results)

	SULLOM VOE	FLOTTA	COW GREEN
Total Number of Impacts Identified	44	500	8
Number Accurate	*	16	4
Number Inaccurate	*	8	3

* Not Yet Available

Details of those predictions audited are given in the written accounts of the case studies, which are still in preparation.

TABLE 2: Breakdown of Impacts for Flotta

Total Number of Impact Predictions	500
Oil Slick Movement Impacts	367
Untestable Impacts	25
Impacts Testable but not yet tested because	(451)
(i) further data to be obtained	1
(ii) appropriate conditions not present	418
(iii) lack of monitoring data	32
Tested Impacts	24

have potentially disastrous effects on fish spawning areas" and "...the likelihood of objectionable odours reaching local inhabitants is small". Such predictions are impossible to audit because of their lack of precision. Secondly, and perhaps more importantly, changes in development design after formulation of certain predictions nullify those predictions relating to the design feature changed. Such predictions, therefore, cannot be audited.

Of the remaining impact predictions, 450 were testable in theory but could not be audited. 32 could not be audited because data were not available to allow any assessment to be made; and 418 could not be audited within the time period of the research because the necessary conditions allowing them to be audited had not yet occurred. For example, most of the oil slick movement impact predictions were based on various assumptions about wind and oil drift rate conditions. If these assumptions had not been fulfilled when an oil spill occurred then the predictions could not be audited because the predictions did not refer to the actual conditions present at the time of spillage.

It has only been possible to audit 25 impact predictions because of all these difficulties. Those audited cover a range of impact types and are not restricted to a particular category of impact, for example, air pollutant or noise impacts. For Flotta, impacts audited cover: oil impacts, effects of the terminal on seal movements and breeding success, and effects on various bird populations. Of the 25 audited impacts, 16 are believed to be accurate, whereas 8 are not. Receipt of monitoring data in the near future will determine the accuracy of the remaining single impact predictions.

At present, results for Cow Green indicate that 4 predictions are accurate while 3 are not. Again these cover a range of impact types, but in this case they are mostly postulated changes in vegetation communities. However, it should be borne in mind that long-term changes may not have been detected by the available monitoring schemes or that the systems concerned may suffer a sudden change in the near or distant future due to the accumulated effects of the reservoir. Therefore, making a judgement on the accuracy of many predictions, at one point in time, is an uncertain affair. It is always possible, especially for ecological predictions, that future events may change a judgement on a particular prediction. This state of uncertainty arises from present lack of knowledge of the impacts of projects on different environmental systems.

Figures available at present for Sullom Voe indicate that approximately 30% of the 44 predictions can be audited. Recently, considerable new monitoring data has been made available and as a result the Sullom Voe study is being reviewed. It is not yet clear what the results of the review will indicate for Sullom Voe.

Project Impacts not Included in EISs

It is easier, for many impacts, to ascertain whether or not an impact has occurred due to a project, than to determine whether a particular impact prediction was accurate. For both Flotta and Sullom Voe a number of impacts thought to arise from the developments were omitted from the EISs. Most of these could be classed as indirect or secondary impacts not directly arising from the development. The majority of such impacts were caused by transport and aggregate extraction. Although not a direct consequence of the terminals, such impacts are related to the projects because they result from activities necessary for the construction and operation of the projects.

There were some impacts known to have been caused by the projects which were omitted from the EISs. The Sullom Voe terminal suffered a bunker fuel oil spill which resulted in the death of approximately 4000 seabirds, 14 otters and 50 sheep. In addition, about 2000 fleeces were damaged. Not only was the possibility of a bunker fuel oil spill not mentioned in the EIS, but otters were not considered to be vulnerable to oil spills.

There have also been environmental changes in the vicinity of the Sullom Voe terminal which may or may not be due to its presence. Changes in the numbers of common/grey seals and in the adult to pup ratio in the vicinity of the terminal have been observed. A decrease in benthic populations near the jetty has been noted. Finally, there has been a recent claim by local farmers that the reproduction potential of sheep inhabiting the area near the terminal has been adversely affected. New evidence to be examined in the final period of the research may help a decision to be made regarding the cause of these changes.

The small sample of EISs which have been examined in depth does not contain explicit mention of particular impact prediction techniques or EIA methods which may have been used to prepare the EISs. Although techniques/methods may have been used it has been necessary in this research to go direct to those involved in preparing the EISs to determine the role of techniques/methods in the EIAs.

In the Flotta case it was possible to identify three specific predictive techniques used in the preparation of the EISs. The oil spill predictions (367) were based on similar predictive techniques, but as the predictions could not be audited, the technique could not be assessed. A dispersion model was used to predict the concentrations of oil in the vicinity of the ballast water effluent pipe. Again this technique could not be assessed because the eventual location of the pipe was different from the alternatives used to make the predictions. Finally, a specific technique was used to determine the zone of visual influence and the nature of the appearance of the terminal. These predictions were tested by an observer using subjective judgement. This subjective assessment indicated that the technique was successful in its predictions. No techniques/methods have yet been identified in the Sullom Voe and Cow Green cases. The Anglesey and Redcar studies, however, will enable other predictive techniques to be tested, for example, noise and air pollutant dispersion models.

ANALYSIS OF RESULTS

A number of preliminary points can be made. First, it is clear that a very low number of impacts were audited. As mentioned in the previous section there are a number of reasons for this finding. One of these, the lack of monitoring data, deserves consideration.

One of the findings of the research has been the lack of integration between impact predictions and monitoring schemes. In the Flotta case there is an extensive marine biological monitoring programme to detect and describe the nature of long-term marine environmental changes and to assess the effects of major oil spills. In addition, there is monitoring of the ballast effluent and of beached bird and seal surveys. Although data from some of these programmes has provided useful information for auditing predictions, they have not been designed for this purpose. Therefore, their ability to aid auditing of impact predictions will always be limited.

Since oil pollution is the main impact concern at the Sullom Voe terminal it is not surprising that most monitoring effort is directed toward that factor. However, the lack of linkage between impact prediction and monitoring is shown clearly at Sullom Voe. Of 25 impacts considered important enough in the EIS to require mitigation, only 7 are monitored. A report on the impacts of the terminal and some of the monitoring work has been published recently (2). As a result it is too early to come to a final conclusion on the Sullom Voe case study as changes may have occurred since the preliminary audit attempt.

The Cow Green reservoir also exemplifies the mismatch between pre-development concerns and post-development monitoring. Before authorization for the reservoir was given it was feared that deleterious changes in rare assemblages of vegetation adjacent to the reservoir would occur as a result of wave erosion, micro-climatic modification and increased visitor pressure once the reservoir was operational. Auditing of Cow Green has shown that monitoring effort has not been directed toward these pre-development predictions. The most systematic, comprehensive and extensive monitoring schemes relating to Cow Green have been established to monitor aquatic ecological effects. The Cow Green literature is dominated by accounts of the results of this programme (for example, (3), (4)). It should be noted that aquatic impacts were of minor importance during the assessment phase, apart from the localized effect of river regulation on downstream plant communities. In comparison, the literature on the terrestrial and atmospheric impacts of the reservoir is sparse. Monitoring schemes exist, but the results are not easily accessible. This has made it difficult to determine their scope and coverage and, subsequently, to audit certain of the impact predictions.

Among the predictions audited it has been found that more are accurate than inaccurate (Table 1), although the limitations of the small sample of case studies should be noted. It seems, however, that no particular type of impact prediction, for example ecological impact predictions, appears to be more accurate than the others. Analysis of impacts omitted from audited EISs, but believed to have occurred, indicates that it might have been beneficial to have used an EIA method such as a basic interaction matrix or network in preparation of the EISs. For example, otters are mentioned in the Sullom Voe EIS but not in the context of oil spills. A simple interaction matrix might have alerted those preparing the EIS to the possibility that otters might be affected adversely by oil spills. A more comprehensive picture of likely impacts could have been obtained because EIA methods, such as matrices, ensure that those assessing a proposal consider the possible effects of project actions or characteristics and environmental components in a systematic manner. As well as enabling direct impacts to be identified, the use of a formal EIA method could have alerted those preparing the Flotta and Sullom Voe EISs to the possibility of indirect or secondary impacts occurring as a result of actions associated with construction and operation of the terminals.

PROBLEMS INVOLVED IN AUDITING

The case studies indicate a number of institutional and technological difficulties which may have to be taken into account before making "cause-effect" linkages between environmental changes and a recently operational project. Predictions of impacts can be based on technology or design layout which is subsequently not used. Should this occur in a particular case and not be recognized, an audit of certain impacts might be invalid, because the predictions tested would have been rendered inoperable, being linked with technology or design which was not in fact used. The fate of the ballast effluent pipe in the Flotta case study shows the problems posed for auditing by technological developments and design layout changes.

Changes in activities associated with a project can have the same effect. Predictions are sometimes made on the basis that a particular impact-causing activity is undertaken in a particular way; for example, that a particular construction activity will not take place during heavy rain to avoid erosion. Unless this is checked, it is possible that a false impression of predictive accuracy will be obtained should the effects of the activity be audited. The activity may be different from that upon which the prediction was made. It is also necessary to check whether mitigating measures implemented as a result of EIA will affect the nature of expected impacts contained in EISs.

Auditing of many impacts can only be undertaken when monitoring data allow statistically valid interpretations of cause-effect relationships to be derived. The case studies show that often monitoring data fall short of this requirement. Also, for projects with a long operational life and causing a number of impacts, the case studies show that monitoring may be required for long periods before trends can be identified.

Assuming that a particular monitoring scheme provides the type of data essential for attempting to make a judgement on the accuracy of an impact prediction, there are two remaining difficulties which have to be overcome before a final judgement can be made. First, there is the problem of sample size. For example, predictions on oil slick movements and destinations are tested by keeping track of actual slick movements and their final landfall. During the period available for auditing it might be that information on only 3 or 4 instances of slick movement is available. Can a judgement on impact accuracy be made on the basis of this sample or should a decision be deferred until a larger sample is available? In theory, a larger sample is required, but in practice there may not be enough time to obtain data from a sufficiently large sample. A decision on whether to use the same predictive model for another development will not await the results from a larger sample.

If a sufficiently large sample did exist, there would still be the problem of determining a criterion or series of criteria for deciding whether or not a prediction is accurate. It may be that in a sample of oil slick beaching events, only 60% of oil slicks beached at the time and place predicted. What does this result signify? Can one assume that the prediction is accurate or does the existence of the 40% failure rate indicate a faulty prediction? Such questions are not of wholly abstract significance. Again, the future use of a particular technique may depend on reaching conclusions to this question which will satisfy the different interest groups involved in the EIA process for specific projects. There needs to be some attempt to generate a consensus so that action can be taken based on an agreed reaction to the "meaning" of past experience.

IMPLICATIONS FOR PRACTICE OF EIA

The research has both produced findings on past EIA practice, and highlighted certain problems involved in auditing which have implications for the future of EIA. In this section, some of the lessons derived from the research will be used to formulate some recommendations for improving EIA practice. The recommendations are offered for discussion as part of the debate on the future of EIA, and cover the following: baseline studies, impact prediction, EIA format, and monitoring. It is not possible to consider these in detail; only a brief discussion of salient points can be given.

Baseline Studies

At present, baseline studies on the existing environment are often carried out in isolation from other EIA activities such as impact identification and prediction. Many EISs still consist of lists of facts or items, such as species lists, which are later "dipped into" when required during the impact prediction phase. The auditing work has shown the need for a higher level of linkage and integration between the various actions involved in EIA than exists at present. The acquisition of baseline data should be iteratively linked with impact prediction and the likely requirements for impact monitoring.

Obtaining baseline data relevant to impact prediction is a very difficult task. In practice, as potential impacts are identified and assessed, the requisite baseline data can also be identified and collected. There is a "feedback" process in which the nature of the impacts determines the baseline data acquisition and baseline information helps aid impact identification and prediction. As more information on impacts is gathered, baseline data requirements become more precise.

As developments are imposed on dynamic environmental systems, impacts can only be considered in terms of the trends of the system(s) they may influence. A static, baseline description is of no real value to EIA prediction. While the iterative process of refining baseline data requirements and the scope and nature of impacts is proceeding, it will become apparent which potential impacts should be monitored.

Assuming a project will proceed, it is necessary to set in motion "reference" station monitoring for sites expected to be unaffected by the particular impacts of concern. This "reference" monitoring will grow out of both the baseline data work and impact identification activities. In fact, "reference" monitoring stations should duplicate, as far as possible, stations established to obtain data. This should save both time and money. Paired "reference" and "affected" stations (measuring those parameters likely to undergo changes due to the development) can be established to aid conclusions on "cause-effect" relationships between a project and an environmental change. Through this procedure it is possible to set up "reference" stations indicating dynamic environmental trends and "affected" stations providing the same type of information (before project construction) as well as data on environmental parameters arising from development.

There is a fundamental problem in the above programme for baseline studies. Since few audits have been implemented little is known about the accuracy of most types of impact predictions. The location of "affected" stations depends on predictions of impacts, for example, area to be affected, or target organisms. However, results from these stations are the only means of testing the accuracy of predictions. If the impact predictions were inaccurate because monitoring stations were located incorrectly, then the monitoring data would provide a false or incomplete picture of actual impacts. For impacts which can be identified visually this problem can be overcome, and monitoring stations adjusted accordingly. On the other hand, when impacts are detectable only by statistical analysis it is possible that they may be missed entirely, giving a false impression of the accuracy of predictions. Should this happen, then future monitoring, based on the existence of misleading information from past experience, might not provide the type of data required, as stations might be sited wrongly.

Impact Prediction

The findings of the research concerning the writing/phrasing of predictions and their lack of specificity have already been mentioned. Support for this result comes from a recent review of 21 EISs prepared in Canada. This study examined the utility and relevance of the ecological impacts in the EISs and concluded that "...prediction amounted to generalizations about the possible occurrence of a problem or the need for some corrective action to be taken" (5).

To overcome existing deficiencies in the way impact predictions are formulated and phrased it seems essential that they be treated as hypotheses which can be tested. Such hypotheses would force those preparing EISs to specifiy a number of aspects relating to time-scale of impacts, probability of occurrence, and geographic extent of impacts. As well as ensuring that attention was paid to these aspects, formulation of hypotheses is essential for the interpretation of monitoring data needed for many types of impacts, for example, ecological effects.

A recent report indicates a type of EIA method which will help greatly to formulate impact predictions in the form of hypotheses which can be checked (6). This method, although developed for aquatic ecological impacts, can be used to produce a hierarchy of predictions in the form of impact hypotheses. These are testable statements on the consequences of a proposed project (see Figure 1). This method also gives guidance on environmental parameters which should be monitored to enable the impact hypotheses to be tested 2).

EIS Format

At present, EISs are produced at a single point in time during the history of a development. Due to this factor EISs only provide a "snapshot" of the design history of a project. All predictions relate to the characteristics of this "snapshot". Unfortunately, projects are dynamic entities which can, and often do, change over time. A "stationary" EIS is of limited value in a dynamic situation. To be able to accommodate project change and to update continually the nature of impacts expected from such an evolving project, it would appear necessary for an EIS to be adaptive to its context. Instead of being a single document, an EIS should perhaps be multi-document, or take a loose-leaf form to enable additions and subtractions to be made easily. Only through such a continual, updating process is it possible to keep track of likely impacts and be sure that the links between project characteristics and impacts are correctly assigned through auditing.

Monitoring

It is not possible here to discuss all the complex issues involved in monitoring. Only a few basic general observations can be made. The case studies have shown that monitoring schemes are not always directly related to expected impacts. Even when they are constituted to keep a track of impacts, changes in sampling design, and sampling equipment, have made interpretation of the different sets of results a difficult task. This was one of the main results from an extensive study in the US into the effects of nuclear power stations (7).

Before hypotheses can be tested it is important that the basic features of sampling design are applied to monitoring. Hypotheses are usually tested by determining whether the variation

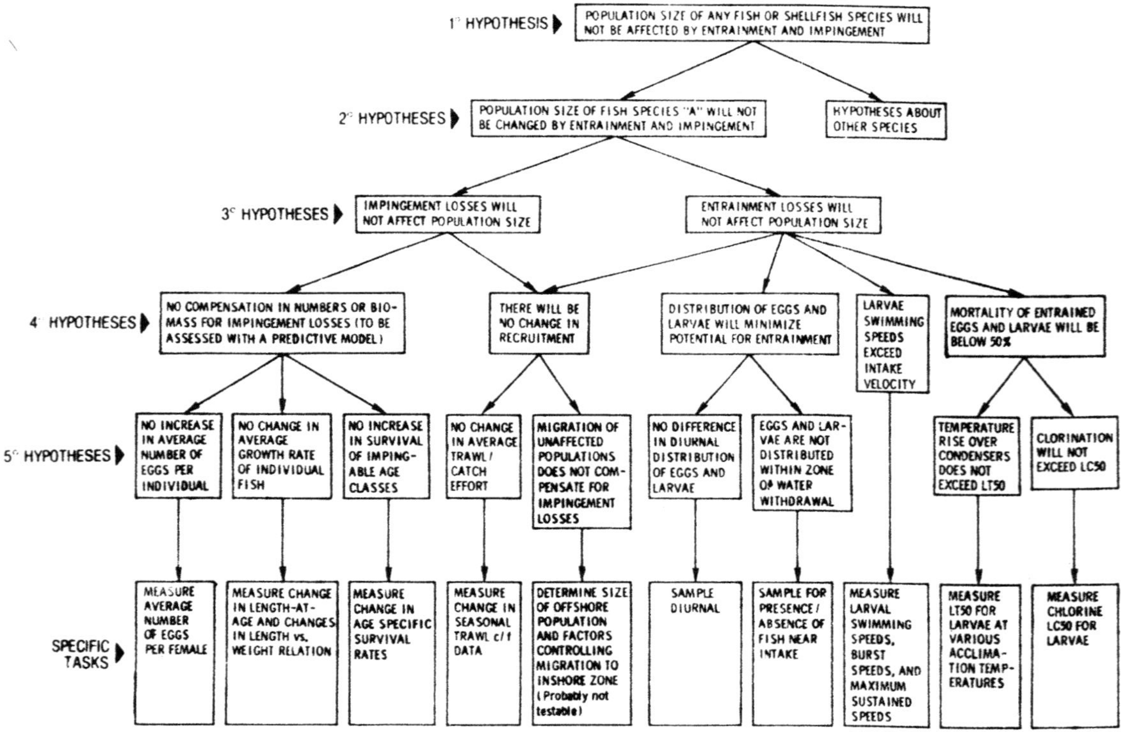

FIGURE 1: Example of a Hierarchy of Hypotheses. Only a Section of Possible Hypotheses to Test the Primary Hypothesis is Shown. (Reproduced by Kind Permission of the US Department of the Interior)

caused by a hypothesized effect in an affected area as compared with an unaffected area is larger than would be expected if the null hypothesis of no change were true. To make this judgement it is necessary to know the error variation, that is the variations within each area.

This approach requires two basic ingredients for success. First, replicate samples are needed within the same location. For each location and for each time of sampling, it is necessary to have, for purposes of an impact study, at least three replicate samples. Replicate samples are needed to determine differences within a location. Differences between locations can be assessed only when differences within locations have been determined. Secondly, these must be randomly distributed within the location. Figure 2 shows a simple example of these factors in a hypothetical study of the effect of an aqueous discharge. A number of text-books describe many of the details of experimental design and statistical analysis (8,9).

The above discussion on hypothesis setting and monitoring is applicable to "biological" impacts on individual organisms or populations. These impacts are often the most difficult to identify and describe. Fortunately, such expensive and time-consuming monitoring programmes may not be required for all impact predictions. In the UK case studies, there are a significant number of impacts which can be verified relatively easily. For example, it was predicted that the effects of the Flotta terminal development would be to cause the movement of common seal populations away from areas of disturbance. Such an impact prediction can be verified by mapping and visual analysis, assuming it is known that there was no trend of movement in the observed direction prior to project operation. The same type of approach can be used for many predictions relating to oil slick movements. It is important that the monitoring requirements for different impact types be borne in mind when devising an audit.

Selection of impacts for auditing can be a problem and there are a number of ways of choosing such impacts. If an audit were being carried out to test specific predictive techniques, for example, then the appropriate impacts would be audited. Such auditing can be undertaken for scientific purposes to improve knowledge on impacts. In most audits it is likely that only impacts of special concern will be studied, to identify deleterious effects early enough to take appropriate action. Such impacts could be chosen with public involvement. If this course were followed it would be possible that certain impacts would be monitored which did not, in the opinion of the experts, require such attention. Alternatively, appropriate experts could be asked to identify those impacts requiring auditing. Finally, impacts could be selected for auditing on the basis of whether they could

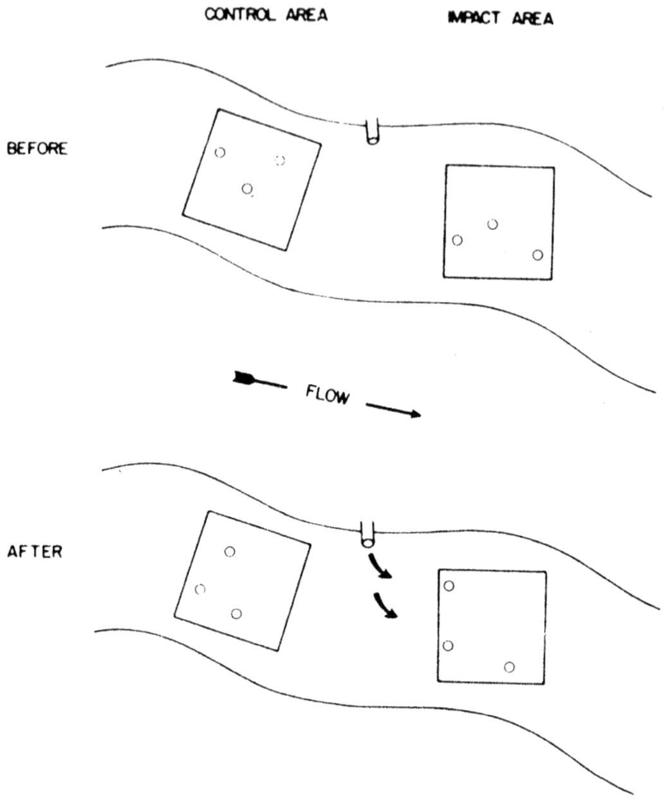

FIGURE 2: A Diagram of a Sampling Design for Impact Monitoring (Reproduced by Kind Permission of Wiley-Interscience)

be mitigated by an appropriate controlling agency or authority. If there were no institutional mechanism for changing the course of a particular impact it could be considered fruitless to monitor its progress unless monitoring were to increase knowledge.

Once it has been decided which impacts to monitor it is useful to identify the type of change which the auditing will be able to identify. This is particularly essential for those impacts which may be economically important to local people and which can be influenced by a controlling authority. Also, there should be a time limit within which these changes must be found. Prior agreement on the magnitude of change to be detected should reduce possible disagreement on whether remedial measures are required. British Petroleum, a major oil company, aim to detect changes in the order of 25% over background natural variation (10). The establishment of changes to be detected should involve all interested parties in, for example, shellfish populations. The changes to be detected should, if possible, be agreed on by local fishermen, fishery experts, pollution control authorities and representatives of the developer.

Monitoring and auditing are long-term commitments and, ideally, require a constant institutional framework. It would be best if personnel establishing a monitoring programme were to remain to carry it out, thereby maintaining consistency. However, staff turnover in both industrial companies and regulatory bodies is unavoidable. Staff changes may result in the invalidation of monitoring data through changes in sampling procedures. These can be altered, but only after calibration has been established through consultation with a statistician. Given the problem of personnel instability, those who initiate monitoring programmes should commit the appropriate details to paper. They can then be used by staff replacements to ensure continuity.

CONCLUSIONS

There are two main conclusions which can be derived from the research. It has shown that it is very difficult, perhaps impossible, to undertake a comprehensive audit of a project and its EIA unless auditing has been kept firmly in mind during baseline studies, formulation of predictions, writing of EISs and implementation of monitoring schemes. Coming to projects once they are operational with the aim of auditing their impacts and comparing them with EIS predictions has been shown to be not very successful.

It would seem appropriate, therefore, that a carefully controlled and co-ordinated limited series of audits for specific selected projects in varying environmental settings be established.

These audits would be established for the purposes of improving knowledge on EIA techniques and impacts. Such a scheme could be initiated by an international organization such as the Commission of the European Communities or the United Nations Environment Programme. EISs should be carried out in a systematic manner using best available methods and techniques, appropriate monitoring schemes devised and checks kept on imponderable and unpredictable factors which might affect the audit.

The recipients of results from such a study, for example governments, would be in a good position to disseminate information relating to the utility of predictive techniques, to those organizations responsible for EIA work. It should not be thought, however, that an international series of audits could avoid all existing problems and provide answers to all questions relating to impact characteristics and predictive accuracy. Gaps in knowledge would still remain, and there would still be a need for monitoring specific impacts to provide an early-warning system against environmental degradation. However, as more comprehensive and impact-specific audits were undertaken the need for such work in the future would decline.

The research on auditing has indicated that EIA should move away from its concentration on baseline studies and impact prediction, toward concern for impact management. It needs to take account of both the pre- and post-development situations if it is to realize its full potential. Auditing of impacts must be considered, in the future, to be a part of EIA so that the detection of harmful impacts can stimulate mitigation measures by changes in project design or operation. EIA should be an iterative process involving a "feedback" loop from initial project design, predictions, auditing of predictions, and impact management, to subsequent changes in project design. If EIA were to evolve in this direction some of the uncertainties would be removed from decision-making on major projects. Further, future EIAs would be better equipped to deal with uncertainty, and the actual impacts arising from projects might be rendered less harmful through the application of EIA containing both a predictive and management component.

FOOTNOTES

1. See papers by Bisset and Canter on EIA methods in this volume.

2. A summary of the role of hypothesis setting in EIA is given in Bisset, R., "Post-Development Audits to Investigate the Accuracy of Environmental Impact Predictions". *Zeitschrift für*

Umweltpolitik (in press).

REFERENCES

1. Johnson, W.C. and S.P. Bratton. "Biological monitoring in UNESCO Biosphere Reserves with special reference to the Great Smoky Mountains National Park". *Biological Conservation* 13 (2) (1978) 105-115.

2. Pearson, T.H. and S.O. Stanley, eds. *Symposium: The Marine Environment of Sullom Voe and the Implications of Oil Development*, Proceedings: Section B (Biological Sciences) vol. 80 parts 1/4 (Edinburgh: Royal Society of Edinburgh, 1981).

3. Armitage, P.D. "A quantitative study of the invertebrate fauna of the River Tees below Cow Green reservoir". *Freshwater Biology* 6 (1976) 229-240.

4. Crisp, D.T. *et al.* "The effects of impoundment and regulation upon the stomach contents of fish at Cow Green, Upper Teesdale". *Journal of Fisheries Biology* 12 (1978) 5-11.

5. Beanlands, G.E. and P.N. Duinker. *Project on the Ecological Basis for Environmental Impact Assessment in Canada: Progress Report* (Halifax, Nova Scotia: Institute of Resource and Environmental Studies, Dalhousie University, 1981).

6. Fritz, E.S. *et al.* *Strategy for Assessing Impacts of Power Plants on Fish and Shellfish Populations*, Report No. FWS/OBS/80/34 (Washington, D.C.: U.S. Fish and Wildlife Service, U.S. Department of the Interior, 1980).

7. Bisset, R. "Problems and issues in the implementation of EIA audits". *Environmental Impact Assessment Review* 1 (4) (1981) 379-396.

8. Sokal, R.R. and R.J. Rohlf. *Biometry* (San Francisco, California: Freeman, 1969).

9. Green, R.H. *Sampling Design and Statistical Methods for Environmental Biologists* (Chichester, England: John Wiley, 1979).

10. Cowell, E. "Ecological monitoring as a management tool in industry". *Ocean Management* 4 (1978) 273-285.

NAMES AND ADDRESSES OF FACULTY

Ronald Bisset,
Research Fellow,
Project Appraisal for Development Control (PADC),
Department of Geography,
University of Aberdeen,
High Street,
Old Aberdeen, AB9 2UF,
Scotland, UK.

Professor L.W. Canter,
School of Civil Engineering & Environmental Science,
University of Oklahoma,
Norman,
Oklahoma,
USA.

Fotis Chatzimikes,
National Council for Physical Planning and the Environment,
Ministry of Co-ordination,
Zalokasta 1,
Athens,
Greece.

Brian D. Clark,
Project Director,
Project Appraisal for Development Control (PADC),
Department of Geography,
University of Aberdeen,
High Street,
Old Aberdeen, AB9 2UF,
Scotland, UK.

Dr Kriton Curi,
School of Engineering,
Department of Civil Engineering,
Boğaziçi University,
PK 2, Bebek,
Istanbul, Turkey.

Dr Jack C. Dacre,
US Army Medical Bioengineering R & D Laboratory,
Fort Detrick,
Frederick, MD 21701,
USA.

Robert R. Everitt,
ESSA Environmental & Social Systems Analysts Ltd.,
678 West Broadway,
Vancouver, BC,
Canada, V52 166.

Murray G. Jones,
Shell Canada,
Environmental Affairs Dept.,
Head Office, Box 400,
Terminal "A",
Toronto, Ontario,
Canada, M5W 1E1.

Dr William V. Kennedy,
Office of Federal Activities
(A-104),
US Environmental Protection Agency,
401 M. St. SW,
Washington DC 20460,
USA.

Xaver Monbailliu,
20 Rue Saint Martin,
75004 Paris,
France.

Professor R. E. Munn,
Institute for Environmental
Studies,
University of Toronto,
Ontario,
Canada, M5S 1A4.

Paul Tomlinson,
Research Assistant,
Project Appraisal for Develop-
ment Control (PADC),
Department of Geography,
University of Aberdeen,
High Street,
Old Aberdeen, AB9 2UF,
Scotland, UK.

Robert G. H. Turnbull,
49 Coillesdene Avenue,
Edinburgh, EH15 2JP,
Scotland, UK.

Professor Geoffrey Wandesford-
Smith,
Division of Environmental
Studies,
University of California,
Davis, California 95616, USA.

Professor C. P. Wolf,
Social Impact Assessment,
Box 587,
Canal Street Station,
New York, NY 10013, USA.

NAMES AND ADDRESSES OF PARTICIPANTS

Mr David J. Anderson,
Environmental Assessment Div.,
Dept. of Environment,
1400 615 2nd St, SE,
Calgary, Alberta, Canada.

Prof. Luis Soares Barreto,
Instituto Superior de Agronómia,
Tapada de Ajuda,
1399 Lisboa Codex,
Portugal.

Dr George H. Bechet,
Unterer Mühlenweg 63,
D-7800 Freiburg,
Federal Republic of Germany.

Dr John F. Benson,
Dept. of Town & Country Plann-
ing,
The University,
Newcastle upon Tyne, NE1 7RU,
England, UK.

Ms Celen Birkan,
Prime Ministry Undersecretariat
for the Environment,
Karanfil Sok No. 8,
Kizilay-Ankara, Turkey.

Mr Eric Børset,
Norsk Institut for by-og
regionforskning,
Postboks 15,
Grefsen, Oslo 4, Norway.

Dr Costas A. Cassios,
3-5 Ippokratous Str.,
Greek Forest Service,
Athens, Greece.

Prof. Antonio Cendrero,
Departamento de Geológia,
Facultad de Ciencias,
Universidad de Santander,
Santander, Spain.

Mr Hendrick C. Christerus,
Ministry of Culture, Recreation
and Social Welfare, Room 1207,
Gebouw H, Sir Winston
Churchillaan 362, Postbox 5406,
2280 Rijswijk, Holland.

Mme Robe Marie Christine,
Commissariat Enérgie Atomique,
SPT/DPR,
91191 Gifsur Yvette Cedex,
France.

Mr Gordon H. Copp,
c/o 1204 Bryan,
Joliet, IL 60435,
USA.

Ms Patricia A. Cross,
2507 P Street, NW,
Washington DC 20007,
USA.

Dr David Curtis,
Dept. of Biology,
Paisley College of Technology,
High Street, Paisley, PA1 2BE,
Scotland, UK.

Mr Carlos E. Da Cunha Dias,
Av: Dr Lourenco Peixinho
-242-2°.
3800 Aveiro,
Portugal.

Mr Rui A. Da Rocha Boaventura,
Dept. Eng. Quimica - Faculdade
De Engenharia,
4099 Porto Codex,
Portugal.

Mr Claudio Da Rocha Miranda,
Rua Fonseca Teles, 121-17°
Andar,
Rio de Janeiro,
Brazil CEP 20940.

Mr Alberto De Caires Vila Nova,
Secretaria De Estado do
Ambiente,
R. Barata Salgueiro N°. 37-5°.
1200 Lisboa, Portugal.

Ms Mario Do Cev Teixeira
Goncalves,
Dept. Eng. Quimica - Faculdade De Engenharia,
4099 Porto Codex, Portugal.

Prof. Wilfred Erbguth,
Zentralinstitut für Raumplanung
an der Universität Münster,
Frauenstr. 39,
4400 Münster,
Federal Republic of Germany.

Mr Edward K. G. Faber,
TNO Complex Apeldoorn,
Postbox 342,
7300 AH Apeldoorn,
The Netherlands.

Mr Ned Farquhar,
Center for Northern Studies,
Wolcott, Vermont 05680,
USA.

Ms Belinda Foster,
Dept. of Geography,
University of Aberdeen,
High Street,
Old Aberdeen, AB9 2UF,
Scotland, UK.

Prof. Charles D. Fowle,
York University,
Dept. of Biology,
4700 Keele Street,
Downsview,
Ontario, Canada, M3J 1P3.

Dr Samir I. Ghabbour,
Dept. of Natural Resources,
Inst. of African Research
and Studies,
Cairo University, Giza (Cairo)
Egypt.

Dr Ir Eric A. Goewie,
DG Wetenschapsbeleid,
Min. Onderwijs en Wetenschappen,
Stadhoudersplantsoen 2,
Den Haag, The Netherlands.

Mr Henri Grisel,
SOGREAH - BP 172 x,
38042, Grenoble, Cedex,
France.

Dr Meir Gross,
107 E Pleasant St.,
Amherst, MA 01002,
USA.

Dr Orcan Gündüz,
1748 Sokak No. 11/1.
Karsiyaka-Izmir,
Turkey.

Ms Gülem Gürses,
Validecesme (Kassambasi),
Armagan Sok No. 31-33/5.
Besiktas, Istanbul,
Turkey.

Ms Maria Heli,
Miraktis 8,
Athens 511,
Greece.

Dr George F. Hepner,
Dept. of Geography,
Western Michigan University,
Kalamazoo, Michigan 49008,
USA.

Dr Stanley M. Hirst,
Environmental and Socio-Economic Services,
BC Hydro,
555 West Hastings Street,
Vancouver, BC, Canada V6B 4T6.

Drs Robert Jeltes,
Ministry of Health and Environmental Hygiene,
Postbus 439,
2260 AK Leidschendam,
The Netherlands.

Mr Tore K. Jenssen,
Det Norske Veritas, FDE,
112 Station Road,
Sidcup, Kent, DA15 7BU,
England, UK.

Mr Gert Johansen,
Søborg Parkalle 26,
DK-2860 Søborg,
Denmark.

John S. Klenavic,
Canadian Government,
Federal Environmental Assessment Review Office,
13th Floor, Fontaine Building,
200 Sacré Coeur,
Hull, Québec, Canada, K1A 0H3.

Mr Stuart Klosinski,
Planning Department,
Council of the Borough of Barrow-in-Furness,
Town Hall,
Barrow-in-Furness, Cumbria, UK.

Prof. C. Gregory Knight,
302 Walker Building,
Dept. of Geography,
Pennsylvania State University,
University Park,
PA 16082, USA.

Ms Günay Kocasoy,
Dept. of Civil Engineering,
Boğaziçi University,
PK 2 Bebek,
Istanbul, Turkey.

Prof. Xenophon G. Kondakis,
Medical School,
University of Patras,
142 Maizonos St.,
Patras, Greece.

Prof. William W. L. Lee,
Dept. of Civil & Urban Engineering,
University of Pennsylvania,
Philadelphia, PA 19104, USA.

Mr Tor Lerstang,
Norsk Institut for by-og regionforskning,
Postboks 15,
Grefsen, Oslo 4, Norway.

Mr Edward B. Liebow,
1334 W. Lynwood,
Phoenix,
Arizona 85007,
USA.

Mr David E. Martin,
British Petroleum,
Britannic House,
Moor Lane,
London EC2Y 9BU, England, UK.

Dr Gregory O. Michalenko,
Dept. of Man-Environment Studies,
University of Waterloo,
Waterloo, Ontario,
Canada, N2L 3G1.

Ms Elizabeth Monosowski,
Av. Professor Frederico Hermann Jr., 345,
05459 São Paulo,
SP Brazil.

Mr Laurence J. Nicholl,
2155 Buchtel Blvd., Apt. B402,
Denver, Colorado 80210,
USA.

Mr Antonio C. S. Pessoa,
Rua Pau de Bandeira, No. 9,
1200 Lisboa, Portugal.

Mr Arif K. Pinarci,
Dept. of Environmental Engineering,
Middle East Technical University,
Ankara,
Turkey.

Prof. Satya Prakash,
Institut für okologische Chemie,
Ingolstädter Landstr. 1,
D-8042 Neuherberg,
West Germany.

Mr David W. Purchon,
Dept. of Environmental Health,
Cleansing & Transport,
Leeds City Council,
Merrion House,
Merrion Centre,
Leeds, LS2 8HL, England, UK.

Dr Mark Reed,
Applied Science Associates,
529 Main Street,
Wakefield,
Rhode Island 02879,
USA.

Roberto A. Sanchez,
Institut fur Raumplanung,
Universitat Dortmund,
Postfach 500 500,
4600 Dortmund 50.

Dr Semra Siber,
Dept. of Civil Engineering,
Water Resources Laboratory,
Middle East Technical University,
Ankara, Turkey.

Mr A. J. O'Sullivan,
Garden Flat, Osborne House,
Seapoint Avenue, Monkstown,
Co. Dublin, Ireland.

Mr Haluk Sur,
c/o Prof. K. Curi,
Dept. Civil Engineering,
Boğaziçi University,
PK 2 Bebek,
Istanbul, Turkey.

Ms Nuran Talu,
Prime Ministry Undersecretariat
for the Environment,
Karanfil Sokak No. 8,
Kizilay, Ankara, Turkey.

Ms Sheila A. Thomson,
The Planning Exchange,
186 Bath Street,
Glasgow, G2 4HG,
Scotland, UK.

Dr David van Rest,
Interdisciplinary Higher
Degrees Scheme (IHD),
University of Aston,
70 Duke Street,
Birmingham, B4 7PJ,
England, UK.

Ms Iara Verocai Dias Moreira,
Rua Fonseca Teles, 121-17º.
Andar,
Rio de Janeiro,
Brazil CEP 20940,

The Hon. Mr Douglas N. Weir,
Creagdhu Lodge,
by Newtonmore,
Inverness-shire,
Scotland, UK.

Dr William Williams,
Nature Conservancy Council,
12 Hope Terrace,
Edinburgh, EH9 2AS,
Scotland, UK.

Ms Anne L. Zoeller,
Argonne National Laboratory,
Bld. 8, 9700 S. Cass Avenue,
Argonne, IL 60439, USA.

Observers

Mr Robin Mabey,
Dept. of the Environment,
Becket House,
1 Lambeth Palace Road,
London SE1 7ER,
England, UK.

Mr Simon Swaffield,
3 Grange Cottages,
Garvald, Haddington,
East Lothian, EH41 4LL,
Scotland, UK.

INDEX

GPSR Compliance

The European Union's (EU) General Product Safety Regulation (GPSR) is a set of rules that requires consumer products to be safe and our obligations to ensure this.

If you have any concerns about our products, you can contact us on ProductSafety@springernature.com

In case Publisher is established outside the EU, the EU authorized representative is:

Springer Nature Customer Service Center GmbH
Europaplatz 3
69115 Heidelberg, Germany

Batch number: 10816188

Printed by Printforce, the Netherlands